Aquatic Medicinal Plants

The use of medicinal plants in herbal and modern medicine has gained popularity over the last few decades due to consumers taking more natural approaches to medicine. Aquatic medicinal plants are rich in bioactive compounds and demonstrate various commercial, nutraceutical, and biological applications. *Aquatic Medicinal Plants* offers the reader a wealth of information on uses of bioactive components of these plants, along with crucial references, and explains their traditional uses, phytochemistry, and pharmacological properties.

Features:

- Provides information on aquatic and semiaquatic medicinal plants and their uses globally.
- Discusses phytochemical components with the known active constituents and their pharmaceutical applications.

This volume in the *Exploring Medicinal Plants* series is appropriate for scientists, experts, and consultants associated with the exploration of aquatic medicinal plant usage. This book is an essential tool for identifying important aquatic medicinal plants and possibilities for the synthesis or preparation of modern drugs.

Exploring Medicinal Plants

Series Editor

Azamal Husen
Wolaita Sodo University, Ethiopia

Medicinal plants render a rich source of bioactive compounds used in drug formulation and development; they play a key role in traditional or indigenous health systems. As the demand for herbal medicines increases worldwide, supply is declining as most of the harvest is derived from naturally growing vegetation. Considering global interests and covering several important aspects associated with medicinal plants, the Exploring Medicinal Plants series comprises volumes valuable to academia, practitioners, and researchers interested in medicinal plants. Topics provide information on a range of subjects including diversity, conservation, propagation, cultivation, physiology, molecular biology, growth response under extreme environment, handling, storage, bioactive compounds, secondary metabolites, extraction, therapeutics, mode of action, and healthcare practices.

Led by Azamal Husen, PhD, this series is directed to a broad range of researchers and professionals consisting of topical books exploring information related to medicinal plants. It includes edited volumes, references, and textbooks available for individual print and electronic purchases.

Phytopharmaceuticals and Biotechnology of Herbal Plants, *Sachidanand Singh, Rahul Datta, Parul Johri, and Mala Trivedi*

Omics Studies of Medicinal Plants, *Ahmad Altaf*

Exploring Poisonous Plants: Medicinal Values, Toxicity Responses, and Therapeutic Uses, *Azamal Husen*

Plants as Medicine and Aromatics: Conservation, Ecology, and Pharmacognosy
Mohd Kafeel Ahmad Ansari, Bengu Turkyilmaz Unal, Munir Ozturk and Gary Owens

Sustainable Uses of Medicinal Plants, *Learnmore Kambizi and Callistus Bvenura*

Medicinal Plant Responses to Stressful Conditions *Arafat Abdel Hamed Abdel Latef*

Aromatic and Medicinal Plants of Drylands and Deserts: Ecology, Ethnobiology and Potential Uses *David Ramiro Aguillón Gutiérrez, Cristian Torres León, and Jorge Alejandro Aguirre Joya*

Secondary Metabolites from Medicinal Plants: Nanoparticles Synthesis and their Applications
Rakesh Kumar Bachheti and Archana Bachheti

Aquatic Medicinal Plants *Archana Bachheti, Rakesh Kumar Bachheti, and Azamal Husen*

Antidiabetic Medicinal Plants and Herbal Treatments *Azamal Husen*

Ethnobotany and Ethnopharmacology of Medicinal and Aromatic Plants: Steps Towards Drugs Discovery *Adnan Mohd, Mitesh Patel and Mejdi Snoussi*

Wild Mushrooms and Health Diversity, Phytochemistry, Medicinal Benefits, and Cultivation
Kamal Ch. Semwal, Steve L. Stephenson, and Azamal Husen

Medicinal Roots and Tubers for Pharmaceutical and Commercial Applications *Rakesh Kumar Bachheti and Archana Bachheti*

Aquatic Medicinal Plants

Edited by
Archana Bachheti, Rakesh Kumar Bachheti, and
Azamal Husen

CRC Press
Taylor & Francis Group
Boca Raton London New York

CRC Press is an imprint of the
Taylor & Francis Group, an **informa** business

Designed cover image: https://www.shutterstock.com/image-photo/pond-landscaping-aquatic-plants-water-lilies-5010268

First edition published 2024
by CRC Press
6000 Broken Sound Parkway NW, Suite 300, Boca Raton, FL 33487-2742

and by CRC Press
4 Park Square, Milton Park, Abingdon, Oxon, OX14 4RN

CRC Press is an imprint of Taylor & Francis Group, LLC

Library of Congress Cataloging-in-Publication Data
Names: Bachheti, Archana, editor. | Bachheti, Rakesh Kumar, editor. | Husen, Azamal, editor.
Title: Aquatic medicinal plants / Edited by Archana Bachheti, Rakesh Kumar Bachheti, and Azamal Husen.
Description: First edition. | Boca Raton, FL : CRC Press, 2023. |
Includes bibliographical references and index. |
Summary: "The use of medicinal plants in herbal and modern medicine has gained popularity over the last few decades due to consumers taking more natural approaches to medicine. Aquatic medicinal plants are rich in bioactive compounds and demonstrate various commercial, nutraceutical, and biological applications. Aquatic Medicinal Plants offers the reader a wealth of information on uses bioactive components of these plants, along with crucial references, and explains their traditional uses, phytochemistry, and pharmacological properties. Features Provides information on aquatic and semiaquatic medicinal plants and their uses globally. Discusses phytochemical components with the known active constituents and their pharmaceutical applications. This volume in the Exploring Medicinal Plants series is appropriate for scientists, experts, and consultants associated with the exploration of aquatic medicinal plant usage. This book is an essential tool for identifying important aquatic medicinal plants and possibilities for the synthesis or preparation of modern drugs"– Provided by publisher.
Identifiers: LCCN 2022058406 (print) | LCCN 2022058407 (ebook) |
ISBN 9781032185408 (hardback) | ISBN 9781032188904 (paperback) | ISBN 9781003256830 (ebook)
Subjects: LCSH: Medicinal plants. | Materia medica, Vegetable. | Botany, Medical. | Aquatic plants.
Classification: LCC QK99 .A88 2023 (print) | LCC QK99 (ebook) | DDC 581.6/34–dc23/eng/20230209
LC record available at https://lccn.loc.gov/2022058406
LC ebook record available at https://lccn.loc.gov/2022058407

ISBN: 9781032185408 (hbk)
ISBN: 9781032188904 (pbk)
ISBN: 9781003256830 (ebk)

DOI: 10.1201/9781003256830

Typeset in Times
by Newgen Publishing UK

Contents

Preface

The prevention and treatment of illnesses have always been accomplished with herbal medicine in every culture. These herbs are made from various plant components, including leaves, stems, blossoms, roots, seeds, and so on. Most of these plants grow inland, i.e. in the terrestrial ecosystem, while some are grown in the aquatic ecosystem. Aquatic and semiaquatic medicinal plants are also vital to human welfare since they provide many of our daily needs. They have shown commercial, nutraceutical, medicinal, and numerous biological applications. Some of the important families of aquatic and semiaquatic medicinal plants are Acanthaceae, Alismataceae, Amaranthaceae, Apiaceae, Araceae, Asteraceae, Boraginaceae, Ceratophyllaceae, Cyperaceae, Fabaceae, Hydrocharitaceae, Lythraceae, Marsileaceae, Menyanthaceae, Nelumbonaceae, Nymphaeaceae, Onagraceae, Plantaginaceae, Poaceae, Polygonaceae, Pontederiaceae, Primulaceae, Scrophulariaceae, and Zingiberaceae. Numerous plants growing in the aquatic ecosystem have a high potential to be used as medicinal plants and are often used by indigenous people. For example, the coastal states in Australia, America, Europe, Sri Lanka, and India utilize aquatic plants for their medicinal, ornamental, and nutritional value. At present, with the help of published data, many researchers have established that numerous aquatic plants have a high medicinal value. They are rich in terms of significant chemical compounds such as alkaloids, flavonoids, terpenoids, phenolics, saponins, tannins, dietary fibre, glycosidic derivatives, carbohydrates, and proteins. These aquatic and semiaquatic medicinal plant-based phytochemicals have been used for their antimicrobial, antioxidant, hepatoprotective, sedative, anticonvulsant, cytotoxic, antiparasitic, and antidiabetic activities. Moreover, several parts of the plants are used as dietary supplements. Several aquatic and semiaquatic plants are used for the green synthesis of metal and metal oxide nanomaterials, which have shown significant applications. They are also known for their commercial value and are used as ingredients in some pharmaceutical and cosmetic industries.

This book provides the reader with a wide spectrum of information on aquatic medicinal plants, including important references. The book consists of 15 chapters, and the vast coverage of diverse aspects of the subject reflects well in the table of contents. We hope that this compendium of chapters will be very useful as a reference book for graduate and postgraduate students and researchers working in the fields of medicinal plants, plant science, economic botany, ecology, chemistry, biotechnology, pharmacognosy, pharmaceuticals, industrial chemistry, nanoscience, and many other interdisciplinary subjects. This book will be helpful in identifying some important aquatic medicinal plants and their future possible use in the synthesis or preparation of modern drugs. Additionally, it will be useful for scientists, experts, and consultants associated with the exploration of medicinal plants for various purposes.

We are highly thankful to all eminent authors who have contributed chapters and provided their valuable time and knowledge for this edited book. We are also pleased to express our gratitude to the reviewer who has contributed specifically to the screening and revision of the articles. We shall be happy to receive comments and criticism, if any, from subject experts and general readers of this book.

Archana Bachheti
Rakesh Kumar Bachheti
Azamal Husen

Editors

Archana (Joshi) Bachheti completed her BSc in 1997 and MSc in 1999 at Hemwati Nandan Bahuguna Garhwal University, Srinagar, India. She received her PhD from the Forest Research Institute, Dehradun, India in 2006. She has carried out research projects and consultancy work in the areas of eco-restoration/development of wasteland and physicochemical properties of *Jatropha curcus* seed oil and its relation with altitudinal variation, and has been a consultant ecologist on a project funded by a government agency. Dr Bachheti is currently a Professor at Graphic Era University, Dehradun, India. She has also served in many capacities in academia within India and provided expertise internationally for more than 15 years where she taught Ecology and Environment, Environmental Science, Freshwater Ecology, Disaster Management, and Bryophytes and Pteridophytes. Her major research interests encompass the broad, interdisciplinary field of plant ecology, with focus on eco-restoration, green chemistry, especially the synthesis of nanomaterial, and medicinal properties of plants. The breadth of her research spans from degraded land ecological amelioration and physical and chemical properties of plant oil to plant-based nanomaterials. She has guided one PhD student and is currently supervising three scholars as well as guiding graduate and undergraduate students in their research projects. While it was the fascination with forest biodiversity that captured her interest, it has been her love for the exploration of values of biodiversity and social upliftment through it that has maintained that passion. Dr Bachheti has published more than 65 research articles in international and national journals along with ten book chapters. She has organized several national seminars/conferences at Graphic Era University, India.

Rakesh Kumar Bachheti graduated from the Hemwati Nandan Bahuguna University, Garhwal, India in 1996. He completed his MSc in Organic Chemistry from Hemwati Nandan Bahuguna University, Garhwal, India in 1998. He undertook a one-year Post Graduate Diploma in Pulp and Paper Technology from the Forest Research Institute, Dehradun, India in 2001. He obtained his PhD in organic chemistry from Kumaun University, Nainital, India in 2007. He is presently working as an Associate Professor of Organic Chemistry in the Department of Industrial Chemistry at the Addis Ababa Science and Technology University (AASTU) of Ethiopia, where he teaches PhD, graduate, and undergraduate students. Before joining AASTU, Dr Bachheti was Dean Project (Assistant) in Graphic Era University (accredited by the National Assessment and Accreditation Council as 'A' grade), Dehradun, India. He has also presented papers at various international (Malaysia, Thailand, and India) and national conferences and been a member of various important committees such as the Internal Quality Assurance Cell (IQAC) and the Antiaging Committee. His major research interests include natural products for industrial application, biofuel and bioenergy, green synthesis of nanoparticles, their application, and pulp and paper technology. He retains a fundamental love for natural products, which permeates all of his research. He has also successfully advised 30 MSc and three PhD students to completion and countless undergraduates have performed research in his laboratory. Dr Bachheti is actively involved in curriculum development for BSc/MSc/PhD programmes. He has over 76 publications dealing with various aspects of natural product chemistry and nanotechnology and has eight book chapters published by Springer, Elsevier, and Nova Science Publishers. He is currently supervising three

PhD students and three Master's students and is also working on two research projects funded by Addis Ababa Science and Technology University, Ethiopia.

Azamal Husen served as Professor and Head of the Department of Biology, University of Gondar, Ethiopia and is a Foreign Delegate at Wolaita Sodo University, Wolaita, Ethiopia. Previously, he was a Visiting Faculty of the Forest Research Institute, and the Doon College of Agriculture and Forest at Dehra Dun, India. His research and teaching experience of 20 years involves studies of biogenic nanomaterial fabrication and application, plant responses to environmental stresses and nanomaterials at the physiological, biochemical, and molecular levels, herbal medicine, and clonal propagation for improvement of tree species.

He has conducted several research projects sponsored by various funding agencies, including the World Bank (FREEP), the National Agricultural Technology Project (NATP), the Indian Council of Agriculture Research (ICAR), the Indian Council of Forest Research Education (ICFRE), and the Japan Bank for International Cooperation (JBIC). He received four fellowships from India and a recognition award from the University of Gondar, Ethiopia, for excellent teaching, research, and community service. Dr Husen has been on the Editorial board and the panel of reviewers of several reputed journals published by Elsevier, Frontiers Media, Taylor & Francis, Springer Nature, RSC, Oxford University Press, Sciendo, The Royal Society, CSIRO, PLOS, MDPI, John Wiley & Sons, and UPM Journals. He is on the advisory board of Cambridge Scholars Publishing, UK. He is a Fellow of the Plantae group of the American Society of Plant Biologists, and a Member of the International Society of Root Research, Asian Council of Science Editors, and INPST. He has over 200 publications to his credit and is Editor-in-Chief of the *American Journal of Plant Physiology*.

He is also working as Series Editor of *Exploring Medicinal Plants*, published by Taylor & Francis Group, USA; *Plant Biology, Sustainability, and Climate Change*, published by Elsevier, USA; and *Smart Nanomaterials Technology*, published by Springer Nature Singapore Pte Ltd. Singapore.

1 Phytochemicals and Medicinal Importance of *Nelumbo nucifera*

Tuyelee Das,[1] Sailky Sau,[1] Samapika Nandy,[1] Mimosa Ghorai,[2] Sayanti Mandal,[2] Abdel Rahman Al-Tawaha,[3] Ercan Bursal,[4] Vartika Jain,[5] Devendra Kumar Pandey,[6] Mallappa Kumara Swamy,[7] Mahipal S. Shekhawat,[8] Tabarak Malik,[9] Arabinda Ghosh,[10] Rahul Bhattacharjee,[11] and Abhijit Dey[1,]*

[1] Department of Life Sciences, Presidency University, Kolkata, West Bengal, India

[2] Institute of Bioinformatics Biotechnology (IBB), Savitribai Phule Pune University (SPPU), Pune, India

[3] Department of Biological Sciences, Al-Hussein Bin Talal University, Maan, Jordon

[4] Department of Biochemistry, Mus Alparslan University, Mus, Turkey

[5] Department of Botany, Government Meera Girls' College, Udaipur, India

[6] Department of Biotechnology, Lovely Professional University, Punjab, India

[7] Department of Biotechnology, East West First Grade College of Science (Bangalore University), Bengaluru, Karnataka, India

[8] Plant Biotechnology Unit, KM Government Institute for Postgraduate Studies and Research, Puducherry, India

[9] Department of Biochemistry, College of Medicine & Health Science, School of Medicine, University of Gondar, Ethiopia

[10] Department of Botany, Gauhati University, Guwahati, Assam, India

[11] Tel Aviv University, Department of Cell and Developmental Biology, Sackler Faculty of Medicine, Tel Aviv University, Israel

[*] Corresponding author (Abhijit Dey): abhijit.dbs@presiuniv.ac.in

1.1 INTRODUCTION

The species *Nelumbo nucifera* Gaertn. belongs to the monocotyledonous family Nelumbonaceae. This is one of the most common aquatic plants distributed and cultivated in Asia, the Americas, and Oceania. Its common parts include leaves, seedpods, rhizomes, flowers, plumules, and stalks. As an ornamental plant, lotus flowers have beautiful colours and a pleasant fragrance. In India, the lotus is normally called kamala or padma, depending on the region. There are two different types of kamalas. In regard to the two varieties, the first one has white flowers and is often referred to as 'pundarika', while the second variety has pink or reddish-pink flowers and is referred to as 'rakta kamala' (Chopra 1958, p.679; Arya et al. 2022). Numerous medicinal properties are associated with *N. nucifera*. Among the main parts of the lotus plant, rhizome and seeds are the most edible. This plant has gained considerable significance due to its high concentration of bioactive secondary metabolites. It

DOI: 10.1201/9781003256830-1

is traditionally used to treat pectoralgias, spermatorrhoeas, pharyngopathies, small poxes, dysentery, coughs, haematemesis, epistaxis, haemoptysis, erythrocyturia, hypermenorrhoea, fevers, cholera, and hyperdipsia. The Ayurvedic system of medicine uses this plant to treat leprosy, headaches, nausea, vomiting, nervous exhaustion, and other ailments. China, India, and Egypt are the three countries honouring this beautiful aquatic flowering plant throughout history. All of these cultures depict this flower as a symbol of perfection, purity, and beauty (Harer 1985; Karki et al. 2012, 2013). In recent years, the leaves of *N. nucifera* have received extensive attention due to them being a potential source of bioactive components. The physicochemical properties of nuciferine have been studied concerning its possible uses in food and bioproducts (Gao et al. 2020). In this chapter, we examine the phytochemistry, bioactive compounds, and pharmacological properties of *N. nucifera* to facilitate its further development and application in the food, health product, and medicine industries, among others.

1.2 PHYSICAL CHARACTERISTICS AND DESCRIPTION

Nelumbo leaves are 20–90 cm in diameter and come in two varieties: aerial and floating. Unlike the aerial leaves, which are cup-shaped and measure 24–30 cm in diameter, the floating leaves are flat and measure 23–30 cm in diameter. These leaves are petiolate and orbicular. There are small brown dots on the petioles of the aerial leaves, which are erect, smooth, greenish or greenish-brown, and sometimes rough. Lotus plants have solitary, ovoid, smooth, bisexual flowers which range in colour from white to rosy pink. *Neolumbo*'s floral arrangement is polysymmetric, with some organs emerging spirally (petals) and others in parallel whorls (stamens and carpels) (Hayes et al. 2000; Bishayee et al. 2022) (Figure 1.1). It is traditionally eaten as a vegetable and used for its medicinal properties. It is known that the roots of tuberculous plants appear yellowish-white to yellowish-brown in colour. The inner part of the roots contains several large air pockets (Paudel and Panth 2015). Fruits are hard,

FIGURE 1.1 Photographs of lotus (*N. nucifera*). (A) Flower, (B) leaves, (C) rhizomes, and (D) natural habitat. (Adapted from Bishayee et al. (2022).)

brown, one-seeded, aggregated, ovoid, round, or oblong. The seeds of this plant live longer than the seeds of any other flowering plant species, despite its extraordinary dormancy (Loewer 2005).

1.3 CHEMICAL CONSTITUENTS OF *NELUMBO NUCIFERA*

The phytochemical analysis of *N. nucifera* revealed the presence of a range of phytochemicals such as alkaloids, flavonoids, glycosides, triterpenoids, vitamins, and minerals. The rhizome, seeds, and leaves of lotus plants have received significant attention due to their medicinal properties in the food and pharmaceutical industries over the past few decades. This aquatic sacred lotus medicinal plant prompted chemists and pharmacologists around the globe to perform chemical and pharmacological studies, especially on alkaloids and flavonoids. By using (high-performance liquid chromatography (HPLC), high-performance liquid chromatography–diode array detection–electro-spray ionization–mass spectrometry (HPLC-DAD-ESI-MS), high-performance thin-layer chromatography (HPTLC), liquid chromatography–electrospray ionization–tandem mass spectrometry (LC-ESI-MS/MS), gas chromatography–mass spectrometry (GC-MS), and nuclear magnetic resonance, many secondary metabolites have been characterized as well as authenticated. Essential secondary metabolites (Figure 1.2) and their nutraceutical values are tabulated (Table 1.1).

In several Asian countries, lotus plants' rhizomes are consumed as vegetables. Lotus plants' rhizomes are considered healthy for various reasons, including their high mineral content. Lotus rhizomes mainly consist of starch and various trace elements. Starch obtained from *N. nucifera* is used as an adjuvant in the preparation of tablets. The methanol extract of the rhizome yields triterpenoid steroids and betulinic acid (Mukherjee et al. 1996a). A new triterpenoid ester, ursane triterpene, has been found in the rhizomes of this plant, along with many other compounds, including linoleic acid, palmitic acid, and α-amyrin (Chaudhuri et al. 2009).

GC-MS has been used to examine the chemical composition of *N. nucifera*. Lotus receptacles contain procyanidins as their main active ingredient (Zheng et al. 2010). The essential oil components obtained from flowers of *N. nucifera* from Wuhan, Hubei province, China included *Z,Z*-10,12-hexadecadienal and *E*-14-hexadecenal. Several different extraction methods were used to extract the essential oil components from flowers, including headspace, steam distillation, and solvent extraction. As a result of the study, the headspace method showed potential to be used as an *in situ* and a simple method for extracting essential oil ingredients from raw materials (Zhang and Guo et al. 2020). Nineteen chemical constituents isolated from *N. nucifera* seeds mainly comprised oxygenated sesquiterpenes. The medicinally active constituents were 1,8-cineole, α-terpeneol, α-asarone, borneol, and γ-gurjunene (Khan et al. 2016).

Among the most important secondary metabolites of lotus plumules are flavonoids. Lotus plumules were analysed by HPLC combined with ESI-MS/MS to identify five flavonoid-*O*-glycosides, i.e. isoquercitrin, kaempferol 3-*O*-robinobioside, isorhamnetin 3-*O*-rutinoside, rutin, and hyperoside (Chen et al. 2012). Zhu et al. (2017) demonstrated that flavonoid *C*-glycosides were more potent antioxidants than flavonoid *O*-glycosides from different lotus extracts. Lotus flavonoids have been relatively little studied for their biological activities due to *C*-glycosylation. A distinguishable difference between flavonoid *O*-glycosides and *C*-glycosides in lotus is that flavonoid *C*-glycosides are preferentially accumulated in the plumules (where more than 70% of total flavonoid content is concentrated). In contrast, leaves accumulate high amounts of flavonoids but exclusively flavonoid *O*-glycosides (Zhu et al. 2017). Among the flavonoids in lotus leaves, hyperoside, isoquercitrin, and astragalin showed significant antioxidative and antiviral effects on rats' visceral adipose tissue, and some flavonoids, such as (+)-catechin, exhibited lipolytic activity, especially in rats' visceral adipose tissue (Ohkoshi et al. 2007). Flavonoids have been detected in the seed oil of lotus plants using Soxhlet's procedure (Arumugam et al. 2012). Zhu et al. (2015) analysed a mixture of flavonoids extracted from lotus leaves by HPLC-MS/MS and bioactivity tests, following their selective enrichment in a microporous resin column. Four new flavonoids and 14 flavonoid glycosides with five

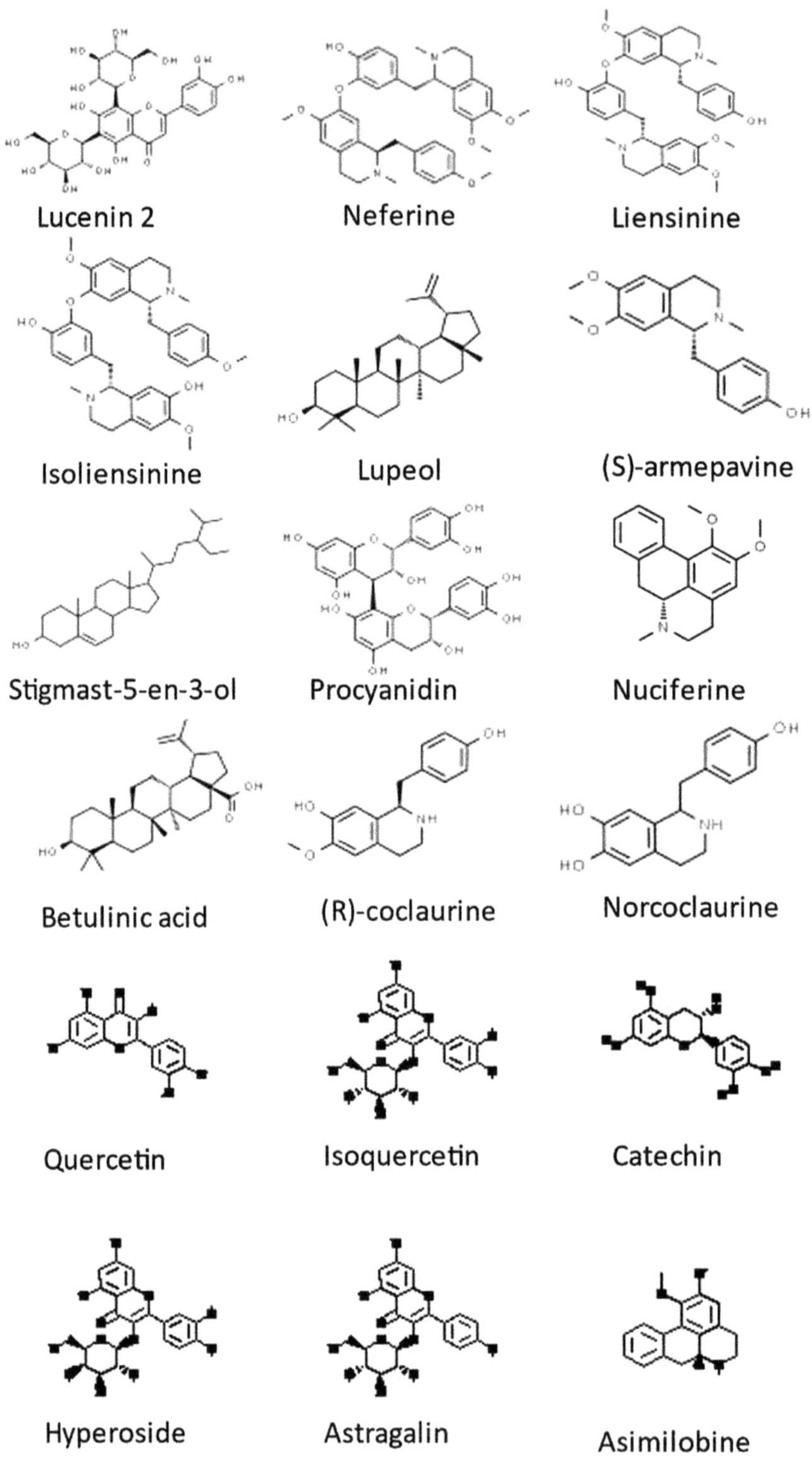

FIGURE 1.2 Essential secondary metabolites of *N. nucifera*.

TABLE 1.1
Identified Secondary Metabolites and Their Medicinal Properties

Plant part	Phytochemicals	Class of chemical	Medicinal properties	References
Embryo	Nuciferine	Aporphine alkaloids	Anti-HIV activity	Kashiwada et al. 2005
	Isoliensinine	Alkaloids	Antiproliferative activity	Xiao et al. 2006
Flower	*Z,Z*-10,12-Hexadecadienal, *E*-14-hexadecenal	Essential oil	–	Zhang and Guo 2020
Leaf	(+)-1(R)-Coclaurine and (−)-1(S)-Norcoclaurine	Alkaloids	Anti-HIV activity	Kashiwada et al. 2005
	7-Hydroxydehydronuciferine	Alkaloids	Antiproliferative activity	Liu et al. 2014
	Quercetin, isoquercitrin, catechin, hyperoside, astragalin	Flavonoids	Anti-obesity activity	Ohkoshi et al. 2007
	Asimilobine, alirinidine	Alkaloids	Serotonin antagonists	Shoji et al. 1987
Seed	Liensinine	Alkaloids	Anti-arrhythmic activity	Wang et al. 1992
	Lucenin 2	Flavonoids	Antibacterial activity	Basile et al. 1999
	Isoliensinine	Benzylisoquinoline alkaloids	Inhibits pulmonary fibrosis	Xiao et al. 2005
			Cytotoxic effect	Zhang et al. 2015
	(S)-Armepavine	Alkaloids	Prevents immune diseases	Liu et al. 2007
	Neferine	Alkaloids	Antifever effects	Sugimoto et al. 2008
	Lupeol	Triterpenes	Anticancer, antiprotozoal, anti-inflammatory, antimicrobial activity	Gallo and Sarachine 2009
	Stigmast-5-en-3-ol	Phytosterols	Antidiabetic property	Sujatha et al. 2010
	1,8-Cineole, α-terpeneol, α-asarone, borneol, γ-gurjunene	–	Inhibits acetylcholinesterase, butyrylcholinesterase	Khan et al. 2016
Seed pod	Procyanidin	Condensed tannins	Superoxide Radical scavenging activity	Ling et al. 2005
Stamen	Myr-3-*O*-Glc, rutin, quer-3-*O*-Glu	Flavonoids	Anti-ageing activity	Tungmunnithum et al. 2022

different types of aglycones were found in lotus leaves via LC-MS/MS (Zhu et al. 2015). In addition, the main active alkaloid in lotus leaves, nuciferine, can alleviate dyslipidaemia and liver steatosis in hamsters by inhibiting the expression of hepatic genes involved in lipid metabolism (Guo et al. 2013). Flavonoids are abundant in lotus epicarps. Cheng et al. (2012) identified 11 flavonoids in lotus epicarps (flavonoids). In different stages of ripening, catechins, epicatechins, hyperosides, and isoquercitrin are present in the epicarps. Flavan-3-ol levels decreased, but quercetin glucoside levels increased as the seeds matured (Liu et al. 2015). Recently, Zhu et al. (2019) performed metabolomic analyses in lotus cultivars with yellow and white petals flowers at five stages of flower colouration. They observed the formation of dihydroflavonols and flavonols opposing significantly between cultivars (Zhu et al. 2019). The plant also possesses phenolics, alkaloids, polysaccharides, saponins, and carbohydrates (Rai et al. 2006). 9-octadecadienoic acid and 2,2,4-trimethyl-3-(3,8,12,16-tetramethyl-heptadeca-3,7,11,15-tetraenyl)-cyclohexanol were obtained from flower receptacles of *N. nucifera* (Krubha et al. 2016). Seed polysaccharides were shown to contain mainly four types of monosaccharides, including D-mannose, D-galactose, L-arabinose, and D-glucose, through acid hydrolysis and methylation (Das et al. 1992). Nuciferine was the first alkaloid isolated from *N. nucifera* leaves. As revealed by leaf extract analysis, the alkaloids identified can be classified into three types: aporphine alkaloids, benzylisoquinoline alkaloids, and bisbenzylisoquinoline alkaloids. A number of aporphine alkaloids have been found to make up most of the active compounds in lotus leaves, including nornuciferine, anonaine, nuciferine, and roemerine (Chen et al. 2013).

1.4 MEDICINAL PROPERTIES OF *NELUMBO NUCIFERA*

1.4.1 Anti-ageing

The sacred lotus seed extract contains anti-ageing ingredients that can help alleviate symptoms such as loss of pliability, fine lines, acne, pores, wrinkles, and flaws. This promotes the appearance of younger skin. Water extract from *N. nucifera* leaf, flower, and the seed was tested for possible functional cosmetic agents and showed an anti-wrinkle effect by inhibiting elastase and adenosine at a dose of 200 μg/ml (Kim et al. 2011). Using a non-invasive device called the Visioscan VC and software to measure the skin surface, we investigated the efficacy of cosmetic formulations containing green tea and lotus extract to reduce facial wrinkles in healthy Asian individuals (Mahmood and Akhtar 2013). In comparison with its whole flower, an ethanolic extract of *N. nucifera* stamens exhibited more significant inhibition of tyrosinase and collagenase (Tungmunnithum et al. 2022).

1.4.2 Antidiabetic Activity

An ethanol extract of *N. nucifera* seed ash (200 mg/kg body weight, taken orally for 30 days) inhibited streptozotocin-induced diabetes in rats. It has been reported that trace elements found in considerable amounts in seeds may have a direct or indirect effect on insulin or act synergistically with insulin (Mani et al. 2010). It has been shown that leaf extract can lower blood insulin levels and glucose levels in pregnant rats with gestational diabetes mellitus (Zeng et al. 2017). Methanolic extract from leaves regulates glucose levels in mice with a high-fat diet. Catechin in the extract significantly enhanced insulin secretion and reversed the glucose intolerance in a dose-dependent manner (Huang et al. 2011).

1.4.3 Anticancer Activity

There is evidence that the anticancer properties of the ethanolic extract of *N. nucifera* are probably due to its ability to quench free radicals. Lisinine, isoliensinine, and neferine, the alkaloids found in the leaves of *N. nucifera*, inhibit cancer cell growth by enhancing reactive oxygen species (ROS) production, activating MAP kinases, and promoting autophagy and apoptosis (Arjun et al. 2012). The ethanolic extract of *N. nucifera* stamens showed dose-dependent anticancer activity (100, 200, and 400 g/ml) against human colon carcinoma HCT116 cells. Several factors associated with apoptosis, including death receptors, fas ligands, caspases (3, 8, and 9), and Bcl-2, also increased (Zhao et al. 2017). Neferine, derived from the embryos of *N. nucifera*, has been reported to exhibit cytotoxicity against hepatocellular carcinoma Hep3B cells by downregulating c-Myc, cyclin D1, D3, CDK4, E2F-1 and upregulating Bim, Bid, Bax, Bak, Puma, caspase-3, -6, -7, -8, and PARP, and the protein expression levels of Bip, calnexin, PDI, calpain-2, and caspase-12 (Yoon et al. 2013).

1.4.4 Antidiarrhoeal Activity

In rats, ethanol extracts of *N. nucifera* rhizomes showed significant antidiarrhoeal properties. In addition, the extract significantly inhibited castor oil-induced diarrhoea and PGE2-induced enteropooling; charcoal meal propulsive movements were also reduced (Mukherjee et al. 1995).

1.4.5 Anti-inflammatory Activity

Leaves, seeds, and seedpods of *N. nucifera* dramatically decreased the protein expression of pro-inflammatory genes such as iNOS and COX-2 and it further inhibits inflammation in LPS-stimulated RAW 264.7 cells via regulating MAPKs, NF-κB, and Nrf2/HO-1 pathways (Lee et al. 2019). The methanolic rhizome extract of *N. nucifera*, at doses of 200 and 400 mg/kg, showed significant anti-inflammatory activity in carrageenin and serotonin-induced rat paw oedema. The effect was comparable to that of the standard drugs phenylbutazone and dexamethasone (Mukherjee et al. 1996c). Further, seed ethanol extract (200, 300, 400 mg/kg) was tested for anti-inflammatory activity in carrageenan-induced male white rats. It seems that 400 mg/kg is the most effective dose for inhibiting inflammation (Fitri et al. 2021).

1.4.6 Anti-obesity Activity

Based on Ono et al.'s (2006) study of leaf extracts, the effects of the extracts on digestion enzyme inhibition and lipid metabolism were observed in high-fat diet-induced obese mice. The leaf extract prevents increased liver triacylglycerol levels, body weight, and fat percentage. This was due to higher UCP3 mRNA expression in skeletal muscle (Ono et al. 2006). The effectiveness of successive ethanol extracts of the root of *N. nucifera* against obesity was investigated by You et al. (2014b). The extracts were studied for their effects on adipogenes. There was a significant reduction in relative body weights in rat fat tissues. This came about because cholesterol levels and triglycerides were reduced significantly in rats with fatty diet (You et al. 2014a).

1.4.7 Diuretic Activity

Methanolic extracts of rhizome of *N. nucifera* induce a noteworthy diuresis in a dose-dependent manner (300, 400, or 500 mg/kg) in rats. Na^+, Cl^-, and K^+ excretion were associated with dose-dependent increases in urinary volume. There was less increase in urine volume when using furosemide (20 mg/kg) as a standard diuretic (Mukherjee et al. 1996b).

1.4.8 Hepatoprotective Activity

The administration of ethanol root extract at 300 mg/kg and 500 mg/kg for 15 days significantly inhibited carbon tetrachloride (CCl_4) hepatotoxicity in rats (Huang et al. 2010). A rat hepatocyte toxicity model using CCl_4 and aflatoxin B1 (AFB1)-induced hepatocyte damage was shown to exhibit antioxidant and antihepatotoxic effects by an ethanol extract with a dose-dependent manner ranging from 10 to 500 mg/ml (Sohn et al. 2003). Lotus leaf extracts have hepatoprotective activity due to their free radical scavenging and antioxidant properties. This may be due to some flavonoids and phenolic compounds found in the leaves. The polyphenol-rich butanolic extract from lotus leaves contains phenolic compounds such as quercetin, catechin, ferulic acid, rutin, and protocatechuic acid, which were tested for hepatoprotective effects, and was found to reduce lipid peroxidation, inhibit ROS formation, and increase glutathione levels in cells when exposed to oxidative stress (Je et al. 2015). Rats were significantly protected against CCl_4 and paracetamol-induced hepatotoxicity by oral administration of a 50% hydroalcoholic extract of *N. nucifera* flowers (200 and 400 mg/kg). Flower extracts possess hepatoprotective effects because of the control of lipid peroxidation, inhibition of cytochrome P450 activity, stabilization of hepatocellular membranes, and enhanced protein synthesis (Rao et al. 2005).

1.4.9 Antimicrobial Activity

The ethanolic extract of *N. nucifera* flowers has significant antibacterial and antifungal activity against five and two important strains with concentrations of 500 μg/ml and 1000 μg/ml, respectively.

Six bacterial strains were inhibited by leaf extracts of *N. nucifera* with minimum inhibitory concentration (MIC) values ranging from 4 mg/ml to 10 mg/ml (Lee et al. 2006). *Staphylococcus aureus* and *Escherichia coli* were inhibited by MICs of 430 μg and 450 μg, respectively (Brindha et al. 2010). LB2 (polysaccharide-protein complex) from a *N. nucifera* decoction inhibited HIV-1 transcription by inhibiting reverse transcriptase and integrase with an IC_{50} value of 5.28 μM through modulation of IL2B, IL4B, and IL-10. LB2 showed anti-HIV-1 activity by downregulating tumour necrosis factor α (Jiang et al. 2011).

1.4.10 Antipyretic Activity

Yeast-induced pyrexia in rats was treated with methanolic extracts of *N. nucifera* rhizome stalks (200 and 400 mg/kg, oral). The results were comparable to those obtained by administering 150 mg/kg of paracetamol intraperitoneally as a standard antipyretic (Sinha et al. 2000). Alcoholic extracts (400 and 600 mg/kg, oral) were evaluated for brewer's yeast-induced pyrexia and showed that 400 and 600 mg/kg are effective in reducing body temperature (Deepa et al. 2009).

1.4.11 CONCLUSION

N. nucifera is gaining popularity due to its historical relevance and its potential use in nutraceuticals. Many parts of *N. nucifera*, including the leaves, rhizomes, seeds, stamens, stalk, and flowers, have been documented as having potential therapeutic properties for treating various diseases worldwide. In addition, *in vitro* and *in vivo* studies have demonstrated that various extracts (mostly flavonoids and alkaloids) isolated from lotus plants possess health-promoting biological and therapeutic properties. Furthermore, it is a natural source that is reported to be far less toxic than those extracted from artificial sources. Many *in vitro* and *in vivo* tests have been conducted using lotus extracts to demonstrate their pharmacological properties. Many chemical groups of bioactive compounds are responsible for the effects, but alkaloids, flavonoids, glycosides, triterpenoids, vitamins, and minerals dominate. In view of this, *N. nucifera* is well researched in terms of its pharmacological effects and active ingredients in its various parts. In the food and pharmaceutical industries, lotus compounds have become increasingly popular for their nutritional properties. As a result of deep processing the lotus products, many residuals are left over (lotus rhizome skins, residues, nodes, seedpods, stems, flowers, stamens, and leaves). This not only pollutes the environment but also contributes to serious resource waste. Considering the use of *N. nucifera* in traditional medicine and its newly determined pharmacological activities, it is imperative to assess its therapeutic value further and develop protocols for the efficient extraction and validation of the active principles for their use in combating different diseases in humans. The cultivation of lotus plants should be promoted at a large scale so that it can be consumed as a low-cost nutritious food and used as a low-cost pharmaceutical to treat a wide range of disorders and diseases.

REFERENCES

Arjun, P., Saranya Sivan, P.S., Mohana Priya, S., Krishnamoorthy, M. and Balasubramanian, K. 2012. Phytochemical analysis and anticancer activity of *Nelumbo nucifera* extracts. *Journal of Scientific and Industrial Research* 1(2):81–85.

Arumugam, A. and Dhailappan, A. 2012. Fatty acid composition and antidermatophytic and antidiarrheal activity of *Nelumbo nucifera* seed oil. *International Journal of Pharmacy and Pharmaceutical Sciences* 4:264–270.

Arya, A.K., Durgapal, M., Bachheti, A., Joshi, K.K., Gonfa, Y.H., Bachheti, R.K. and Husen, A. 2022. Ethnomedicinal use, phytochemistry, and other potential application of aquatic and semiaquatic medicinal plants. *Evidence-Based Complementary and Alternative Medicine* 2022:4931556.

Basile, A., Giordano, S., López-Sáez, J.A. and Cobianchi, R.C. 1999. Antibacterial activity of pure flavonoids isolated from mosses. *Phytochemistry* 52(8):1479–1482.

Bishayee, A., Patel, P.A., Sharma, P., Thoutireddy, S. and Das, N. 2022. Lotus (*Nelumbo nucifera* Gaertn.) and its bioactive phytocompounds: a tribute to cancer prevention and intervention. *Cancers* 14(3):529.

Brindha, D. and Arthi, D. 2010. Antimicrobial activity of white and pink *Nelumbo nucifera* gaertn flowers. *Asian Journal of Pharmaceutical Research and Health Care* 2(2):147–155.

Chaudhuri, P.K. and Singh, D. 2009. A new lipid and other constituents from the rhizomes of Nelumbo nucifera. *Journal of Asian natural products research* 11(7):583–587.

Chen, S., Zhang, H., Liu, Y., Fang, J. and Li, S. 2013. Determination of lotus leaf alkaloids by solid phase extraction combined with high performance liquid chromatography with diode array and tandem mass spectrometry detection. *Analytical Letters* 46(18):2846–2859.

Cheng, L., Li, Y., Yin, J., Chen, X. and Li, L. 2012. Expression of genes related to starch synthesis during rhizome swelling of lotus root (Nelumbo nucifera Gaerfn). *Scientia Agriculture Sinica* 45(16):3330–3336.

Chopra RN. *Indigenous Drugs of India*, 2nd edn. Calcutta: U.N. Dhur and Sons Pvt, 1958.

Das, S., Ray, B. and Ghosal, P.K. 1992. Structural studies of a polysaccharide from the seeds of *Nelumbo nucifera. Carbohydrate Research* 224:331–335.

Deepa, P.K., Usha, P.T.A., Nair, A.C. and Prasannakumari, K.T. 2009. Antipyretic activity of seeds from red and white type of lotus (*Nelumbo nucifera*) in Albino rat. *Veterinary World* 2(6):213–214.

Fitri, K., Khairani, T.N., Sianturi, K.T., Leny, L. and Hafiz, I. 2021. Anti-inflammatory activity of ethanol extract of Lotus (*Nelumbo nucifera* G.) seed against white male rats using paw edema method. *Journal of Drug Delivery and Therapeutics* 11(4):1–4.

Gallo, M.B. and Sarachine, M.J. 2009. Biological activities of lupeol. *International Journal of Biomedical and Pharmaceutical Sciences* 3(1):46–66.

Gao, H., Zhang, L., Zhu, A., Liu, X., Wang, T., Wan, M., Yang, X., Zhang, Y. and Zhang, Y. 2020. Metabolic profiling of nuciferine in vivo *and in vitro. Journal of Agricultural and Food Chemistry* 68(48):14135–14147.

Guo, F.C., Yang, X., Li, X.X., Feng, R.N., Guan, C.M., Wang, Y.W., Li, Y., and Sun, C.H. 2013. Nuciferine prevents hepatic steatosis and injury induced by a high-fat diet in hamsters. *PLoS One* 8(5):e63770.

Harer, W.B. 1985. Pharmacological and biological properties of the Egyptian lotus. *Journal of the American Research Center in Egypt* 22:49–54.

Hayes, V., Schneider, E.L. and Carlquist, S. 2000. Floral development of *Nelumbo nucifera* (Nelumbonaceae). *International Journal of Plant Sciences* 161(S6):S183–S191.

Huang, B., Ban, X., He, J., Tong, J., Tian, J. and Wang, Y. 2010. Hepatoprotective and antioxidant activity of ethanolic extracts of edible lotus (*Nelumbo nucifera* Gaertn.) leaves. *Food Chemistry* 120(3):873–878.

Huang, C.F., Chen, Y.W., Yang, C.Y., Lin, H.Y., Way, T.D., Chiang, W. and Liu, S.H. 2011. Extract of lotus leaf (*Nelumbo nucifera*) and its active constituent catechin with insulin secretagogue activity. *Journal of Agricultural and Food Chemistry* 59(4):1087–1094.

Je, J.Y., and Lee, D.B. 2015. *Nelumbo nucifera* leaves protect hydrogen peroxide-induced hepatic damage via antioxidant enzymes and HO-1/Nrf2 activation. *Food & Function* 6(6):1911–1918.

Jiang, Y., Ng, T.B., Liu, Z., Wang, C., Li, N., Qiao, W. and Liua, F. 2011. Immunoregulatory and anti-HIV-1 enzyme activities of antioxidant components from lotus (*Nelumbo nucifera* Gaertn.) rhizome. *Bioscience Reports* 31(5):381–390.

Karki, R., Jeon, E.R. and Kim, D.W. 2013. *Nelumbo nucifera* leaf extract inhibits neointimal hyperplasia through modulation of smooth muscle cell proliferation and migration. *Nutrition* 29(1):268–275.

Karki, R., Jung, M.A., Kim, K.J. and Kim, D.W. 2012. Inhibitory effect of *Nelumbo nucifera* (Gaertn.) on the development of atopic dermatitis-like skin lesions in NC/Nga mice. *Evidence-Based Complementary and Alternative Medicine* 2012:153568.

Kashiwada, Y., Aoshima, A., Ikeshiro, Y., Chen, Y.P., Furukawa, H., Itoigawa, M., Fujioka, T., Mihashi, K., Cosentino, L.M., Morris-Natschke, S.L. and Lee, K.H. 2005. Anti-HIV benzylisoquinoline alkaloids and flavonoids from the leaves of *Nelumbo nucifera*, and structure-activity correlations with related alkaloids. *Bioorganic & Medicinal Chemistry* 13(2):443–448.

Khan, S., Khan, H., Ali, F., Ali, N., Khan, F.U. and Khan, S.U. 2016. Antioxidant, cholinesterase inhibition activities and essential oil analysis of *Nelumbo nucifera* seeds. *Natural Product Research* 30(11):1335–1338.

Kim, T., Kim, H.J., Cho, S.K., Kang, W.Y., Baek, H., Jeon, H.Y., Kim, B. and Kim, D. 2011. *Nelumbo nucifera* extracts as whitening and anti-wrinkle cosmetic agent. *Korean Journal of Chemical Engineering* 28(2):424–427.

Krubha, A. and Vasan, P.T. 2016. Phytochemical analysis and anticancer activity of *Nelumbo nucifera* floral receptacle extracts in MCF-7 Cell Line. *Journal of Academia and Industrial Research (JAIR)* 4(12):251.

Lee, E.J., Seo, Y.M., Kim, Y.H., Chung, C., Sung, H.J., Sohn, H.Y., Park, J.Y. and Kim, J.S. 2019. Anti-inflammatory activities of ethanol extracts from leaf, seed, and seedpod of *Nelumbo nucifera. Journal of Life Science* 29(4):436–441.

Lee, K.S., Oh, C.S. and Lee, K.Y. 2006. Antimicrobial effect of the fractions extracted from a lotus (*Nelumbo nucifera*) leaf. *Journal of the Korean Society of Food Science and Nutrition* 35(2):219–223.

Ling, Z.Q., Xie, B.J. and Yang, E.L. 2005. Isolation, characterization, and determination of antioxidative activity of oligomeric procyanidins from the seedpod of *Nelumbo nucifera* Gaertn. *Journal of Agricultural and Food Chemistry* 53(7):2441–2445.

Liu, C.M., Kao, C.L., Wu, H.M., Li, W.J., Huang, C.T., Li, H.T. and Chen, C.Y. 2014. Antioxidant and anticancer aporphine alkaloids from the leaves of *Nelumbo nucifera* Gaertn. cv. Rosa-plena. *Molecules* 19(11):17829–17838.

Liu, C.P., Kuo, Y.C., Shen, C.C., Wu, M.H., Liao, J.F., Lin, Y.L., Chen, C.F. and Tsai, W.J. 2007. (S)-armepavine inhibits human peripheral blood mononuclear cell activation by regulating Itk and PLCγ activation in a PI-3K-dependent manner. *Journal of Leukocyte Biology* 81(5):1276–1286.

Liu, Y., Ma, S.S., Ibrahim, S.A., Li, E.H., Yang, H. and Huang, W. 2015. Identification and antioxidant properties of polyphenols in lotus seed epicarp at different ripening stages. *Food Chemistry* 185:159–166.

Loewer, P. 2005. *Seeds: The Definitive Guide to Growing, History and Lore.* Portland, OR: Timber Press.

Mahmood, T. and Akhtar, N. 2013. Combined topical application of lotus and green tea improves facial skin surface parameters. *Rejuvenation Research* 16(2):91–97.

Mani, S.S., Subramanian, I.P., Pillai, S.S. and Muthusamy, K. 2010. Evaluation of hypoglycemic activity of inorganic constituents in *Nelumbo nucifera* seeds on streptozotocin-induced diabetes in rats. *Biological Trace Element Research* 138(1):226–237.

Mukherjee, K., Das, J.B.R.S., Balasubramanian, R., Kakali, S., Pal, M. and Saha, B.P. 1995. Antidiarrhoeal evaluation of *Nelumbo nucifera* rhizome extract. *Indian Journal of Pharmacology* 27(4):262.

Mukherjee, P.K., Giri, S.N., Saha, K., Dutta, M.S., Pal, M. and Saha, B.P. 1996a. Pharmaceutical application of starch isolated from *Nelumbo nucifera* Gaertn (Fam. Nymphaeaceae). *Indian Journal of Pharmaceutical Sciences* 58(2):59.

Mukherjee, P.K., Pal, M., Saha, K., Saha, B.P. and Das, J. 1996b. Diuretic activity of extract of the rhizomes of *Nelumbo nucifera* Gaertn. (Fam. Nymphaeaceae). *Phytotherapy Research* 10(5):424–425.

Mukherjee, P.K., Saha, K., Balasubramanian, R., Pal, M. and Saha, B.P. 1996c. Studies of psychopharmacological effects of *Nelumbo nucifera* Gaertn. rhizome extract. *Journal of Ethnopharmacology* 54(2–3):63–67.

Ohkoshi, E., Miyazaki, H., Shindo, K., Watanabe, H., Yoshida, A. and Yajima, H. 2007. Constituents from the leaves of *Nelumbo nucifera* stimulate lipolysis in the white adipose tissue of mice. *Planta Medica* 73(12):1255–1259.

Ono, Y., Hattori, E., Fukaya, Y., Imai, S. and Ohizumi, Y. 2006. Anti-obesity effect of *Nelumbo nucifera leaves* extract in mice and rats. *Journal of Ethnopharmacology* 106(2):238–244.

Paudel, K.R. and Panth, N. 2015. Phytochemical profile and biological activity of *Nelumbo nucifera. Evidence-Based Complementary and Alternative Medicine* 2015:789124.

Rai, S., Wahile, A., Mukherjee, K., Saha, B.P. and Mukherjee, P.K. 2006. Antioxidant activity of *Nelumbo nucifera* (sacred lotus) seeds. *Journal of Ethnopharmacology* 104(3):322–327.

Rao, G.M.M., Pushpangadan, P. and Shirwaikar, A. 2005. Hepatoprotective activity of *Nelumbo nucifera* Geartn. flower: an ethnopharmacological study. *ACTA Pharmaceutica Sciencia* 47(1):79–88.

Shoji, N., Umeyama, A., Saito, N., Iuchi, A., Takemoto, T., Kajiwara, A. and Ohizumi, Y. 1987. Asimilobine and lirinidine, serotonergic receptor antagonists, from *Nelumbo nucifera. Journal of Natural Products* 50(4): 773–774.

Sinha, S., Mukherjee, P.K., Mukherjee, K., Pal, M., Mandal, S.C. and Saha, B.P. 2000. Evaluation of antipyretic potential of *Nelumbo nucifera* stalk extract. *Phytotherapy Research* 14(4):272–274.

Sohn, D.H., Kim, Y.C., Oh, S.H., Park, E.J., Li, X. and Lee, B.H. 2003. Hepatoprotective and free radical scavenging effects of *Nelumbo nucifera. Phytomedicine* 10(2–3):165–169.

Sugimoto, Y., Furutani, S., Itoh, A., Tanahashi, T., Nakajima, H., Oshiro, H., Sun, S. and Yamada, J. 2008. Effects of extracts and neferine from the embryo of *Nelumbo nucifera* seeds on the central nervous system. *Phytomedicine* 15(12):1117–1124.

Sujatha, S., Anand, S., Sangeetha, K.N., Shilpa, K., Lakshmi, J., Balakrishnan, A. and Lakshmi, B.S. 2010. Biological evaluation of (3β)-STIGMAST-5-EN-3-OL as potent anti-diabetic agent in regulating glucose transport using *in vitro* model. *International Journal of Diabetes Mellitus* 2(2):101–109.

Tungmunnithum, D., Drouet, S. and Hano, C. 2022. Validation of a high-performance liquid chromatography with photodiode array detection method for the separation and quantification of antioxidant and skin anti-aging flavonoids from *Nelumbo nucifera* Gaertn. stamen extract. *Molecules* 27(3):1102.

Wang, J.L., Nong, Y. and Jing, M.X. 1992. [Effects of liensinine on haemodynamics in rats and the physiologic properties of isolated rabbit atria.] *Yao xue xue bao = Acta Pharmaceutica Sinica* 27(12):881–885.

Xiao, J.H., Zhang, J.H., Chen, H.L., Feng, X.L. and Wang, J.L. 2005. Inhibitory effects of isoliensinine on bleomycin-induced pulmonary fibrosis in mice. *Planta Medica* 71(3):225–230.

Xiao, J.H., Zhang, Y.L., Feng, X.L., Wang, J.L. and Qian, J.Q. 2006. Effects of isoliensinine on angiotensin II-induced proliferation of porcine coronary arterial smooth muscle cells. *Journal of Asian Natural Products Research* 8(3):209–216.

Yoon, J.S., Kim, H.M., Yadunandam, A.K., Kim, N.H., Jung, H.A., Choi, J.S., Kim, C.Y. and Kim, G.D. 2013. Neferine isolated from *Nelumbo nucifera* enhances anti-cancer activities in Hep3B cells: molecular mechanisms of cell cycle arrest, ER stress-induced apoptosis and anti-angiogenic response. *Phytomedicine* 20(11):1013–1022.

You, J.S., Lee, Y.J., Kim, K.S., Kim, S.H. and Chang, K.J. 2014a. Anti-obesity and hypolipidaemic effects of *Nelumbo nucifera* seed ethanol extract in human pre-adipocytes and rats fed a high-fat diet. *Journal of the Science of Food and Agriculture* 94(3):568–575.

You, J.S., Lee, Y.J., Kim, K.S., Kim, S.H. and Chang, K.J. 2014b. Ethanol extract of lotus (*Nelumbo nucifera*) root exhibits an anti-adipogenic effect in human pre-adipocytes and anti-obesity and antioxidant effects in rats fed a high-fat diet. *Nutrition Research* 34(3):258–267.

Zeng, Z., Xu, Y. and Zhang, B. 2017. Antidiabetic activity of a lotus leaf selenium (Se)-polysaccharide in rats with gestational diabetes mellitus. *Biological Trace Element Research* 176(2):321–327.

Zhang, C.Y. and Guo, M. 2020. Comparing three different extraction techniques on essential oil profiles of cultivated and wild Lotus (*Nelumbo nucifera*) flower. *Life* 10(9):209.

Zhang, X., Wang, X., Wu, T., Li, B., Liu, T., Wang, R., Liu, Q., Liu, Z., Gong, Y. and Shao, C. 2015. Isoliensinine induces apoptosis in triple-negative human breast cancer cells through ROS generation and p38 MAPK/JNK activation. *Scientific Reports* 5(1):1–13.

Zhao, X., Feng, X., Wang, C., Peng, D., Zhu, K. and Song, J.L. 2017. Anticancer activity of *Nelumbo nucifera* stamen extract in human colon cancer HCT-116 cells *in vitro*. *Oncology Letters* 13(3):1470–1478.

Zheng, Z., M. Wang, D. Wang, W. Duan, X. Wang, and C. Zheng. 2010. Preparative separation of alkaloids from *Nelumbo nucifera* leaves by conventional and pH-zone-refining counter-current chromatography. *Journal of Chromatography B* 878(19):1647–1651.

Zhu, M., Liu, T., Zhang, C. and Guo, M. 2017. Flavonoids of lotus (*Nelumbo nucifera*) seed embryos and their antioxidant potential. *Journal of Food Science* 82(8):1834–1841.

Zhu, M.Z., Wu, W., Jiao, L.L., Yang, P.F. and Guo, M.Q. 2015. Analysis of flavonoids in lotus (*Nelumbo nucifera*) leaves and their antioxidant activity using macroporous resin chromatography coupled with LC-MS/MS and antioxidant biochemical assays. *Molecules* 20(6):10553–10565.

Zhu, H.H., Yang, J.X., Xiao, C.H., Mao, T.Y., Zhang, J. and Zhang, H.Y. 2019. Differences in flavonoid pathway metabolites and transcripts affect yellow petal colouration in the aquatic plant *Nelumbo nucifera*. *BMC Plant Biology* 19(1):1–18.

2 Traditional Uses, Important Phytochemicals, and Therapeutic Profile of *Persicaria hydropiper*

Alam Zeb[1,*] *and Muhammad Ayaz*[2]
[1] Department of Biochemistry, University of Malakand, Khyber Pakhtunkhwa, Pakistan
[2] Department of Pharmacy, University of Malakand, Khyber Pakhtunkhwa, Pakistan
[*] Corresponding author (Alam Zeb): Azebuom@gmail.com

2.1 INTRODUCTION

Persicaria hydropiper is commonly known as water pepper. It is an important plant that belongs to the family Polygonaceae. It is usually found in temperate zones of Asia, Europe, North America, Australia, and New Zealand. In East Asia, it is also cultivated for food or medicinal purposes. The habitat of the plant is in damp places and shallow water.

Different members of the family Polygonaceae have recently been studied for their pharmacological relevance. Polygonaceae represents flowering plants, commonly referred to as knotweed and smartweed. The name of the family Polygonaceae comes from the *Polygonum* genus, first coined by Antoine Laurent, a French botanist. The words 'poly' and 'gonum' mean 'many' and 'joints', respectively (Ayaz et al. 2020). Polygonaceae comprises ~1200 species distributed in about 50 genera (Mabberley 2008; A. Rahman and Gondha 2014). The family is distributed worldwide and represented by various genera including *Eriogonum*, *Rumex*, *Coccoloba*, *Polygonum*, *Persicaria*, and *Calligonum*. Several species of Polygonaceae are used as ornamental plants, while some are considered very important from a therapeutic perspective. One such plant is *Persicaria hydropiper* (L.) Delarbre (synonym: *Polygonum hydropiper* L.) (Figure 2.1) which has gained tremendous attention due to its diverse phytochemistry and medicinal uses. *P. hydropiper* is also known as water/marsh pepper in English, and Bishkatali in Bengali and has a cosmopolitan distribution (Oany et al. 2017; Arya et al. 2022).

A careful review of the ethnomedicinal literature indicates various reported uses of *P. hydropiper* in epilepsy, oedema, inflammation, headache, rheumatoid arthritis, chill, colic pain, fever, joint pain, and other infectious diseases. It is also reported for use as a central nervous system stimulant, diuretic, anthelmintic, and in the treatment of kidney diseases, hypertension, haemorrhoids, diarrhoea, piles, bleeding, parasitic worms, colic pain, and angina (Ayaz et al. 2014a, 2014b; Kimura et al. 2014; Said et al. 2015; Oany et al. 2017; Xiao et al. 2017). Various species of Polygonaceae, especially *P. hydropiper*, are also reported to have a neuroprotective behaviour.

Although used widely in traditional medicines and studied for its chemical constituents and biological activities, Ayaz et al. (2020) comprehensively reviewed *P. hydropiper* regarding its

DOI: 10.1201/9781003256830-2

FIGURE 2.1 Photographs of *P. hydropiper*. (Reproduced with kind permission of Elsevier; Ayaz et al. (2020).)

traditional uses, bioactive compounds, and mechanisms underlying various pharmacological activities. Similarly, Nasir et al. (2021) also reviewed the phytochemical and pharmacological aspects of the plant. Bairagi et al. (2022) reviewed the ethnopharmacology, phytochemistry, and pharmacology of *P. hydropiper*. This chapter presents the updated literature on the traditional uses of this plant, its bioactive constituents, and its pharmacological activities with a particular focus on neurological disorders.

2.2 TRADITIONAL USES AS FOOD AND MEDICINE

P. hydropiper is used as food and medicine all over the world. The detailed traditional uses are given in Table 2.1. Whole plants or individual parts (roots, seeds, aerial parts) are used in different traditional medicine systems. The most common uses in many Asian countries is as food because it has a strong spicy taste. Other uses include using juice or decoction/extracts of plants administered orally to treat many diseases. One study also reported using steam from boiling leaves to treat haemorrhoids (Ningthoujam et al. 2013). A few topical applications for snakebites and insect bites were also observed.

2.3 BIOACTIVE CHEMICAL CONSTITUENTS

Polygodial is the active pungent component in *P. hydropiper*. Several classes of secondary metabolites have been isolated and identified from different parts of *P. hydropiper*. A detailed list of these secondary metabolites and their chemical classes is given in Table 2.2.

Among them, flavonoids are one of the most studied chemical classes (Haraguchi et al. 1992). Persicarin was the first sulphated flavonoid isolated from natural sources (Kawaguchi and Kim 1937, 1940). Several other sulphated flavonoids are also reported further. For example, Haraguchi et al. (1996) isolated five sulphated flavonoids, including persicarin, quercetin-3-sulphate,

TABLE 2.1
Traditional Uses of *P. hydropiper*

Country	Plant part used	Ethnopharmacological uses	References
Azerbaijan	Whole plant	Plant is used as astringent, analgesic, and haemostatic agent. It is also used for the treatment of kidney stones, oedema, gastric ulcers, eczema, asthma, and thyroid diseases	Shiraliyeva 2017
Bangladesh	Leaves, seeds	Used as sedative, insecticide, and colic	Rahmatullah et al. 2010
Bangladesh	Leaves	Juice from the leaves is applied in case of insect bites	Uddin et al. 2012; A.H.M.M. Rahman 2013
Bangladesh	Leaves	Leaves are crushed and the juice is taken thrice daily as antiseptic and for the treatment of bites from snakes and worms	Khan et al. 2011
China	Aerial parts, leaves	Used for the spice taste in traditional Chinese dishes	Lee et al. 2011
China	Leaves	Juice of the leaves is used locally or orally for the treatment of snakebites Juice of the leaves is also used locally for the treatment of itching and swelling Decoction is used for the treatment of neck and shoulder pain	Sanghai Scientific and Technical Publisher 1985
India	Leaves	Leaves are used in treatment of abdominal pain, high blood pressure, dysuria, hair regeneration, intestinal worms, jaundice, toothache, and skin allergies	Gairola et al. 2014
India	Leaves	Leaves are boiled in water and the steam obtained is directed at the anal region for the treatment of haemorrhoids	Ningthoujam et al. 2013
India	Whole plant	Used as sedative, antiseptic, and antidote	Ganesan and Xu 2017
India	–	Used as diuretic, and for the treatment of dysentery, menstrual and other gynaecological problems, and cardiac problems	Choudhary et al. 2011
Japan	Young shoot	Young shoots are consumed as spice with raw fish	Haraguchi et al. 1992
Japan	Aerial parts, leaves	Smashed aerial parts are mixed with an equal amount of ginger and used for the treatment of food poisoning Juice of the leaves is used for the treatment of insect bites	Okada and Mitsuhashi 2002; Watanabe et al. 2018
Japan	Whole plant	Whole plant is used for the treatment of rheumatism, beriberi, and heatstroke Seeds are used for the treatment of gastroenteritis, stomach ache and facial swelling	Suzuki 2005
Japan	Sprouts	Used as hot-tasting spice. Sprouts are used as garnish for 'Sashimi' due to their bright red-purplish colour and pungency	Miyazawa and Tamura 2007
Malaysia	–	Along with other aromatic herbs, it is well known in Malaysia and commonly used in Malaysian dishes to give unique flavour and aroma to food such as 'asam pedas', curry, and 'laksa'	Noor Hashim et al. 2012; Aziman et al. 2014

(*continued*)

TABLE 2.1 (Continued)
Traditional Uses of *P. hydropiper*

Country	Plant part used	Ethnopharmacological uses	References
Nepal	Seeds	Seeds are used to relieve stomach ache	A.R. Joshi and Edington 1990; Rajbhandari 2001
Nepal	Aerial parts	Plant is boiled in water and filtered water is used as anthelmintic. Plant is also used as fish poison	Manandhar 2000
Nepal	Roots	Juice and decoction from the roots are used as a drink for removing stones from the urinary bladder	K. Joshi 2004
Nepal	Whole plant	Whole plant parts are used to prevent moths and other insects from infesting clothes and furniture	A.R. Joshi and Joshi 2004
Nepal	Whole plant	About 10 tablespoons of paste of plant are used as an anthelmintic veterinary medicine	Manandhar 2001
Pakistan	Whole plant	Powder of whole plant is used for the treatment of respiratory diseases	Kayani et al. 2014
Vietnam, Cambodia, Laos	Whole plant	Used for diuretic and homeostatic properties. Used in hypertension, metrorrhagia, oedemas, and snakebites	Van Duong 1993

Reproduced with kind permission of Elsevier (Ayaz et al. 2020).

TABLE 2.2
Chemical Constituents Reported from *P. hydropiper*

Chemical class	Compounds	Plant parts	References
Phenolic acids	4-Caffeoylquinic acid	Whole plant	Mahnashi et al. 2022
	5-Feruloylquinic acid Chlorogenic acid Ferulic acid	Leaves	Aziman et al. 2021
Sulphated flavonoids	Persicarin (isorhamnetin 3-sulphate)	Leaves	Kawaguchi and Kim 1937, 1940; Haraguchi et al. 1996
	Quercetin 3-sulphate	Leaves	Yagi et al. 1994; Haraguchi et al. 1996
	Isorhamnetin 3,7-disulphate	Leaves	Yagi et al. 1994; Haraguchi et al. 1996
	Tamarixetin 3-glucoside-7-sulphate	Leaves	Yagi et al. 1994; Haraguchi et al. 1996
Flavonoids	Quercetin		Yusif and Blinova 1984; Haraguchi et al. 1992; Haraguchi et al. 1996; Peng et al. 2003; X. Yang et al. 2011; Noor Hashim et al. 2012; Mahnashi et al. 2022
	7,4'-Dimethylquercetin	Leaves	Haraguchi et al. 1992
	7,3'-Dimethylquercetin (rhamnazin)	Leaves	Yusif and Blinova 1984
	Isorhamnetin (3'-methylquercetin)	Leaves	Kawaguchi and Kim 1937, 1940; Yusif and Blinova 1984; Haraguchi et al. 1992; Lee et al. 2011; X. Yang et al. 2011; Xiao et al. 2017b
	Rhamnetin (7-methylquercetin)	Leaves	Noor Hashim et al. 2012

TABLE 2.2 (Continued)
Chemical Constituents Reported from *P. hydropiper*

Chemical class	Compounds	Plant parts	References
	Isoquercitrin (quercetin 3-*O*-glucoside)	Leaves	Fukuyama et al. 1983; Haraguchi et al. 1992; Haraguchi et al. 1996; Lee et al. 2011; X. Yang et al. 2011; Noor Hashim et al. 2012
	Galloyl quercetin-3-*O*-glucoside		Noor Hashim et al. 2012
	Hyperin (quercetin 3-*O*-galactoside)		X. Yang et al. 2011; Takabe et al. 2018
	Quercetin 3-*O*-glucuronide	Leaves	Peng et al. 2003
	Cyanidin 3-*O*-galactoside		Takabe et al. 2018
	Quercitrin (quercetin 3-*O*-rhamnoside)		Fukuyama et al. 1983; Peng et al. 2003; X. Yang et al. 2011; Noor Hashim et al. 2012
	Quercetin-3-hexoside		Mahnashi et al. 2022
	Qercetin-3-glucoronide		
	Kaemferol-3-(*p*-coumaroyl-diglucoside)-7-glucoside		
	3-*O*-(6-Galloyl) glucosylquercetin	Leaves	Peng et al. 2003; Noor Hashim et al. 2012
	3,7-Dihydroxy-5,6-dimethoxyflavone		Xiao et al. 2017b
	Isalpinin		Xiao et al. 2017b
	Kaempferol	Leaves	X. Yang et al. 2011
	Kaempferol 3-*O*-glucoside	Leaves	Peng et al. 2003; Noor Hashim et al. 2012
	Kaempferol 3-*O*-rutinoside	Leaves	Noor Hashim et al. 2012
	Scutillarein 7-*O*-glucoside	Leaves	Peng et al. 2003
	6-Hydroxyapigenin	Leaves	Peng et al. 2003
	Apigenin 7-*O*-glucoside	Leaves	Noor Hashim et al. 2012
	3-*O*-(6-galloyl) glucosylkaempferol	Leaves	Peng et al. 2003; Noor Hashim et al. 2012
	Scutillarein	Leaves	Peng et al. 2003
	6-Hydroxyluteolin	Leaves	Peng et al. 2003
	6-Hydroxyluteolin 7-*O*-glucoside	Leaves	Peng et al. 2003
	(+)-Catechin	Callus and cell suspension-cultured cells from hypocotyls of seedlings	Ono et al. 1998; X. Yang et al. 2011
	(−)-Epicatechin	Callus and cell suspension-cultured cells from hypocotyls of seedlings	Ono et al. 1998; X. Yang et al. 2011
	(−)-Epicatechin 3-*O*-gallate	Callus and cell suspension-cultured cells from hypocotyls of seedlings	Ono et al. 1998
Chalcone derivative	Cardamonin	Leaves	Xiao et al. 2017b
	Pinosylvin	Leaves	Xiao et al. 2017b
Phenylpropanoid derivative	Hydropiperoside	Roots	Fukuyama et al. 1983
	Vanicoside A	Leaves	Noor Hashim et al. 2012

(continued)

TABLE 2.2 (Continued)
Chemical Constituents Reported from *P. hydropiper*

Chemical class	Compounds	Plant parts	References
	Vanicoside B	Leaves	Noor Hashim et al. 2012; Xiao et al. 2017b
	Vanicoside D	Leaves	Noor Hashim et al. 2012
	Vanicoside E	Leaves	Xiao et al. 2017b
	Vanicoside F	Leaves	Xiao et al. 2017b
Simple phenolics	3,5-Dihydroxy-5-methoxybenzoic acid	Leaves	Noor Hashim et al. 2012
	Gallic acid	Roots	Fukuyama et al. 1983
	Elagic acid 3,3'-dimethyl ether	Roots	Fukuyama et al. 1983
Anthraquinone	Anthraquinone	Roots	Fukuyama et al. 1983
Isocoumarin	Polygonolide	Roots	Furuta et al. 1986
Terpenoids	2-Desoxy-4-epi-pulchellin		Takabe et al. 2018
Monoterpenes	Bornyl acetate	Leaves, flowers	Prota et al. 2014
	Limonene	Leaves, flowers	Prota et al. 2014
Sesquiterpene	Polygodial	Leaves, flowers	Barnes and Loder 1962; Starkenmann et al. 2006; Prota et al. 2014
	9-Epi-polygodial	Leaves, flowers	Prota et al. 2014
	Isodrimenin	Leaves, flowers	Prota et al. 2014
	Drimenin	Leaves, flowers	Prota et al. 2014
	Drimenol	Flowers	Prota et al. 2014
	Caryophyllene	Leaves, flowers, sprouts	Starkenmann et al. 2006; Prota et al. 2014
	α-Humulene	Leaves, flowers	Prota et al. 2014
	(*E*)-β-Farnesene	Leaves, flowers	Prota et al. 2014
	α-Bisabolol	Leaves, flowers	Prota et al. 2014
	α-Muurolene	Leaves, flowers	Prota et al. 2014
	Aristolone	Leaves, flowers	Prota et al. 2014
	Caryophyllene oxide	Leaves, flowers	Prota et al. 2014
	(*E*)-β-Bergamotene	Leaves, flowers	Prota et al. 2014
	β-Bisabolene	Leaves, flowers	Prota et al. 2014
	(*E*)-Nerolidol	Leaves, flowers	Prota et al. 2014
	(*E*)-α-Bergamotene	Leaves, flowers	Prota et al. 2014
	β-Elemene	Leaves, flowers	Prota et al. 2014
	Humulene epoxide II	Leaves, flowers	Prota et al. 2014
	δ-Cadinene	Flowers	Prota et al. 2014
	β-Bisabolol	Flowers	Prota et al. 2014
	β-Selinene	Leaves	Prota et al. 2014
	Warburganal	Leaves	Fukuyama et al. 1980
	Polygodial	Leaves	Fukuyama et al. 1980
	Isopolygodial	Leaves	Fukuyama et al. 1980
	Isodrimeninol	Leaves	Fukuyama et al. 1980
	Drimenol	Leaves	Fukuyama et al. 1980
	Confertifolin	Leaves	Fukuyama et al. 1980
	3β-Angeloyloxy-7-epi-futronolide	Whole plant	Sultana et al. 2011
	Polygonumate	Whole plant	Sultana et al. 2011
	Dendrocarbin L	Whole plant	Sultana et al. 2011
	(+)-Winterin	Whole plant	Sultana et al. 2011
	(+)-Fuegin	Whole plant	Sultana et al. 2011

TABLE 2.2 (Continued)
Chemical Constituents Reported from *P. hydropiper*

Chemical class	Compounds	Plant parts	References
	Changweikangic acid	Whole plant	Sultana et al. 2011
	Futronolide	Whole plant	Sultana et al. 2011
	7-Ketoisodrimenin	Whole plant	Sultana et al. 2011
Norsesquiterpene	Polygonal	Leaves	Fukuyama et al. 1980
Diterpenes	Neophytadiene isomer I	Leaves, flowers	Prota et al. 2014
	Neophytadiene isomer II	Leaves, flowers	Prota et al. 2014
	Neophytadiene isomer II	Leaves, flowers	Prota et al. 2014
Steroids and triterpenoids	β-Sitosterol	Leaves	Xiao et al. 2017b; Ayaz et al. 2019
	Stigmasterol	Leaves	Ayaz et al. 2019
Lignan	(+)-Ketopinoresinol	Leaves	Xiao et al. 2017b
Other compounds	5,6-Dehydrokawain	Leaves	Xiao et al. 2017b
	Aniba-dimer-A	Leaves	Xiao et al. 2017b
	6,6'-[(1α,2α,3β,4β)-2,4-Diphenylcyclobutane-1,3-diyl] bis(4-methoxy-2*H*-pyran-2-one)	Leaves	Xiao et al. 2017b
	Heneicosane	Leaves, flowers	Prota et al. 2014
	Dihydro-α-ionone	Leaves, flowers	Prota et al. 2014
	Nonanal	Leaves	Prota et al. 2014
	Confertifolin	Essential oil of fresh leaves	Duraipandiyan et al. 2010

Updated and reproduced with kind permission of Elsevier (Ayaz et al. 2020).

isorhamnetin-3,7-disulphate, rhamnzain-3-sulphate, and tamarixetin-3-glucoside-7-sulphate from the leaves (Figure 2.2). Various other flavonoid derivatives, including quercetin, kaempferol, apigenin, luteolin, and catechins derivatives, are also widely reported (Figure 2.3). Takabe et al. (2018) isolated cyanidin 3-*O*-galactoside and quercetin 3-*O*-galactoside from *P. hydropiper* from Japan. However, the authors mention the Japanese name of this plant as 'Benitade'. According to the YList online database (http://ylist.info/ylist_detail_display.php?pass=1358), 'Benitade' refers to a different species, *Persicaria orientalis* (L.) Spach, which should be clarified further in future studies. Apart from isolation from a plant obtained from a natural source, Ono et al. (1998) reported (+)-catechin, (−)-epicatechin, and (−)-epicatechin-3-*O*-gallate from the callus and cell suspension-cultured cells from hypocotyls of seedlings of *P. hydropiper*. Various phenylpropanoids derivatives such as hydropiperoside (Figure 2.4) (Fukuyama et al. 1983) and its derivatives vanicosides A, B, C, D, E, and F are also reported (Noor Hashim et al. 2012; Xiao et al. 2017). Mahnashi et al. (2021) reported the isolation and purification of two bioactive compounds, i.e. 4-methyl-5-oxo-tetrahydrofuran-3-yl acetate and 4-hydroxy-3-methoxybenzoate, from the active fractions. Recently, Mahnashi et al. (2022a) reported kaemferol-3-(*p*-coumaroyl-diglucoside)-7-glucoside (275.4 mg/g), *p*-coumaroylhexose-4-hexoside (96.5 mg/g), quercetin-3-glucuronide (76.0 mg/g), 4-caffeoylquinic acid (58.1 mg/g), quercetin (57.9 mg/g), 5,7,3-trihydroxy-3,6,4,5-tetramethoxyflavone (55.5 mg/g), 5-feruloylquinic acid (45.8 mg/g), cyanidin-3-glucoside (26.8 mg/g), delphinidin-3-glucoside (24 mg/g), and quercetin-3-hexoside (20.7 mg/g) were highly abundant compounds. In Malaysian samples, chlorogenic acid, ferulic acids, rutin, myristin, and quercetin were reported in the ethanolic extract (Aziman et al. 2021).

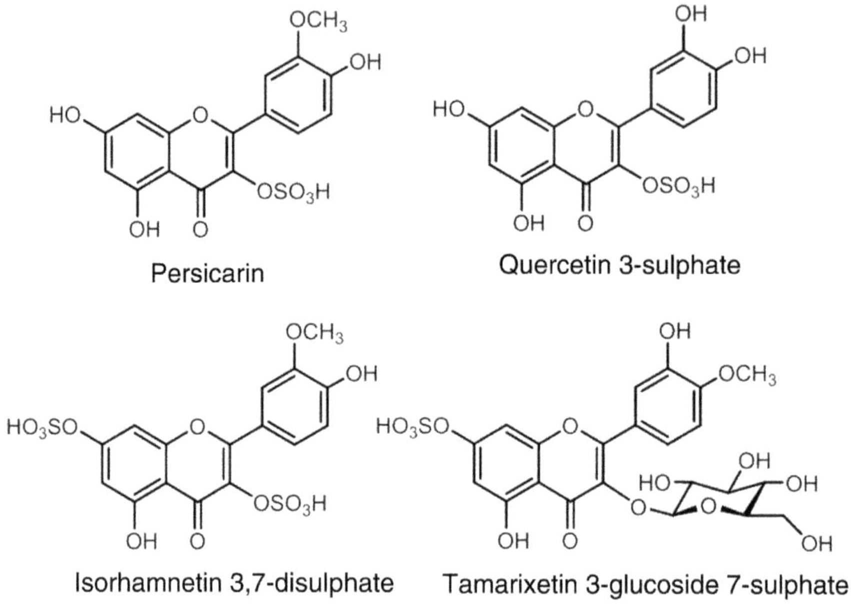

FIGURE 2.2 Structures of sulphated flavonoids.

Polygonal, also known as tadeonal, a sesquiterpene dialdehyde, is the active pungent component in *P. hydropiper* (Barnes and Loder 1962). Many other sesquiterpenoid derivatives are also reported (Figure 2.5) (Prota et al. 2014; Starkenmann et al. 2006).

There have also been studies on volatile constituents. Prota et al. (2014) analysed the chemical constituents of leaves and flowers of three species of *Persicaria* (*P. hydropiper*, *P. maculosa*, and *P. minor*) by gas chromatography–mass spectrometry and compared the results. The main constituents were sesquiterpenoids. *P. hydropiper* contained polygodial in a high amount (6.2 mg/g fresh weight), but the other two species had a very low abundance of it. Similarly, Starkenmann et al. (2006) also performed a comparison between the volatile constituents of *P. odorata* (Lour.) Sojak and *P. hydropiper*. Miyazawa and Tamura (2007) analysed the components in essential oil obtained from the sprouts collected in Japan. They identified a total of 53 compounds where the main constituents were sesquiterpenoids such as (*E*)-β-farnesene, phytol, (*E*)-caryophyllene, and (*E*)-neridol. Mahnashi et al. (2022b) reported 141 compounds in the essential oils of *P. hydropiper*, among which the most abundant compounds were β-elemene, dihydro-α-ionone, *cis*-geranyl acetone, α-bulnesene, bicyclo[4.1.0]heptane,-3-cyclopropyl,-7-hydroxymethyl, trans, nerolidol, bicyclo[2.2.2]oct-2-ene, 1,2,3,6-tetramethyl, (1R,5S,8R,9R)-4,4,8-trimethyltricyclo [6.3.1.0(1,5)] dodeca-2-en-9-ol, β-caryophyllene epoxide, and decahydronaphthalene. These compounds showed important bioactivities. *P. hydropiper* samples from Malaysia showed the presence of dodecanal, caryophyllene, caryophyllene oxide, decanal, α-caryophyllene, citronellol, heptadecanal, linalool, and phytol in its essential oil (Aziman et al. 2021). Ayaz et al. (2022) showed the presence of phytosterol and β-sitosterol in *P. hydropiper* and their potential biological effects.

2.4 PHARMACOLOGICAL STUDIES

Due to the severe side effects of allopathic medicines, researchers have focused on exploring medicinal plants to treat various diseases. For decades, plants belonging to the Polygonaceae family have been used as traditional medicine. One of the important family members is *P. hydropiper* which

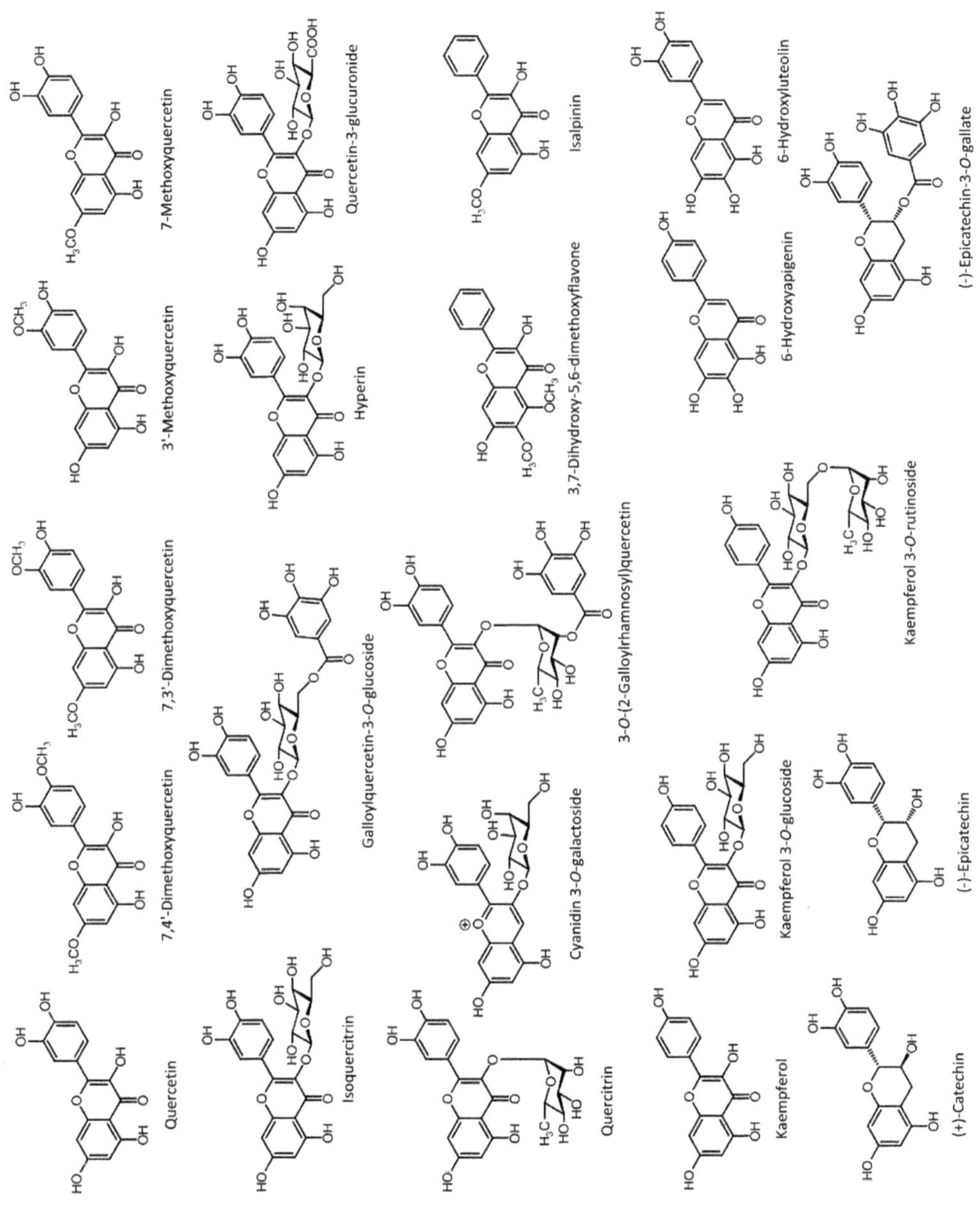

FIGURE 2.3 Structures of other flavonoids reported from *P. hydropiper*.

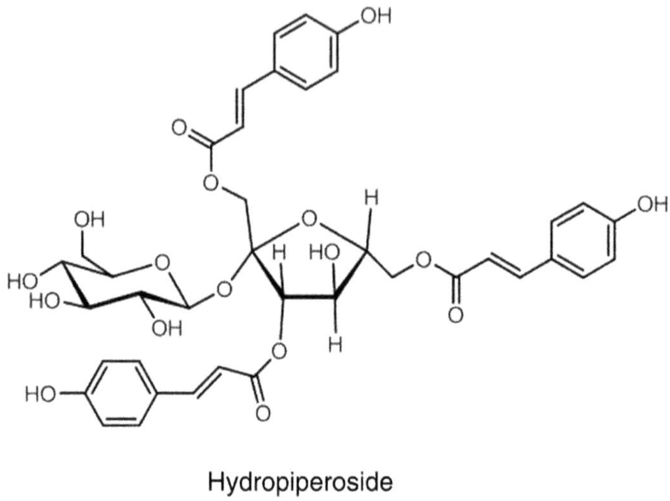

FIGURE 2.4 Structure of hydropiperoside.

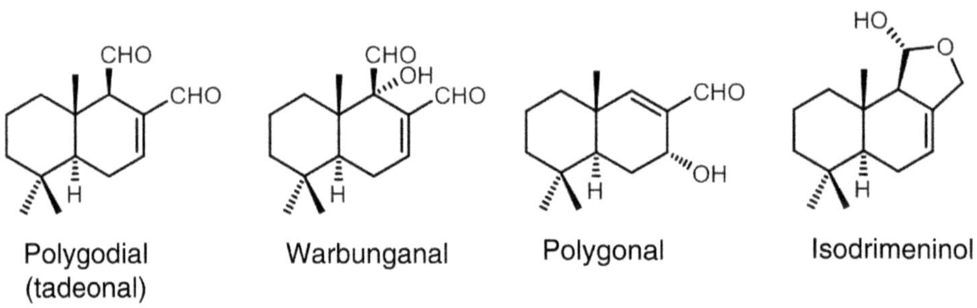

FIGURE 2.5 Structures of sesquiterpenoids reported from *P. hydropiper*.

contains plentiful chemical substances, including flavonoids, phenolic compounds, and glycosides (Narasimhulu et al. 2014). The current literature summarizes its potential anti-inflammatory, antineoplastic, antiulcer, anti-Alzheimer, and other miscellaneous activities (Table 2.3). Currently, several studies have been reported on its potential pharmacological activities which have been discussed below.

2.4.1 Anti-inflammatory Activity

P. hydropiper contains a significant amount of phenolic compounds, including flavonol glycosides and flavonol sulphates. Its methanolic extract was reported as an effective xanthine oxidase (XO) inhibitor (Noor Hashim et al. 2012). Gout, which is caused through accumulation of uric acid in the bloodstream (hyperuricaemia), is responsible for joint inflammation. The XO inhibitors can be used for gout treatment as they block uric acid formation in the body (Mohamed Isa et al. 2018). The flavonoids in the butanol fraction of *P. hydropiper* extract were reported to possess anti-inflammatory potential prompted by lipopolysaccharides. During *in vivo* and *in vitro* studies, its anti-inflammatory mechanism was mediated by suppression of c-Jun N-terminal kinase, extracellular signal-regulated kinase, and mitogen-activated protein kinase signalling pathways (Tao et al. 2016). In a recent study, *P. hydropiper* (125, 250, and 500 mg/kg) was orally administered for one

TABLE 2.3
Reported Biological Activities of *P. hydropiper* Extracts and Compounds

Activity	Plant part	Activity/studies	*In vivo* or *in vitro*	Tested extracts/active metabolites	Mechanism	References
Anti-inflammatory	Flower	Displayed enhanced DPPH, ferric reducing antioxidant power, metal chelating, hydrogen peroxide, nitric oxide superoxide radical scavenging activity while inhibited XO	*In vitro*	Methanol extract	Inhibiting XO and scavenged superoxide radicals	Mohamed Isa et al. 2018
–	Whole plant	Improved the levels of T-SOD, T-AOC, GSH-PX, and GSH while reduced the levels of TNF-α, MPO, and MDA	*In vivo* and *in vitro*	Flavonoids from butanol fraction	Inhibiting the phosphorylation of JNK, ERK and c-JUN in MAPKs signalling pathways	Tao et al. 2016
–	Stalks	Enhanced TNBS-induced symptoms like macroscopic score and histological examination, improved MPO activity and GSH content	*In vivo* and *in vitro*	Water extract	Downregulated the TNBS-induced increase in the activity of iNOS and levels of Cox-2, TNF-α, and IL-1β while unregulated the protein expression of NF-κB	Zhang et al. 2018
Antioxidant activity	Whole plant	Exhibited DPPH & ABTS scavenging activity while inhibited AChE & BChE activity	*In vitro*	Crude methanolic extract, *n*-hexane, chloroform, ethyl acetate, *n*-butanol and saponins	Scavenging free radicals whereas inhibiting AChE & BChE	Ayaz et al. 2014a
–	Whole plant	Showed antioxidant and anticholinesterase activity	*In vitro*	Flavonoids, flavonoid glycoside, and phenylpropanoid glycoside	Scavenging ferric thiocyanate & DPPH	Noor Hashim et al. 2012
–	Whole plant	Exhibited DPPH scavenging activity	*In vitro*	Methanol extract contains hydropiperoside A, hydropiperoside B, & vanicosides	Scavenging free radicals	Kiem et al. 2008
Anti-Alzheimer's activity	Whole plant	Inhibitory potential against AChE & BChE while improving the memory, behaviour, and motor coordination in mice	*In vivo* and *in vitro*	β-sitosterol	Showed robust anticholinesterase activities; improved the memory & behaviour	Ayaz et al. 2017
–	Leaves & flowers	Inhibitory potential against AChE & BChE while scavenging free radicals	*In vivo* and *in vitro*	Caryophylene oxide & decahydronaphthalene	Displayed anticholinesterase & antioxidant activities	Ayaz et al. 2015

(*continued*)

TABLE 2.3 (Continued)
Reported Biological Activities of *P. hydropiper* Extracts and Compounds

Activity	Plant part	Activity/studies	*In vivo* or *in vitro*	Tested extracts/active metabolites	Mechanism	References
Antitumour activity	Whole plant	Characterization of 9 epipolygodial, possesses 20-fold potency vs polygodial against five cancer cell lines	*In vitro*	Polygodial (terpenenoid dialdehyde)	TRPV1 agonist & anticancer activities	Dasari et al. 2015a; Mahnashi et al. 2022
–	Whole plant	Employed as antiproliferative due to its cytostatic potential against apoptosis-resistant cancer cells	*In vitro*	C12-Wittig derivatives of polygodial	Cytostatic agent with action against drug-resistant cancer cells & capable of pyrrolylation of primary amines	Dasari et al. 2015b
–	Whole plant	Antiproliferative potential against oral squamous cell carcinoma	*In vitro*	Polygodial analogues P3 and P27	Prompted a substantial concentration-dependent decrease in the mitochondrial transmembrane efficacy & subsequent apoptosis	De La Chapa et al. 2018
Antinociceptive activity	Leaves	Inhibited the glutamate & cinnamaldehyde-induced pain in mice. Actively involved in cGMP & ATP-sensitive K^+ channel pathways	*In vitro*	Methanol extract	Possess antinociceptive activity both in peripheral & central pain models of mice	Khatun et al. 2015
–	Whole plant	Displayed a substantial activity in acetic acid-induced writhing in mice	*In vitro*	Hexane, ethyl acetate & methanol extracts	Reduced the amount of writhing induced by acetic acid in mice	E. Rahman et al. 2002
Antihyperglycaemic activity	Leaves & stem	Significantly reduced the blood glucose levels in glucose-loaded mice	*In vivo* and *in vitro*	Ethanolic extract	Possess antihyperglycaemic activity	E. Rahman et al. 2002
–	Whole plant	Inhibited the formation & accumulation of advanced glycation end products	*In vivo* and *in vitro*	Cyanidin 3-*O*- & quercetin 3-*O*-galactoside	Inhibited glycation	Takabe et al. 2018
Antimicrobial activity	Stem, leaves, & flowers	Showed broad-spectrum activity against bacterial & fungal strains	*In vitro*	Various crude extracts	Possess significant antimicrobial activity	Ayaz et al. 2016
Sedative & anxiolytic activity	Leaves	Showed sedative & anxiolytic effects on the central nervous system	*In vivo* and *in vitro*	Methanol extracts	Possess significant sedative & anxiolytic activity	Shahed-Al-Mahmud and Lina 2017

Reproduced with kind permission of Elsevier (Ayaz et al. 2020).

week to rats having inflammation in their intestine triggered by 2,4,6-trinitrobenzene sulphonic acid. The *P. hydropiper* extract significantly protected intestinal inflammation in rats. It has shown a potential role in inhibiting the nuclear factor kappa-light-chain-enhancer of activated B-cell signalling pathways (Zhang et al. 2018).

2.4.2 Antioxidant Activity

The leaves and flowers of *P. hydropiper* contain various flavonoids which retain a strong antioxidant potential as determined by using 1,1-diphenyl-2-picrylhydrazyl (DPPH) and 2,2'-azinobis 3-ethylbenzothiazoline-6-sulphonic acid (ABTS) free radicals (Ayaz et al. 2014a). Among the antioxidants, phenolic compounds are well known for their antioxidant action (Zeb 2020). The antioxidant activity was assessed by measuring H_2O_2 (hydrogen peroxide) free radicals due to linoleic acid oxidation by using the ferric thiocyanate including DPPH scavenging capability with half-maximal inhibitory concentration (IC_{50}) (13.30 μg/ml). Similarly, 3,5-dihydroxy-4-methoxy benzoic acid (IC_{50} 8.08 μg/ml), quercetin (IC_{50} 11.14 μg/ml), and quercetin-3-*O*-rhamnoside (IC_{50} 18.46 μg/ml) showed enhanced activity as compared to vitamin C (IC_{50} 6.80 μg/ml) (Noor Hashim et al. 2012). The methanol leaf extract of *P. hydropiper* contains hydropiperoide B and vanicoside A that revealed antioxidant activity by using a DPPH assay with scavenging concentration (SC_{50}) values of 23.4 and 26.7 μg/ml as compared to ascorbic acid (SC_{50} 22.0 μg/ml) (Kiem et al. 2008). Hussain et al. (2021) showed the highest antioxidant activity for leaves, followed by stem and root, suggesting the presence of high amounts of phenolic compounds in leaves.

2.4.3 Antitumour Studies

Polygodial (bicyclic sesquiterpene) was isolated from *P. hydropiper* and is a recognized transient receptor potential vanilloid 1 receptor (TRPV1) agonist. Several polygodial analogues have been reported for TRPV1 agonistic and anticancer activities. In this regard, 9-epipolygodial was investigated for its 20-fold potency compared to polygodial against five cancer cell lines comprising apoptosis-resistant human glioblastoma, human A549 non-small cell lung cancer, human SKMEL-28 melanoma, apoptosis-sensitive human Hs683 anaplastic oligodendroglioma, and human MCF-7 breast cancer (Dasari et al. 2015a). A derivative of polygodial (C12-Wittig designated as P3) was synthesized and evaluated for its anticancer activity. C12-Wittig was found to be an interesting chemotype due to its antiproliferative mechanism generally over its cytostatic properties, clarifying their action alongside apoptosis-resistant cancer cells as compared to parent polygodial, which shows fixative and broad-spectrum cytotoxic features against human cells (Dasari et al. 2015b). A more effective derivative of polygodial known as P27 has been synthesized which showed greater antiproliferative properties as compared to P3 during *in vitro* studies against oral squamous cell carcinoma (De La Chapa et al. 2018). More recently, phytosterols isolated from the plant were reported to possess considerable cytotoxicity against HeLa, MCF-7, and NIH/3T3 cells (Ayaz et al. 2019). Similarly, Mahnashi et al. (2021) reported the cytotoxicity property of *P. hydropiper* against cancer cells. The anti-angiogenic and antitumour results suggested additional tumour-suppressive properties.

2.4.4 Antinociceptive Studies

P. hydropiper leaves possess substantial antinociceptive potential reported in peripheral and central pain models of mice. It was helpful in trivial dose-dependent inhibition of paw licking during neurogenic and inflammatory pain produced by formalin injection. Additionally, it has considerably reduced glutamate and cinnamaldehyde-induced pain in mice. Antinociception was considerably reversed due to an effective opioid system during the pretreatment with naloxone in the hot plate and tail immersion test (Khatun et al. 2015). Another study has reported that the hexane, ethyl acetate, and methanol extracts of PHL given at doses of 250 and 500 mg/kg to mice exhibited an enhanced action on acetic acid-induced writhing. However, among the tried extracts, ethyl acetate has shown much higher activity (E. Rahman et al. 2002).

2.4.5 Antihyperglycaemic Studies

The antihyperglycaemic activity was reported in the ethanolic leaf extract of *P. hydropiper* during oral glucose tolerance tests in mice. Various leaf extract doses (50, 100, 200, and 400 mg/kg) of body weight have significantly reduced the level of blood sugar by 48.8%, 51.5%, 54.1%, 58.2% (p <0.05) among the tested mice compared to the control mice. However, the blood glucose level decreased by 42.1%, when glibenclamide (an antihyperglycaemic drug) was given at a dose of 10 mg/kg body weight to the mice (Oany et al. 2017). Hyperglycaemia is interlinked with glycation, a non-enzymatic reaction between proteins and a reducing agent, ultimately forming advanced glycation end products. Two antiglycative compounds (cyanidin 3-*O*-galactoside and quercetin 3-*O*-galactoside) have been isolated and characterized from *P. hydropiper* mutant (Benitade) (Takabe et al. 2018). Recently, Mahnashi et al. (2022a) showed that phenolic compounds might be responsible for both *in vitro* antidiabetic potentials of the plant samples. Chloroform, saponins, and crude methanolic extract were the most potent fractions in α-amylase inhibitory activity and displayed IC_{50} values of 90, 100, and 200 μg/ml, respectively. In another study, Mahnashi et al. (2022b) reported that essential oils from the leaves of *P. hydropiper* have strong inhibitory activity toward α-glucosidase and amylase enzymes. Inhibition of these enzymes is the key target in reducing postprandial hyperglycaemia.

2.4.6 Antimicrobial Studies

The *P. hydropiper* extract has broad-spectrum antimicrobial activity, which implies its prospective therapeutic efficacy against bacterial and fungal infections. Different *P. hydropiper* samples containing various types of crude extracts were tried against a broad range of pathogenic bacterial and fungal species. Minimum inhibitory concentrations (MICs) of the crude extract against *Enterococcus faecalis* and *Staphylococcus aureus* were 32.00 and 64.00 μg/ml, respectively. The minimum fungicidal concentration of *P. hydropiper* extract against *Aspergillus niger* was 125.00 μg/ml and against *Fusarium oxysporum* was 46.66 μg/ml (Ayaz et al. 2016). Confertifolin (an antimicrobial agent) was isolated from *P. hydropiper* leaves using column chromatography. The lowest MIC was observed against *E. faecalis* (31.25 μg/ml); however, the highest MICs were observed against the fungal strains *Magnaporthe grisea* and *Trichophyton simii* (62.5 and 125 μg/ml, respectively) (Duraipandiyan et al. 2010). Aziman et al. (2021) showed that the ethanolic extract and essential oil of *P. hydropiper* possessed antibacterial activity against the bacteria *Staphylococcus aureus* only at different concentrations, with MIC values of 0.625 and 5 mg/ml, respectively; and minimum bactericidal concentration values of 5 and 40 mg/ml, respectively.

2.4.7 Antihypertensive Studies

In a study, young leaf powder and young leaf extract of *P. hydropiper* L. were evaluated for chymase inhibitory and antihypertensive potentials in a salt-induced hypertension model (Devarajan et al. 2018). *P. hydropiper* young leaf powder and extract significantly reduced the development of hypertension in the 2% saltwater-induced hypertension model. As *P. hydropiper* therapy was stated at the pre-hypertensive stage, so authors concluded that the plant might be a prophylactic regimen against the development of salt-induced hypertension. Further, as chymase upregulation occurs in salt-induced hypertensive crises, being an effective inhibitor of chymase, *P. hydropiper* might be a helpful option in managing salt-induced hypertension (Devarajan et al. 2018).

2.4.8 Other Pharmacological Studies

Nowadays, obesity has become a global threat connected with metabolic syndrome and liver disease (Imran et al. 2019). It has been confirmed that the Wnt/β-catenin pathway has a close link with obesity and the small molecules that control the Wnt/β-catenin pathway are actively involved in regulating the adipogenesis of obesity. HEK 293-TOP cells with the Wnt/β-catenin signalling reporter gene are activated when screened against 11 *P. hydropiper* extracts. The extracts contain isoquercitrin that overwhelms the adipogenesis of 3T3-L1 cells via the inhibition of Wnt/β-catenin signalling (Lee et al. 2011). In animal models, *P. hydropiper* crude extracts, essential oils, and saponins were reported to possess urease inhibitory, anti-proteus, and gastroprotective potentials. Pretreatment with samples exhibited considerable anti-ulcerogenic potentials at 100–400 mg/kg doses in the pyloric ligation ulceration model and normalized biochemical parameters in animal models (Ayaz et al. 2017).

2.5 EFFECTIVENESS IN NEUROLOGICAL DISORDERS

2.5.1 Alzheimer's Disease

Recently, β-sitosterol was purified from *P. hydropiper* to check its potential against Alzheimer's disease. *In vitro* and *in vivo* molecular docking and behavioural studies were conducted using shallow water maze (SWM), Y-maze, and balance beam tests (Figures 2.6 and 2.7). The β-sitosterol-treated groups displayed an IC_{50} value of 55 and 50 µg/ml against acetylcholinesterase (AChE) and butyrylcholinesterase (BChE), respectively. While studying the behavioural studies in the transgenic mice, steady progress was observed in memory, unplanned fluctuation in behaviour, and motor coordination (Ayaz et al. 2017).

Similarly, the leaf and flower essential oils of *P. hydropiper* contained caryophyllene oxide (41.4%) and decahydronaphthalene (38.2%). Leaf oil and flower oil showed IC_{50} values of 120 and 220 µg/ml in AChE, whereas IC_{50} values of 130 and 225 µg/ml were shown in BChE inhibitory assays, respectively (Ayaz et al. 2015).

Solvent-based fractions, including crude methanolic extract, *n*-hexane, chloroform, ethyl acetate, and saponins, were reported to possess considerable AChE inhibitory potentials with IC_{50} values of 330, 35, 55, 310, and 108 µg/ml respectively. Moreover, some fractions, including *n*-hexane, chloroform, ethyl acetate, and aqueous, also exhibited significant inhibitory effects against BChE with IC_{50} values of 40, 215, 81.01, and 03 µg/ml, respectively (Ayaz et al. 2014a). A Chinese herbal formula containing *P. hydropiper* was reported to possess intestinal endocrine stimulatory and neuroprotective potentials via stimulation of 5HT, GLP-1, CCK, and hTGR5 pathways. Noor Hashim et al. (2012) reported the antioxidant, cholinesterase inhibitory potentials of *P. hydropiper* and quantified constituents in the most active fraction via liquid chromatography– diode array detection–electrospray ionization–tandem mass spectrometry (LC-DAD-ESI-MS/MS). Various bioactive compounds, including 3,5-dihydroxy-4-methoxybenzoic acid, vanicoside A, B, D, quercetin, galloyl quercetin-3-*O*-glucoside, quercetin-3-*O*-glucoside, hydropiperoside, kaempferol rutinoside, apigenin-7-*O*-glucoside, galloyl kaempferol-3-*O*-glucoside, rhamnetin, and quercetin-3-*O*-rhamnoside, may be responsible for the overall cholinesterase inhibitory potentials of the plant. Among the isolated compounds, quercetin, galloyl quercetin-3-*O*-glucoside, quercetin-3-*O*-rhamnoside, and rhamnetin, 3,5-dihydroxy-4-methoxybenzoic acid exhibited IC_{50} values of 9.30, 28.06, 15.95, 23.94, 33.38 and 6.22 µg/ml respectively (Noor Hashim et al. 2012). In another study, Sultana et al. (2011) reported the isolation of drimane-type sesquiterpenes from *P. hydropiper* and

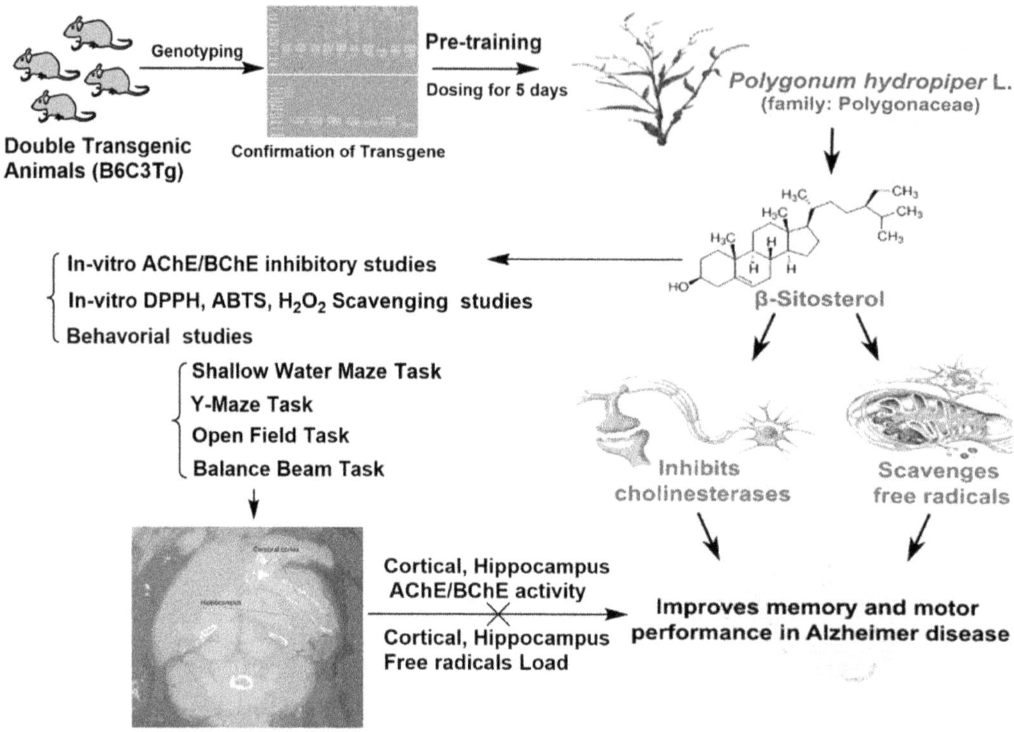

FIGURE 2.6 Graphical representation of anti-Alzheimer and neuroprotective studies on β-sitosterol isolated from *P. hydropiper*. (Reproduced with kind permission of Elsevier; Ayaz et al. (2020).)

evaluated them for ACh/BChE inhibitory potentials. Among the isolated compounds, weikangic acid A exhibited moderated anticholinesterase activities.

2.5.2 Neuroprotective Studies

Persicarin, a major component of *P. hydropiper* and *Oenanthe javanica*, has been reported to exhibit neuroprotective properties via inhibition of intracellular calcium influx, cellular peroxide production, and intracellular nitric oxide production in a glutamate-mediated neurotoxicity model of cortical cells. Moreover, persicarin also augments the anti-radical activities of enzymes, including glutathione reductase, superoxide dismutase, and glutathione peroxidase (Haraguchi et al. 1996; Ma et al. 2010). *P. hydropiper* extracts were reported to possess strong antioxidant and anti-ageing potentials mediated via scavenging of free radicals, inhibition of elastase, tyrosinase, and photo-haemolysis of red blood cells (H. J. Yang et al. 2008). Tong et al. (2021) showed that essential oils from *P. hydropiper* exhibit significant neuroprotective potential due to their anti-radical and cholinesterase inhibitory activities.

2.5.3 Sedative and Anxiolytic Studies

The methanol extract of *P. hydropiper* leaves has shown sedative and anxiolytic activities in mice. The sedative effect was investigated by giving different doses (50, 200, 350, and 500 mg/kg) in various conditions such as open field, rota-rod, and thiopental sodium-induced sleeping time tests in mice. In contrast, the anxiolytic activity was assessed in the elevated plus-maze (EPM), light–dark box, hole-board, and marble-burying tests. The sedative and anxiolytic activities were

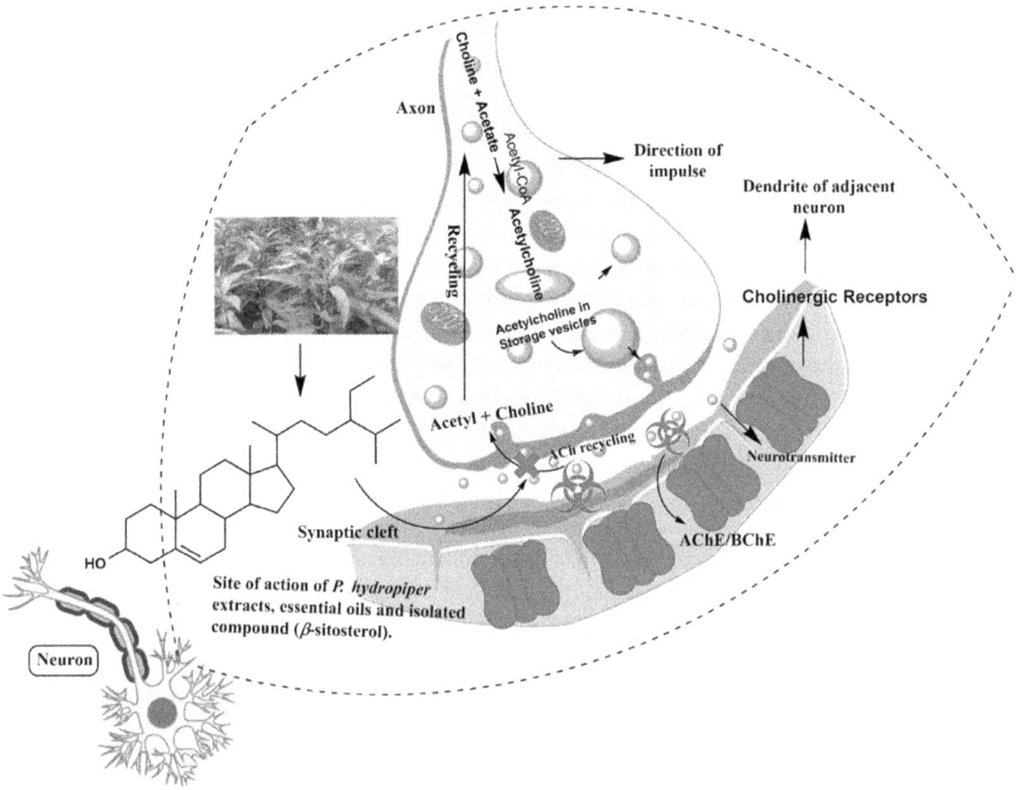

FIGURE 2.7 Acetylcholine synthesis, action, termination via AChE/BChE, and site of action of *P. hydropiper*-derived extracts, essential oils, and compounds. (Reproduced with kind permission of Elsevier; Ayaz et al. (2020).)

compared to diazepam as a positive control, including an acute toxicity test (Shahed-Al-Mahmud and Lina 2017).

Open field is a widely used behavioural task for the assessment of the locomotor performance of rodents. The open field apparatus is usually made of wood with dimensions of 72 × 72 × 36 cm, and the floor of the apparatus is divided into 16 squares with 18 × 18 cm dimensions. The number of squares crossed in time intervals of 0, 30, 60, 90, and 120 minutes are observed in all groups of animals administered with test samples and standard drugs. In this study, the sedative effect of the test samples increased significantly in a dose and time-dependent manner, with the highest sedative effect observed at 500 mg/kg after 120 minutes. Other behavioural tasks for sedative potentials of test compounds include the hole cross test, which is used to analyse the hyperemotional response after exposure to a new environment. The rota-rod test to assess motor coordination and performance of animal models post drug therapy was used in the study. The thiopental sodium-induced sleeping time paradigm was also used for the sedative effect of samples. The number of holes crossed significantly increased in test and standard treated group animals after 120 minutes of drug administration. Likewise, rota-rod test treatment with *P. hydropiper* samples caused a significant increase in the number of animals falls in a time and dose-dependent manner. At the same time, the muscle relaxant effect of *P. hydropiper* was comparable with the positive control (diazepam 1 mg/kg) group. In the thiopental-induced sleeping test, *P. hydropiper* significantly reduced the onset of sleep and increased the duration of sleep (Ayaz et al. 2020).

An extensively used anxiolytic assessment apparatus, the EPM, was used in anxiolytic assays. The EPM apparatus consists of two open and two closed arms with an open roof of 50 cm elevated from the floor to observe the anxiolytic potentials of the test samples. Various parameters were recorded post drug administration, including time spent at open and closed arms and the number of animal entries to both closed and open arms. Another behavioural tool, the light–dark apparatus, which predicts animal exploratory/anxiolytic behaviour using time spent in each compartment, was applied. Other anxiolytic behavioural tools, including the hole-board test and marble-burying test, were also used for all groups of animals. In the EPM test, *P. hydropiper* methanolic extract (MEPH)-treated group animals spent more time in open arms, and the number of entries to open arms was significantly higher than in another group of animals. This group of animals exhibited a significant anxiolytic effect comparable to the diazepam-treated group of animals.

Moreover, in the light–dark test, *P. hydropiper*-treated animals spent more time in the light compartment, and a significant decline in head dips was observed in the hole-board test. Finally, a significant decline in the number of marbles buried comparable with the diazepam group was observed. The current study suggests many sedative and anxiolytic potentials of MEPH and warrants the isolation of bioactive compounds for further studies (Ayaz et al. 2020).

2.5.4 Antidepressant Studies

Phytosterols, including beta-sitosterol which is abundant in *P. hydropiper* (Ayaz et al. 2019) and other plants (*Sargassum horneri*), are reported to exhibit considerable antidepressant potential. *P. hydropiper* extracts were reported to possess antidepressant and anxiolytic potentials using a force swing test and open field test (Sharif et al. 2013). Another study reported that beta-sitosterol considerably reduced immobility times in force swimming tests and animal suspension tests and did not affect the locomotor activity of animal models. Furthermore, beta-sitosterol norepinephrine, 5-hydroxytryptamine, and its metabolite 5-hydroxy indoleacetic acid are in animal brain tissues (Zhao et al. 2016). As these neurotransmitters are implicated in the aetiology of depression, so the antidepressant effect of beta-sitosterol might be linked with the modulation of these neurotransmitters in the brain.

2.6 TOXICOLOGICAL ASPECTS

2.6.1 Heavy Metal Content and Phytotoxicity

Ayaz et al. (2014b) reported contents of the heavy metals, including lead (Pb), cadmium (Cd), copper (Cu), zinc (Zn), iron (Fe), and nickel (Ni) in crude dried powder of the whole *P. hydropiper* plant, crude methanolic extract, various fractions including *n*-hexane, chloroform, ethyl acetate, *n*-butanol, and aqueous fractions as well as soil samples via atomic absorption spectrophotometer. The heavy metal concentrations were Pb (not detected), Cd [0.35 parts per million (ppm)], Cu (7.21 ppm), Zn (0.199 ppm), Fe (95.0 ppm), and Ni (0.199 ppm) in the crude powder material of the plant. Moreover, the concentrations of Cd were 22.8, 20.2, 16.0, 9.55, 37.1, 23.1, and 55.2 ppm in crude methanolic extract, *n*-hexane, chloroform, ethyl acetate, *n*-butanol, aqueous, and soil samples respectively. Likewise, the Pb concentrations in crude methanolic extract, *n*-hexane, chloroform, ethyl acetate, *n*-butanol, aqueous, and soil samples were 79.3, 67.3, 55.4, 70.3, and 90.15 ppm, respectively. Pb and Cd, being non-essential heavy metals, are associated with several malfunctions in the body. The concentration of essential heavy metals was in a good range for human use. According to the World Health Organization's (WHO's) recommended limits for heavy metals in humans, the concentrations of all the heavy metals were in the recommended range in crude powder. In contrast, the concentrations were relatively high in other fractions. WHO recommends assessing the heavy metal content of plant materials before use in humans to avoid health hazards associated with heavy metals. According to the authors, the plant was collected from a marshy place, and the high metal contents in the soil sample indicate that heavy metals might come from drain water. The heavy metal contents of medicinal plants vary from place to place depending on the irrigation system and environment.

The phytotoxic effects of the same samples on radish seeds were also analysed (Ayaz et al. 2014b). Plant methanolic, ethyl acetate, and water-soluble saponins exhibited 86.6%, 89.3%, and 89.2% inhibitory effects on radish seed germination. The same fractions also showed considerable inhibitory effects on the root lengths of the germinated plants. The authors suggested that the phytotoxic effect of the plant samples might be related to heavy metal content and that the plant can be a potential source of natural herbicidal agents against unwanted weeds.

2.6.2 Acute and Sub-chronic Toxicity

In an acute toxicity study, the acute toxicological effects of the plant methanolic extract were tested at 20–600 mg/kg in Swiss Webster albino mice. No mortality was observed at 400 mg/kg of the sample after 24 hours of administration. However, 100% of animals died at 600 mg/kg, suggesting a median lethal dose (LD_{50}) of 500 mg/kg of the crude methanolic extract after intraperitoneal administration (Raihan et al. 2012). A subacute study showed the toxicity of plant ethanolic extract in F344/DuCrj rats for 13 weeks. The sample caused no clinical adverse effects in experimental animals at 1000 ppm, and no considerable changes in serum biochemistry, haematology, body weight, and histopathology of the vital organs were observed (Kuroiwa et al. 2006).

2.6.3 Topical Toxicity

Plant leaves cause skin blisters during handling due to excessive polygodial, an irritant compound. Polygodial is reported to cause an increase in extracellular glutamate concentrations via inhibition of glutamate uptake by astrocytes. Polygodial exhibits neurotoxic effects by increasing glutamate release by synaptosomes, thus increasing its concentration in the cortex (Martini et al. 2006).

2.6.4 Genotoxicity

Kuroiwa et al. (2006) reported the genotoxic effects of plant ethanolic extracts in *Salmonella typhimurium* TA 100, TA 98, and chromosomal changes in Chinese hamster-derived CHL/IU cells. Samples exhibited positive mutagenicity in the Ames test whereas they displayed a negative effect in the micronuclei test of bone marrow cells. The polygodial component of the plant is reported to have no mutagenic effects in the Ames test using TA 100, TA 98, TA 2637 strains of and mammalian cells in the V79/HGPRT test (Morales et al. 1992; Lajter et al. 2013).

2.7 CONCLUSION AND FUTURE PROSPECTS

P. hydropiper is used as a traditional medicine worldwide and as a spice in food recopies. Various compounds have been reported as active compounds, including phenolic acids, flavonoids, phenylpropanoid derivatives, and main sesquiterpenoids. The extracts and compounds from *P. hydropiper* show diverse biological activities, including anti-inflammatory, antioxidant, cytotoxic, and antimicrobial. Although various research reports show diverse biological activities, very few studies have focused on the mechanism of action. Also, as it is widely used in treating snakebites and insect bites, the effects of extracts and compounds still need to be well explored. Future studies should also establish the link between traditional uses, active compounds, and reported pharmacological activities.

REFERENCES

Arya, A.K., Durgapal, M., Bachheti, A., Joshi, K.K., Gonfa, Y.H., Bachheti, R.K. & Husen, A. (2022). Ethnomedicinal use, phytochemistry, and other potential application of aquatic and semiaquatic medicinal plants. *Evidence-Based Complementary and Alternative Medicine*, 2022, 4931556. https://doi.org/10.1155/2022/4931556

Ayaz, M., Junaid, M., Ahmed, J., Ullah, F., Sadiq, A., Ahmad, S., & Imran, M. (2014a). Phenolic contents, antioxidant and anticholinesterase potentials of crude extract, subsequent fractions and crude saponins from *Polygonum hydropiper* L. *BMC Complementary and Alternative Medicine*, 14, 145. https://doi.org/10.1186/1472-6882-14-145

Ayaz, M., Junaid, M., Subhan, F., Ullah, F., Sadiq, A., Ahmad, S., ... Shah, S.M. (2014b). Heavy metals analysis, phytochemical, phytotoxic and anthelmintic investigations of crude methanolic extract, subsequent fractions and crude saponins from *Polygonum hydropiper* L. *BMC Complementary and Alternative Medicine*, 14, 465. https://doi.org/10.1186/1472-6882-14-465

Ayaz, M., Junaid, M., Ullah, F., Sadiq, A., Khan, M.A., Ahmad, W., ... Ahmad, S. (2015). Comparative chemical profiling, cholinesterase inhibitions and anti-radicals properties of essential oils from *Polygonum hydropiper* L: a preliminary anti-Alzheimer's study. *Lipids in Health and Disease*, 14(1), 141. https://doi.org/10.1186/s12944-015-0145-8

Ayaz, M., Junaid, M., Ullah, F., Sadiq, A., Ovais, M., Ahmad, W., ... Zeb, A. (2016). Chemical profiling, antimicrobial and insecticidal evaluations of *Polygonum hydropiper* L. *BMC Complementary and Alternative Medicine*, 16(1), 502. https://doi.org/10.1186/s12906-016-1491-4

Ayaz, M., Junaid, M., Ullah, F., Subhan, F., Sadiq, A., Ali, G., ... Ahmad, S. (2017). Anti-Alzheimer's studies on ß-sitosterol isolated from *Polygonum hydropiper* L. *Frontiers in Pharmacology*, 8, 697. https://doi.org/10.3389/fphar.2017.00697

Ayaz, M., Ahmad, I., Sadiq, A., Ullah, F., Ovais, M., Khalil, A.T. & Devkota, H.P. (2020). *Persicaria hydropiper* (L.) Delarbre: a review on traditional uses, bioactive chemical constituents and pharmacological and toxicological activities. *Journal of Ethnopharmacology*, 251, 112516.

Ayaz, M., Wadood, A., Sadiq, A., Ullah, F., Anichkina, O., & Ghufran, M. (2022). In-silico evaluations of the isolated phytosterols from *Polygonum hydropiper* L against BACE1 and MAO drug targets. *Journal of Biomolecular Structure and Dynamics*, 40(20), 10230–10238.

Ayaz, M., Sadiq, A., Junaid, M., Ullah, F., Subhan, F., & Ahmed, J. (2017). Neuroprotective and anti-aging potentials of essential oils from aromatic and medicinal plants. *Frontiers in Aging Neuroscience*, 9, 168. https://doi.org/10.3389/fnagi.2017.00168

Ayaz, M., Sadiq, A., Wadood, A., Junaid, M., Ullah, F., & Zaman Khan, N. (2019). Cytotoxicity and molecular docking studies on phytosterols isolated from *Polygonum hydropiper* L. *Steroids*, 141, 30–35. https://doi.org/10.1016/j.steroids.2018.11.005

Aziman, N., Abdullah, N., Bujang, A., Mohd Noor, Z., Abdul Aziz, A., & Ahmad, R. (2021). Phytochemicals of ethanolic extract and essential oil of *Persicaria hydropiper* and their potential as antibacterial agents for food packaging polylactic acid film. *Journal of Food Safety*, 41(1), e12864.

Aziman, N., Abdullah, N., Noor, Z.M., Kamarudin, W.S.S.W., & Zulkifli, K.S. (2014). Phytochemical profiles and antimicrobial activity of aromatic Malaysian herb extracts against food-borne pathogenic and food spoilage microorganisms. *Journal of Food Science*, 79(4), M583–M592. https://doi.org/10.1111/1750-3841.12419

Barnes, C.S., & Loder, J.W. (1962). The structure of polygodial: a new sesquiterpene dialdehyde from *Polygonum hydropiper* L. *Australian Journal of Chemistry*, 15(2), 322–327. https://doi.org/10.1071/CH9620322

Bairagi, J., Saikia, P.J., Boro, F., & Hazarika, A. (2022). A review on the ethnopharmacology, phytochemistry and pharmacology of *Polygonum hydropiper* Linn. *Journal of Pharmacy and Pharmacology*, 74(5), 619–645.

Choudhary, R.K., Oh, S., & Lee, J. (2011). An ethnomedicinal inventory of knotweeds of Indian Himalaya. *Journal of Medicinal Plants Research*, 5(10), 2095–2103. https://doi.org/10.5897/JMPR.9000068

Dasari, R., De Carvalho, A., Medellin, D.C., Middleton, K.N., Hague, F., Volmar, M.N.M., ... Kornienko, A. (2015a). Synthetic and biological studies of sesquiterpene polygodial: activity of 9-epipolygodial against drug-resistant cancer cells. *ChemMedChem*, 10(12), 2014–2026. https://doi.org/10.1002/cmdc.201500360

Dasari, R., De Carvalho, A., Medellin, D.C., Middleton, K.N., Hague, F., Volmar, M.N.M., … Kornienko, A. (2015b). Wittig derivatization of sesquiterpenoid polygodial leads to cytostatic agents with activity against drug-resistant cancer cells and capable of pyrrolylation of primary amines. *European Journal of Medicinal Chemistry*, 103, 226–237. https://doi.org/10.1016/j.ejmech.2015.08.047

De La Chapa, J., Singha, P.K., Sallaway, M., Self, K., Nasreldin, R., Dasari, R., … Gonzales, C.B. (2018). Novel polygodial analogs P3 and P27: efficacious therapeutic agents disrupting mitochondrial function in oral squamous cell carcinoma. *International Journal of Oncology*, 53(6), 2627–2636. https://doi.org/10.3892/ijo.2018.4585

Devarajan, S., Yahiro, E., Uehara, Y., Kuroda, R., Hirano, Y., Nagata, K., Miura, S., Saku, K., & Urata, H. (2018). Depressor effect of the young leaves of *Polygonum hydropiper* Linn. in high-salt induced hypertensive mice. *Biomedicine & Pharmacotherapy*, 102, 1182–1187.

Duraipandiyan, V., Indwar, F., & Ignacimuthu, S. (2010). Antimicrobial activity of confertifolin from *Polygonum hydropiper*. *Pharmaceutical Biology*, 48(2), 187–190. https://doi.org/10.3109/13880200902902471

Fukuyama, Y., Sato, T., Asakawa, Y., & Takemoto, T. (1980). A potent cytotoxic warburganal and related drimane-type sesquiterpenoids from *Polygonum hydropiper*. *Phytochemistry*, 21(12), 2895–2898. https://doi.org/10.1016/0031-9422(80)85063-1

Fukuyama, Y., Sato, T., Miura, I., Asakawa, Y., & Takemoto, T. (1983). Hydropiperoside, a novel coumaryl glycoside from the root of *Polygonum hydropiper*. *Phytochemistry*, 22(2), 549–552. https://doi.org/10.1016/0031-9422(83)83043-X

Furuta, T., Fukuyama, Y., & Asakawa, Y. (1986). Polygonolide, an isocoumarin from *Polygonum hydropiper* possessing anti-inflammatory activity. *Phytochemistry*, 25(2), 517–520. https://doi.org/10.1016/S0031-9422(00)85513-2

Gairola, S., Sharma, J., & Bedi, Y.S. (2014). A cross-cultural analysis of Jammu, Kashmir and Ladakh (India) medicinal plant use. *Journal of Ethnopharmacology*, 155(2), 925–986. https://doi.org/10.1016/j.jep.2014.06.029

Ganesan, K., & Xu, B. (2017). Ethnobotanical studies on folkloric medicinal plants in Nainamalai, Namakkal District, Tamil Nadu, India. *Trends in Phytochemical Research*, 1(3), 153–168.

Haraguchi, H., Hashimoto, K., & Yagi, A. (1992). Antioxidative substances in leaves of *Polygonum hydropiper*. *Journal of Agricultural and Food Chemistry*, 40(8), 1349–1351. https://doi.org/10.1021/jf00020a011

Haraguchi, H., Ohmi, I., Sakai, S., Fukuda, A., Toihara, Y., Fujimoto, T., … Yagi, A. (1996). Effect of *Polygonum hydropiper* sulfated flavonoids on lens aldose reductase and related enzymes. *Journal of Natural Products*, 59(4), 443–445. https://doi.org/10.1021/np9601622

Hussain, F., Hadi, F. and Rongliang, Q. (2021). Effects of zinc oxide nanoparticles on antioxidants, chlorophyll contents, and proline in *Persicaria hydropiper* L. and its potential for Pb phytoremediation. *Environmental Science and Pollution Research*, 28(26), 34697–34713.

Imran, M., Khan, N., Shah, A.A., & Ahmad, I. (2019). Overweight and obesity prevalence pattern and associated risk factors among the people of Malakand Division, Khyber Pakhtunkhwa Pakistan. *Arabian Journal for Science and Engineering*, 44(1), 65–74. https://doi.org/10.1007/s13369-018-3457-y

Joshi, A.R., & Edington, J.M. (1990). The use of medicinal plants by two village communities in the central development region of Nepal. *Economic Botany*, 44, 71–83. https://doi.org/10.1007/BF02861069

Joshi, A.R., & Joshi, K. (2004). Insecticidal plants of the Bagmati Watershed, Nepal: ethnobotany and traditional uses. *Bionotes*, 6(2), 37–39.

Joshi, K. (2004). Documentation of medicinal plants and their indigenous uses in Likhu sub watershed, Nepal. *Journal of Non-Timber Forest Products*, 112(2), 86–93.

Kawaguchi, R., & Kim, K.W. (1937). Chemical constituents of *Persicaria hydropiper* (Part 2). *Journal of the Pharmaceutical Society of Japan*, 57(8), 767–769.

Kawaguchi, R., & Kim, K.W. (1940). Chemical constituents of *Persicaria hydropiper* (Part 3). *Journal of the Pharmaceutical Society of Japan*, 60, 352–353.

Kayani, S., Ahmad, M., Zafar, M., Sultana, S., Khan, M.P.Z., Ashraf, M.A., … Yaseen, G. (2014). Ethnobotanical uses of medicinal plants for respiratory disorders among the inhabitants of Gallies – Abbottabad, Northern Pakistan. *Journal of Ethnopharmacology*, 156, 47–60. https://doi.org/10.1016/j.jep.2014.08.005

Khan, M.A.S.A., Sultana, F., Rahman, M.H., Roy, B., & Anik, S.I. (2011). Status and ethno-medicinal usage of invasive plants in traditional health care practices: a case study from northeastern Bangladesh. *Journal of Forestry Research*, 22(4), 649–658. https://doi.org/10.1007/s11676-011-0174-8

Khatun, A., Imam, M.Z., & Rana, M.S. (2015). Antinociceptive effect of methanol extract of leaves of *Persicaria hydropiper* in mice. *BMC Complementary and Alternative Medicine*, 15, 63. https://doi.org/10.1186/s12906-015-0558-y

Kiem, P. Van, Nhiem, N.X., Cuong, N.X., Hoa, T.Q., Huong, H.T., Huong, L.M., … Kim, Y.H. (2008). New phenylpropanoid esters of sucrose from *Polygonum hydropiper* and their antioxidant activity. *Archives of Pharmacal Research*, 31(11), 1477–1482. https://doi.org/10.1007/s12272-001-2133-y

Kimura, H., Ishihara, T., Michida, M., Ogawa, S., Akihiro, T., & Yokota, K. (2014). Short communication: identification and quantitative analysis of polyphenolic compounds from the indigo plant (*Polygonum tinctorium* Lour). *Natural Product Research*, 28(7), 492–495. https://doi.org/10.1080/14786419.2013.871719

Kuroiwa, K., Shibutani, M., Inoue, K., Lee, K.Y., Woo, G.H., & Hirose, M. (2006). Subchronic toxicity study of water pepper extract in F344 rats. *Food and Chemical Toxicology*, 44(8), 1236–1244. https://doi.org/10.1016/j.fct.2006.01.020

Lajter, I., Zupkó, I., Molnár, J., Jakab, G., Balogh, L., Vasas, A., & Hohmann, J. (2013). Antiproliferative activity of Polygonaceae species from the Carpathian Basin against human cancer cell lines. *Phytotherapy Research*, 27(1), 77–85. https://doi.org/10.1002/ptr.4690

Lee, S.H., Kim, B., Oh, M.J., Yoon, J., Kim, H.Y., Lee, K.J., … Choi, K.Y. (2011). *Persicaria hydropiper* (L.) Spach and its flavonoid components, isoquercitrin and isorhamnetin, activate the Wnt/-βcatenin pathway and inhibit adipocyte differentiation of 3T3-L1 cells. *Phytotherapy Research*, 25(11), 1629–1635. https://doi.org/10.1002/ptr.3469

Ma, C.J., Lee, K.Y., Jeong, E.J., Kim, S.H., Park, J., Choi, Y.H., … Sung, S.H. (2010). Persicarin from water dropwort (*Oenanthe javanica*) protects primary cultured rat cortical cells from glutamate-induced neurotoxicity. *Phytotherapy Research*, 24(6), 913–918. https://doi.org/10.1002/ptr.3065

Mabberley, D. (2008). *Mabberley's Plant-Book*, 3rd ed. Cambridge: Cambridge University Press.

Mahnashi, M.H., Alqahtani, Y.S., Alyami, B.A., Alqarni, A.O., Ullah, F., Wadood, A., … Ayaz, M. (2021). Cytotoxicity, anti-angiogenic, antitumor and molecular docking studies on phytochemicals isolated from *Polygonum hydropiper* L. *BMC Complementary Medicine and Therapies*, 21(1): 239.

Mahnashi, M.H., Alqahtani, Y.S., Alyami, B.A., Alqarni, A.O., Ayaz, M, Ghufran, M., … Murthy, H.C.A. (2022b). Phytochemical analysis, α-glucosidase and amylase inhibitory, and molecular docking studies on *Persicaria hydropiper* L. leaves essential oils. *Evidence-based Complementary and Alternative Medicine*, 2022, 7924171.

Mahnashi, M.H., Alqahtani, Y.S., Alyami, B.A., Alqarni, A.O., Alqahl, S.A., Ullah, F., … Ayaz, M. (2022a). HPLC-DAD phenolics analysis, α-glucosidase, α-amylase inhibitory, molecular docking and nutritional profiles of *Persicaria hydropiper* L. *BMC Complementary Medicine and Therapies*, 22(1), 26.

Manandhar, N.P. (2000). *Plants and People of Nepal*. Portland, OR: Timber Press, Inc.

Manandhar, N.P. (2001). Herbal veterinary practices in Nepal. *Nepal Journal of Science and Technology*, 3, 65–68.

Martini, L.H., Cereser, L., Junior, I.Z., Jardim, F.M.A., Vendite, D.A., Frizzo, M.E.D.S., … Souza, D.O. (2006). The sesquiterpenes polygodial and drimanial in vitro affect glutamatergic transport in rat brain. *Neurochemical Research*, 31(3), 431–438. https://doi.org/10.1007/s11064-005-9033-3

Miyazawa, M., & Tamura, N. (2007). Components of the essential oil from sprouts of *Polygonum hydropiper* L. ('Benitade'). *Flavour and Fragrance Journal*, 22(3), 188–190. https://doi.org/10.1002/ffj.1779

Mohamed Isa, S.S.P., Ablat, A., & Mohamad, J. (2018). The antioxidant and xanthine oxidase inhibitory activity of plumeria rubra flowers. *Molecules*, 23(2). https://doi.org/10.3390/molecules23020400

Morales, P., Andersson, M., Lewan, L., & Sterner, O. (1992). Structure-activity relationships for unsaturated dialdehydes 6. The mutagenic activity of 11 compounds in the V79/HGPRT assay. *Mutation Research – Fundamental and Molecular Mechanisms of Mutagenesis*, 268(2), 315–321. https://doi.org/10.1016/0027-5107(92)90237-V

Narasimhulu, G., Reddy, K.K., & Mohamed, J. (2014). The genus *Polygonum* (Polygonaceae): an ethnopharmacological and phytochemical perspectives – review. *International Journal of Pharmacy and Pharmaceutical Sciences*, 6(2), 21–45.

Nasir, A., Khalil, A.A.K., Bhatti, M.Z., Rehman, A.U., Li, J. and Parveen, Z. (2021). Review on pharmacological and phytochemical prospects of traditional medicinal plant: *Persicaria hydropiper* (smartweed). *Current Topics in Medicinal Chemistry*, 21(12), 1027–1036.

Ningthoujam, S.S., Das Talukdar, A., Potsangbam, K.S., & Choudhury, M.D. (2013). Traditional uses of herbal vapour therapy in Manipur, North East India: an ethnobotanical survey. *Journal of Ethnopharmacology*, 147(1), 136–147. https://doi.org/10.1016/j.jep.2012.12.056

Noor Hashim, N.H., Abas, F. Shaari K. and Lajis, N.H. (2012). LC-DAD-ESIMS/MS characterization of antioxidant and anticholinesterase constituents present in the active fraction from *Persicaria hydropiper*. *LWT – Food Science and Technology*, 46(2), 468–476.

Oany, A.R., Siddikey, A.A., Hossain, M.U., Islam, R., & Emran, A.-A. (2017). A preliminary evaluation of cytotoxicity, antihyperglycemic and antinociceptive activity of *Polygonum hydropiper* L. ethanolic leaf extract. *Clinical Phytoscience*, 2, 2. https://doi.org/10.1186/s40816-016-0016-5

Okada, M., & Mitsuhashi, H. (eds.). (2002). *Newly Revised Illustrated Medicinal Plants of The World*. Tokyo: Hokuryukan Co., Ltd.

Ono, K., Nakao, M., Toyota, M., Terashi, Y., Yamada, M., Kohno, T., & Asakawa, Y. (1998). Catechin production in cultured *Polygonum hydropiper* cells. *Phytochemistry*, 49(7), 1935–1939. https://doi.org/10.1016/S0031-9422(98)00426-9

Peng, Z.F., Strack, D., Baumert, A., Subramaniam, R., Goh, N.K., Chia, T.F., … Chia, L.S. (2003). Antioxidant flavonoids from leaves of *Polygonum hydropiper* L. *Phytochemistry*, 62(2), 219–228. https://doi.org/10.1016/S0031-9422(02)00504-6

Prota, N., Mumm, R., Bouwmeester, H.J., & Jongsma, M.A. (2014). Comparison of the chemical composition of three species of smartweed (genus *Persicaria*) with a focus on drimane sesquiterpenoids. *Phytochemistry*, 108, 129–136. https://doi.org/10.1016/j.phytochem.2014.10.001

Rahman, A., & Gondha, R. (2014). Taxonomy and traditional medicine practices on Malvaceae (mallow family) of Rajshahi, Bangladesh. *Open Journal of Botany*, 1(2), 19–24.

Rahman, A.H.M.M. (2013). Traditional medicinal plants used in the treatment of different skin diseases of Santals at Abdullahpur Village under Akkelpur Upazilla of Joypurhat District, Bangladesh. *Biomedicine and Biotechnology*, 1(2), 17–20. https://doi.org/10.12691/bb-1-2-4

Rahman, E., Goni, S.A., Rahman, M.T., & Ahmed, M. (2002). Antinociceptive activity of *Polygonum hydropiper*. *Fitoterapia*, 73(7–8), 704–706. https://doi.org/10.1016/S0367-326X(02)00239-3

Rahmatullah, M., Ferdausi, D., Mollik, M.A.H., Jahan, R., Chowdhury, M.H., & Haque, W.M. (2010). A survey of medicinal plants used by Kavirajes of Chalna area, Khulna district, Bangladesh. *African Journal of Traditional, Complementary and Alternative Medicines*, 7(2), 91–97. https://doi.org/10.4314/ajtcam.v7i2.50859

Raihan, M.O., Khalequeuzzaman, M., Brishti, A., Tareq, S.M., Hossain, A., & Rana, S. (2012). Anthelmintic and antiproliferative activity of aerial parts of *Persicaria hydropiper*. *Der Pharmasia Sinica*, 3, 104–110.

Rajbhandari, K.R. (2001). *Ethnobotany of Nepal*. Kathmandu, Nepal: Ethnobotanical Society of Nepal.

Said, M.S., Chinchansure, A.A., Nawale, L., Durge, A., Wadhwani, A., Kulkarni, S.S., … Joshi, S.P. (2015). A new butenolide cinnamate and other biological active chemical constituents from *Polygonum glabrum*. *Natural Product Research*, 29(22), 2080–2086. https://doi.org/10.1080/14786419.2015.1004674

Sanghai Scientific and Technical Publisher (Ed.). (1985). *Chinese Materia Medicine Dictionary*. Tokyo: Shogakukan.

Shahed-Al-Mahmud, M., & Lina, S.M.M. (2017). Evaluation of sedative and anxiolytic activities of methanol extract of leaves of *Persicaria hydropiper* in mice. *Clinical Phytoscience*, 3, 20. https://doi.org/10.1186/s40816-017-0056-5

Sharif, S., Shahriar, M., Haque, A., Chowdhury, Z.S., Islam, R., & Bhuiyan, M.A. (2013). In-vitro antioxidant activities, antinociceptive and neuropharmacological activities of *Polygonum hydropiper*. *Journal of Biology*, 3(19), 61–71.

Shiraliyeva, G. (2017). *Persicaria* species in flora of Azerbaijan and ethnobiology of their use. *International Journal of Current Microbiology and Applied Sciences*, 6(1), 527–531.

Starkenmann, C., Luca, L., Niclass, Y., Praz, E., & Roguet, D. (2006). Comparison of volatile constituents of *Persicaria odorata* (Lour.) Soják (*Polygonum odoratum* Lour.) and *Persicaria hydropiper* L. Spach (*Polygonum hydropiper* L.). *Journal of Agricultural and Food Chemistry*, 54(8), 3067–3071. https://doi.org/10.1021/jf0531611

Sultana, R., Hossain, R., Adhikari, A., Ali, Z., Yousuf, S., Choudhary, M.I., … Zaman, M.S. (2011). Drimane-type sesquiterpenes from *Polygonum hydropiper*. *Planta Medica*, 77(16), 1848–1851. https://doi.org/10.1055/s-0030-1271178

Suzuki, H. (2005). *Encyclopedia of Traditional Oriental Drugs*. Tokyo: Ishiyaku Publishers, Inc.

Takabe, W., Yamaguchi, T., Hayashi, H., Sugimura, N., Yagi, M., & Yonei, Y. (2018). Identification of antiglycative compounds in Japanese red water pepper (red leaf variant of the *Persicaria hydropiper* sprout). *Molecules*, 23(9), 2319. https://doi.org/10.3390/molecules23092319

Tao, J., Wei, Y., & Hu, T. (2016). Flavonoids of *Polygonum hydropiper* L. attenuates lipopolysaccharide-induced inflammatory injury via suppressing phosphorylation in MAPKs pathways. *BMC Complementary and Alternative Medicine*, 16, 25. https://doi.org/10.1186/s12906-016-1001-8

Tong, X., Li, X. Ayaz, M. Ullah, F. Sadiq, A. Ovais, M. Shahid, M. Khayrullin, M. and Hazrat, A.(2021). Neuroprotective studies on *Polygonum hydropiper* L. essential oils using transgenic animal models. *Frontiers in Pharmacology*, 11, 580069.

Uddin, M.Z., Abul Hassan, M.D., Rahman, M., & Arefin, K. (2012). Ethno-medico-botanical study in Lawachara National Park, Bangladesh. *Bangladesh Journal of Botany*, 41(1), 97–104. https://doi.org/10.3329/bjb.v41i1.11087

Van Duong, N. (1993). *Medicinal Plants of Vietnam, Cambodia and Laos*. Vietnam: N.V.D. Press.

Watanabe, M., Devkota, H.P., Sugimura, K., & Watanabe, T. (2018). *A guidebook of Medicinal Plant Park*. Kumamoto, Japan: School of Pharmacy, Kumamoto University.

Xiao, H., Rao Ravu, R., Tekwani, B.L., Li, W., Liu, W.B., Jacob, M.R., ... Wang, W. (2017). Biological evaluation of phytoconstituents from *Polygonum hydropiper*. *Natural Product Research*, 31(17), 2053–2057. https://doi.org/10.1080/14786419.2016.1269094

Yagi, A., Uemura, T., Okamura, N., Haraguchi, H., Imoto, T., & Hashimoto, K. (1994). Antioxidative sulphated flavonoids in leaves of *Polygonum hydropiper*. *Phytochemistry*, 35(4), 885–887. https://doi.org/10.1016/S0031-9422(00)90632-0

Yang, H.J., Kim, E.H., & Park, S.N. (2008). Antioxidative activity and component analysis of *Cayratia japonica* extract. *Journal of the Society of Cosmetic Scientists of Korea*, 34(2), 117–127.

Yang, X., Wang, B.C., Zhang, X., Yang, S.P., Li, W., Tang, Q., & Singh, G.K. (2011). Simultaneous determination of nine flavonoids in *Polygonum hydropiper* L. samples using nanomagnetic powder three-phase hollow fibre-based liquid-phase microextraction combined with ultrahigh performance liquid chromatography-mass spectrometry. *Journal of Pharmaceutical and Biomedical Analysis*, 54(2), 311–316. https://doi.org/10.1016/j.jpba.2010.08.026

Yusif, K.S., & Blinova, K.F. (1984). Flavonoid aglycones of *Polygonum hydropiper*. *Chemistry of Natural Compounds*, 20(5), 625–626. https://doi.org/10.1007/BF00580088

Zeb, A. (2020). Concept, mechanism, and applications of phenolic antioxidants in foods. *Journal of Food Biochemistry*, 44(9), e13394.

Zhang, W., Pan, Y., Qu, S., Wang, D., Cheng, S., & Liu, X. (2018). Anti-inflammatory effects of an extract of *Polygonum hydropiper* stalks on 2,4,6-trinitrobenzenesulphonic acid-induced intestinal inflammation in rats by inhibiting the NF-κ B pathway. *Mediators of Inflammation*, 2018, 1–10. https://doi.org/10.1155/2018/6029135

Zhao, D., Zheng, L., Qi, L., Wang, S., Guan, L., Xia, Y., & Cai, J. (2016). Structural features and potent antidepressant effects of total sterols and β-sitosterol extracted from *Sargassum horneri*. *Marine Drugs*, 14(7), 123. https://doi.org/10.3390/md14070123

3 Ethnobotany, Phytochemistry, and Pharmacological Activity of *Marsilea minuta*

Senjuti Banerjee,[1,2] Kasturi Sarkar,[1] and Parames C. Sil[3,]*
[1] Department of Microbiology, St. Xavier's College, Kolkata, India
[2] Department of Botany, Bhirab Ganguly College, Kolkata, India
[3] Department of Molecular medicine Bose Institute, Kolkata, India
[*] Corresponding author (Parames C. Sil): parames@jcbose.ac.in

3.1 INTRODUCTION

Nutrition and healthcare remain strongly interconnected in traditional societies. Many plants are used as food as well as for their medicinal properties. Researchers are looking for the active ingredients present in these traditional medicinal plants. Identifying the active molecules and their mode of action can lead to discovering new drugs. The vital bioactive components found in medicinal plants are tannins, alkaloids, carbohydrates, terpenoids, steroids, and flavonoids, which are secondary metabolites and are chemically diverse. These molecules are widely used as pharmaceuticals and have been found to have antimicrobial, antioxidant, and many other activities (Parihar and Parihar 2006; Sen et al. 2008, 2011). The phenolic compounds act as scavengers of reactive oxygen species and electrophiles and as metal chelators. Consumption of a diet rich in phenolic compounds is associated with a low incidence of coronary heart disease, atherosclerosis, certain types of cancer, and stroke. Similar properties are associated with flavonoids and flavones. More than 4,000 flavonoids have been identified, purified, and analysed for their properties and structures. Alkaloids are also known for their antimicrobial properties (Mahesh and Satish 2008).

Tannins are polyphenols having antioxidant and immunological properties and are found in fruits, cortexes, leaves, roots, and other parts of plants. India is considered one of the mega biodiversity countries in the global panorama. In India, the history of herbal medicine dates to ancient human civilization, and valuable information on traditional medical practices is available (Vincent et al. 2012). Natural products, either as pure compounds or as standardized plant extracts, provide unlimited opportunities to discover new drugs because of the unmatched availability of chemical diversity. Plant products and related drugs treat 87% of commonly categorized diseases (Grover et al. 2002). Besides the phytochemicals or secondary metabolites (Okwu et al. 2005), many other chemical components have yet to be discovered (Mallikharjuna et al. 2007). These active components provide the plant's defence against microbial attacks and exhibit medicinal properties for treating several diseases. *Marsilea minuta* is an aquatic medicinal plant with phytochemicals that possess various properties such as antimicrobial, anti-aggressive, antidepressant, antipyretic, analgesic, antidiabetic, antitumour, and antitussive activities. In the present chapter, the aquatic fern *M. minuta* will be discussed in light of its composition, traditional uses, and future prospects in pharmacy.

M. minuta and *M. quadrifolia* (Ripa et al. 2009) are closely related and belong to the same group along with some other species, like *M. angustifolia*, *M. drummondii*, *M. fadeniana*, and *M. crenata*. *M. crenata* is the synonym of *M. minuta.*

DOI: 10.1201/9781003256830-3

3.1.1 Habitat and Distribution

3.1.1.1 Habitat

M. minuta is a small, soft herb, light green in colour, having a creeping rhizome that produces erect leaves alternately at intervals. Most members of the Marsileaceae family are either aquatic or semi-aquatic in nature. *M. minuta* is found in wet lowlands like shallow ponds and ditches in submerged conditions. It grows in paddy fields readily and is invasive.

3.1.1.2 Distribution

The genus *Marsilea* is a common water fern distributed worldwide with 60–65 species. They grow in tropical and temperate regions, but most species are distributed in tropical Africa and Australia. About ten species have been reported in India, *M. minuta* being an important species among them (Figure 3.1).

3.1.2 Morphology, Anatomy, and Reproduction

3.1.2.1 Morphology

The sporophytic plant body of *M. minuta* shows the differentiation of stem, leaves, and roots. The stem is a long, slender, and freely branched rhizome of indefinite growth observed on or just below the soil surface. There are nodes and internodes. The leaves arise from the nodes and arrange alternately in two rows on the upper side of the rhizomatous stem. The leaves are palmately compound with four leaflets and have long petioles. Young leaves show circinate vernation and, at maturity, the pinnae extend perpendicularly to the petiole. The venation is a closed reticulate type. The short-lived primary roots are replaced by adventitious roots, which arise from the nodes of the lower side of the rhizomatous stem (Figure 3.2 A,B).

3.1.2.2 Anatomy

Stem: a transverse section of the rhizomatous stem of *M. minuta* shows the epidermis, cortex, and stele from the periphery to the centre. The epidermis is composed of thick-walled cells arranged in

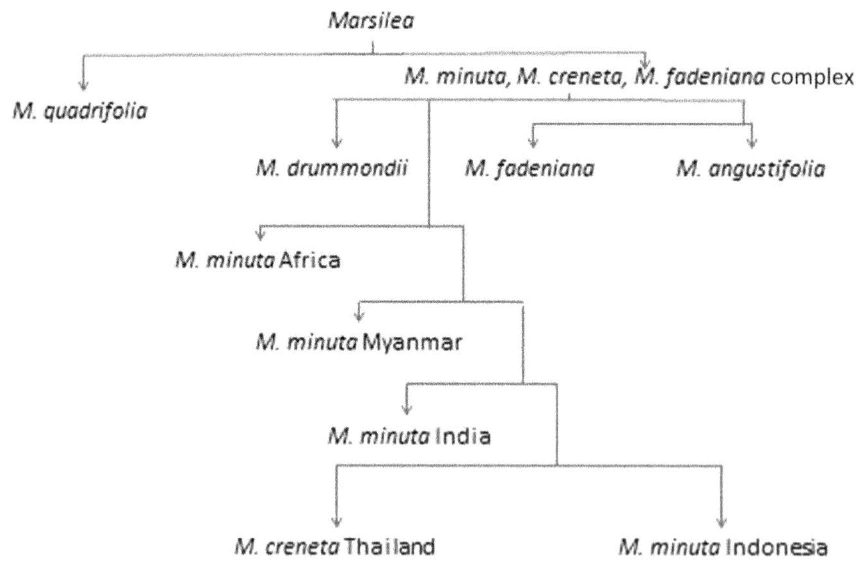

FIGURE 3.1 Hierarchical representation of *M. minuta* complex.

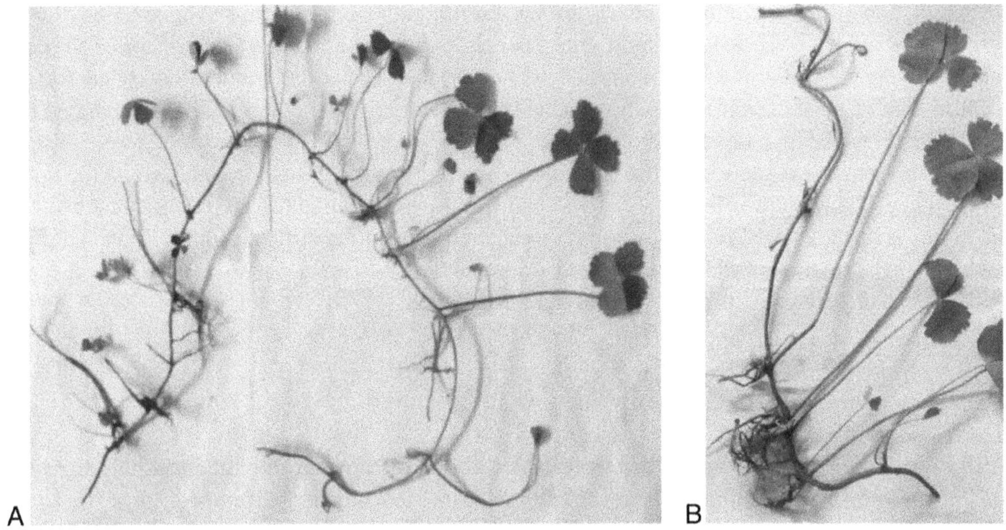

FIGURE 3.2 Photographs of *M. minuta*.

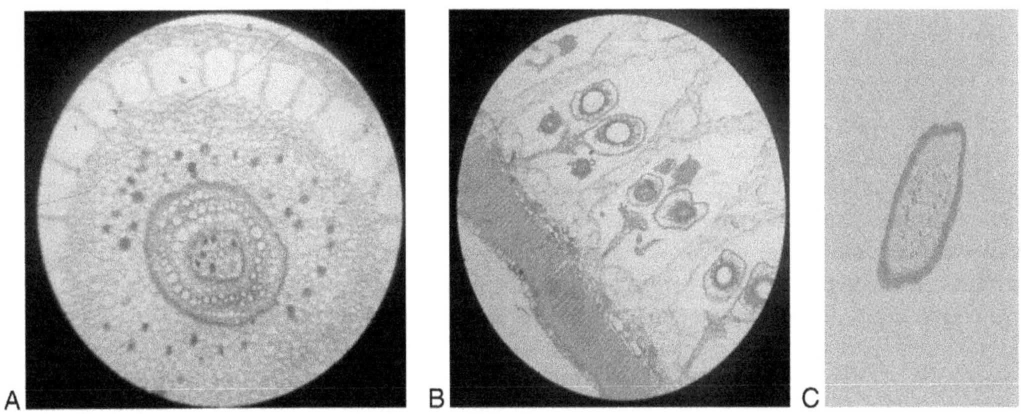

FIGURE 3.3 (A) Transverse section of rhizomatous stem of *M. minuta* under light microscope (10× objective lens). (B) Vertical section of spore-bearing structure (sporocarp) showing megaspores under light microscope (10× objective lens). (C) Vertical section of a single sporocarp under simple microscope (4× eye piece lens).

a compact manner. The cortex is broad and differentiated into three layers: outer, middle, and inner. The outer cortex comprises parenchymatous cells with large air spaces, the middle cortex is sclerenchymatous, and the inner cortex is again parenchymatous but without air space. At the centre, there is an amphiphloic solenostele, i.e. xylem occurs in the form of a ring and the phloem surrounds either side of the xylem ring. The stele is bounded externally by outer endodermis and outer pericycle and internally by inner endodermis and inner pericycle. The centre of the stele is occupied by parenchymatous pith (Figure 3.3 A).

Leaves: the transverse section of the petiole is differentiated into the epidermis, cortex, and stele. The epidermis is single layered and composed of cutinized, rectangular cells. The cortex is differentiated into the outer cortex and inner cortex, while the stele is protostelic, mostly triangular in outline with an exarch xylem. The vertical section of the leaflet shows the epidermis, stomata, mesophyll, and vascular bundle. Stomata are present in the lower epidermis. The mesophyll tissue with vertical airspaces is differentiated into palisade parenchymatous tissue, the upper epidermis and spongy parenchymatous tissue, and the lower epidermis. The vascular bundle is concentric with a central core of the mesarch xylem surrounded by phloem. Each vascular bundle is separated from the mesophyll tissue by a layer of endodermis.

Root: the transverse section of the root shows three distinct regions, the epidermis, the cortex, and the stele. The epidermis is single layered and composed of cuticularized parenchymatous cells. The cortex is differentiated into three zones: aerenchymatous outer cortex, radially separated by air spaces, parenchymatous middle cortex, and sclerenchymatous inner cortex. Successive layers of endodermis and pericycle internally bound the inner cortex.

3.1.2.3 Reproduction

M. minuta reproduces both vegetatively and sexually.

Vegetative reproduction: the subterranean branches of the rhizome develop tubers, which store food in the form of oil globules under certain conditions. The tubers again germinate to give rise to new plants under favourable conditions.

Reproduction by spores: *M. minuta* produces both microspores and megaspores in microsporangia and megasporangia, respectively, which are enclosed in a bean-shaped structure called a sporocarp (Figure 3.3B,C). The sporocarps can withstand desiccation and can germinate even after 20–25 years (Nagalingum et al. 2006).

3.2 CHEMICAL COMPOSITION OF *MARSILEA MINUTA*

Gas chromatography–mass spectrometry analysis of *M. minuta* leaf extract in 50% hexane and 50% methanol identified 12 compounds present in the leaves. The major compounds are benzoic acid-4-ethoxy-, ethyl ester (43.39%), monoester of benzoic acid and farnesol acetate (18.42%), and sesquiterpene compounds. Apart from phenol, flavonoids, saponins, quinones, tannins, terpenoids, coumarins, sugar, and anthroquinones (Table 3.1), *M. minuta* also contains marsiline, marsileagenin-A, quercetin-3-*O*-glucoside, chalcone-*O*-glucoside, kaempferol-3-*O*-glucoside, quercetin-3-*O*-galactoside, and quercetin-3-rutinoside (rutin). Flavonoids like naringenin-7-*O*-glucoside and phytosterols like β-sitosterol and hentriacontane have also been reported (de Britto et al. 2013; Revathi and Catharin 2014).

The methanolic extract of the leaves of *M. minuta* showed the presence of around 36 phytocompounds including *n*-hexadecanoic acid (44.41%) ($C_{16}H_{32}O_2$); (*Z*)6,(*Z*)9-pentadecadien-1-ol (35.49%) ($C_{15}H_{28}O$); phytol (5.10%) ($C_{20}H_{40}O$); 2-cyclohexane-1-one,4-hydroxy-3,5,6-trimethyl-4

TABLE 3.1
Quantitative Analysis of Phytochemicals Present in *M. minuta* (in mg/ml) in Different Solvents

Phytochemical	Petroleum	Benzene	Chloroform	Methanol	Aqueous
Alkaloids (mg/ml)	09.25 ± 0.18	11.20 ± 0.35	10.33 ± 0.29	15.10 ± 0.15	08.90 ± 0.25
Flavonoids (mg/ml)	11.85 ± 0.51	13.80 ± 0.55	12.35 ± 0.10	16.45 ± 0.20	10.32 ± 0.15
Phenolic compounds (mg/ml)	08.53 ± 0.10	09.25 ± 0.30	07.55 ± 0.45	13.45 ± 0.20	06.13 ± 0.24
Saponins (mg/ml)	05.92 ± 0.04	07.22 ± 0.44	06.50 ± 0.35	10.20 ± 0.55	04.70 ± 0.35

TABLE 3.2
Chemical Constituents and Their PubChem Compound Identification Numbers Found in Different Parts of *M. minuta*

Parts of the plant	Names of the compounds present
Leaves	Marsilin (PubChem CID: 84258)
	Phytol (PubChem CID: 5280435)
	Kaempferol-3-*O*-glucoside (PubChem CID: 5282102)
	Chalcone-3-*O*-glucoside (PubChem CID: 5318659)
	Rutin (PubChem CID: 5280805)
	Octadecanoic acid (PubChem CID: 5281)
	10-Octadecanoic acid methyl ester (PubChem CID: 8201)
Stem	1,19-Eicosadiene (PubChem CID 519006)
	N-Hexadecanoic acid (PubChem CID 985)
	β-Sitosterol (PubChem CID 222284)
	Hentriacontane (PubChem CID 12410)
Root	Quercetin-3-*O*-galactoside (PubChem CID 6453392)
	Naringenin-7-*O*-glucoside (PubChem CID 92794)
	Quercetin-3-*O*-glucoside (PubChem CID 5280804)

(3-oxo-1-butenyl) (2.25%) ($C_{13}H_{18}O_3$); 9,12,15-octadecatrienoic acid (Z,Z,Z) (2.0%) ($C_{18}H_{30}O_2$); 3,7,11,15-tetramethyl-2-hexadecan-1-ol (1.99%) ($C_{20}H_{40}O$); and benzofuran, 2,3-dihydro- (1.1%) (C_8H_8O) (Table 3.2) (Arokiyaraj et al. 2018).

The methanolic extract of the stem of *M. minuta* showed the presence of 27 bioactive compounds including 1,19-eicosadiene (40.94%) ($C_{20}H_{38}$); *n*-hexadecanoic acid (20.13%) ($C_{16}H_{32}O_2$); 2(*R*)3(*S*)-1,2,3,4-butanetetrol (16.31%) ($C_4H_{10}O_4$); glycerine (6.70%) ($C_3H_8O_3$); octadecanoic acid (4.24%) ($C_{18}H_{36}O_2$); benzofuran, 2,3-dihydro- (1.73%) (C_8H_8O); 3,7,11,15-tetramethyl-2-hexadecen-1-ol (1.55%) ($C_{20}H_{40}O$); *n*-hexadecanoic acid methyl ester (1.15%) ($C_{17}H_{34}O_2$) and 10-octadecenoic acid methyl ester (1.42%) ($C_{19}H_{36}O_2$) (Table 3.2) (Arokiyaraj et al. 2018).

The considerable amount of carbohydrates, protein, and amino acids reveals the nutritional value of *M. minuta*. Quantitative analysis of biochemical constituents of *M. minuta* reveals carbohydrates are present in the highest amount, followed by proteins, amino acids, flavonoids, saponins, and fats. Saponins have antidiarrhoeal properties as they inhibit histamine release *in vitro*, anticancer properties as they possess membrane permeabilizing properties, and antihelminthic properties as they lead to vacuolization and disintegration of teguments. The flavonoids are antimicrobial as they bind to the cell wall of bacteria and antidiarrhoeal; they inhibit the release of autocoids and prostaglandins, and inhibit contractions caused by spasmogens (Prashant et al. 2011). Juice made from the leaves of *M. minuta* is a diuretic and a febrifuge. It is also used to treat snakebites and abscesses. The medicinal value of the floral species may be transferred to the human system through the production of food items, preferably desserts, which can be easily prepared with the extracts of the leaves (Revathi and Catharin 2014).

3.3 ETHNOBOTANICAL USES

The knowledge of medicinal plants is traditional and age-old. The ethnomedicinal aspects of medicinal plants used by local tribes have been studied and documented (Soni and Singh 2012). *M. minuta* is consumed as a whole, or parts can be used for specific purposes.

3.3.1 Uses of Cooked Plant

The whole plant is consumed as a vegetable and has extensive applications in treating several disorders such as psychopathy, diarrhoea, respiratory diseases, and skin diseases.

3.3.2 Uses of Raw Plant

The juice extracted from the fresh leaves and shoots is used to treat cough and other respiratory troubles by the rural people in West Bengal, India, and Bangladesh (Soni and Singh 2012). The extract of *M. minuta* has been found to have an antitussive effect by prolonging the latency and reducing cough in *in vivo* experiments. It also shows dose-dependent expectorant activity (Chakraborty et al. 2013). The raw juice or paste of the fresh plant is taken to fight against cough, with and without garlic, for adults and babies, respectively (Marndi and Kumar 2017). Regular consumption of whole plant or the young fronds has beneficial effects on lowering hypertension and treating sleeping disorders, insomnia, and mental problems (Bhattamisra et al. 2007). *M. minuta* has been found to treat epilepsy and spastic conditions of the leg and muscle (Chakraborty et al. 2013). It is believed that fresh plant paste taken with curd can cure epilepsy. Two drops of crushed leaf juice in the nostrils twice a day can treat migraine, whereas squeezed juice from the leaves if massaged on the tendons twice daily for a few days, can treat rheumatism (Rahmatullah et al. 2010). There are reports that *M. minuta* can reduce cholesterol, triglycerides, diabetes, anxiety, depression, aggressiveness, and amnesia (Upreti et al. 2009; Soni and Singh 2012). The leaves cooked with rice are used to treat indigestion and gastrointestinal tract-related disorders. The leaf juice stops nose bleeding (Sajini et al. 2019).

3.3.3 Uses in Combination with Other Plants

M. minuta leaves rolled in *Shorea robusta* leaves can be applied to treat gum-related problems (Soni and Singh 2012). Regular consumption of petioles and leaves cooked in oil with salt and spices relieves hypertension, sleep disorder, and headache. The plant cooked with enough garlic, red chilli, cumin, and coriander, covered with a bamboo basket, and consumed in the evening, is believed to help in curing some nervous disorders (Tiwari et al. 2009). The crushed plant with sugar candy or honey can cure an infant's diarrhoea. Ground spores mixed with flour during bread preparation can enrich its starch content (Sen et al. 2008).

3.4 PHARMACOLOGICAL ACTIVITIES

Based on the ethnobotanical knowledge of *M. minuta*, researchers are now trying to elucidate the role of different parts of this plant in various pharmacological activities. The extracts of different parts of the plant have been found to possess beneficial properties such as antioxidant, 2,2-diphenyl-1-picrylhydrazyl (DPPH) scavenging, antimicrobial, antitussive, anti-aggressive, antidepressant, antipyretic, analgesic, antidiabetic, antitumour, and hepatoprotective activities. On the other hand, detrimental effects such as the antifertility activity of the plant extract have also been documented.

3.4.1 Antioxidant Activity

The methanolic extract of *M. minuta* leaves has been shown to have significant antioxidant activity. The crude leaf extract (50 μg/ml) could reduce 60% of ferric reducing antioxidant power activity. The half-maximal inhibitory concentration (IC_{50}) value of *M. minuta* leaf extracts was found to be 37.48 μg/ml compared to the value of 7.42 μg/ml of the standard ethylenediaminetetraacetic acid (EDTA). The antioxidant ability of *M. minuta* extract might be due to the presence of phenolic compounds, which chelate metal ions and prevent the development of free radicals (Arokiyaraj et al.

2018). In another study, Chowdhury et al. (2017) showed the antioxidant property of the methanolic extract of *M. minuta* with the highest free radical scavenging activity at a concentration of 200 µg/ml.

3.4.2 DPPH Scavenging Activity

The methanolic extract of the leaves of *M. minuta* exerted significant dose-dependent inhibition of DPPH. The IC_{50} value of *M. minuta* extracts was reported to be 8.94 µg/ml compared to the value of 5.77 µg/ml of the standard vitamin C (Gopalakrishnan and Udayakumar 2014).

3.4.3 Antimicrobial Activity

The aqueous extract of *M. minuta* leaves was found to have antibacterial activity against *Bacillus subtilis*, *Staphylococcus aureus*, *Enterococcus faecalis*, *Klebsiella pneumoniae*, and *Pseudomonas aeruginosa* (Mahesh and Satish 2008). The minimum inhibitory concentrations of the extract of *M. minuta* leaves against the bacteria lie within the concentration range of 125–250 µg/ml. The extract has maximum inhibitory activity against *P. aeruginosa* (Arokiyaraj et al. 2018). The phenolic compounds, aromatic acid ester, oxygenated sesquiterpene, sesquiterpene, and fatty acids present in the leaves might be responsible for the antibacterial activity (Arokiyaraj et al. 2018).

The ethanolic and DMSO extracts of the rhizome of *M. minuta* also possess antibacterial activity (Arokiyaraj et al. 2018). Sajini et al. (2019) proposed that gold nanoparticles produced with water extract from leaves can serve better antibacterial purposes against *E. coli* and *S. aureus*.

3.4.4 Antitussive Activity

The antitussive activity of *M. minuta* was compared in different solvents and was evaluated on animal models of cough in rats. The methanol extract was found to have the highest activity among all. It showed 59.5% and 55.8% inhibition in coughing induced by ammonium liquor and sulphur dioxide at the dose of 500 mg/kg body weight (Chakraborty et al. 2013).

3.4.5 Anti-Aggressive Activity

The ethanolic extract of *M. minuta* was found to have dose-dependent anti-aggressive activity comparable to that of diazepam (2.5 mg/kg, orally). The maximum anti-aggressive effect of *M. minuta* was observed at a concentration of 400 mg/kg (Tiwari et al. 2009).

3.4.6 Antidepressant Activity

Chatterjee and colleagues reported in 1963 the presence of marsiline, a hexacosanoic acid triacontylester in *M. minuta*, and commented on its role as a sedative and anticonvulsant molecule (Chatterjee et al. 1963). In a study, Bhattamisra et al. (2008) showed the antidepressant activity of marsiline using a forced swimming test (FST), tail suspension test (TST), learned helplessness test (LHT), and 5-hydroxytryptophan (5-HTP)-induced head twitches response in rodents. They used ethanolic extract of *M. minuta* standardized for marsiline (1.15%, w/w) in doses of 100, 200, and 400 mg/kg body weight once daily for three consecutive days and the effects were evaluated on the last day, one hour after administration of the extract. The results were compared with imipramine, the standard antidepressant drug given at a dose of 15 mg/kg/day, intraperitoneally. Results showed that the ethanolic extract of *M. minuta* significantly reduced the immobility time in the FST and TST. The number of escape failures in the LHT decreased, and the head twitch response induced by 5-HTP was significantly attenuated by *M. minuta* (400 mg/kg, orally). 5-HTP, the precursor of serotonin, and 5-HT2A receptors or serotonin receptors are associated with many neurological processes, including controlling mood, cognition, anxiety, and learning and are found more in the frontal cortex of depressed patients. Radioligand receptor binding assays in the *M. minuta*-treated group showed downregulation of the 5-HT2A receptor in the frontal cortex without affecting the benzodiazepine receptor (Bhattamisra et al. 2008).

Oral administration of ethanolic extract of *M. minuta* at dosages of 200 and 400 mg/kg/day for three days significantly reversed amnesia. The extract significantly inhibited acetylcholinesterase activity in the frontal cortex and hippocampus compared to amnesic rats ($p < 0.05$). Ethanolic extract of *M. minuta* at the same dose also increased the binding of 3*H*(*R*)-3-quinuclidinyl benzilate (3*H*-QNB) in the frontal cortex, indicating the upregulation of the muscarinic receptors, suggesting the involvement of central cholinergic system in the process (Bhattamisra et al. 2012).

3.4.7 Antipyretic and Analgesic Activity

Antipyretic activity and analgesic activity of ethanolic leaf extract of *M. minuta* were found at a dose of 400 mg/kg (dried leaves extracted in ethanol) compared to standard antipyretics, paracetamol (150 mg/kg, orally), and diclofenac sodium (9 mg/kg, intraperitoneally) (Madhu et al. 2015).

3.4.8 Antidiabetic Activity

The antidiabetic property of ethanolic extract of the leaves of *M. minuta* was evaluated on albino rats in a diabetes model. Administration of the extract (250 and 500 mg/kg) for three weeks checked the rise in blood glucose concentration compared to the untreated group with improved lipid profile (Chakraborty et al. 2013). The fasting blood glucose, cholesterol, and serum triglyceride levels were reduced while serum high-density lipoprotein and cholesterol levels were elevated in *M. minuta*-treated rats compared to normal control (Madhu et al. 2012). Further studies with the isolated components of *M. minuta* present in ethanol will reveal the mode of action in reducing blood sugar levels.

3.4.9 Antitumour Activity

The ethanolic extract of *M. minuta* was found to have antitumorous effect on *Agrobacterium tumefaciens*-induced potato disc tumour assay. In a study, the antitumorous activity of *M. minuta* was compared with two other pteridophytes viz. *Selaginella ciliaris* (Retz.) and *Thelypteris prolifera* (Retz.). Among these three, *M. minuta* (82.32%) showed maximum inhibition in tumour development followed by *S. ciliaris* (80%) and *T. prolifera* (75.68%). *M. minuta* was found to inhibit tumorigenesis at a concentration of 10 parts per million (ppm) compared to 100 and 1000 ppm of *S. ciliaris* and *T. prolifera* indicating the presence of tumour inhibitor metabolites. Isolation of these metabolites and elucidating their mechanism of action might serve an important role in developing antitumour drugs for human beings, as there lies a similarity between human and plant tumour formation mechanisms (Sarker et al. 2011).

3.4.10 Hepatoprotective Activity

The hepatoprotective activity of *M. minuta* was evaluated by Balne et al. (2013).

They checked the activity of different fractions of methanolic extract of *M. minuta* (MMME) on toxin-induced (carbon tetrachloride, paracetamol, and ethanol) acute liver damage in rats by monitoring the levels of serum parameters, microsomal drug-metabolizing enzymes, histology, and prothrombin time. Results showed that the detrimental effects caused by the toxins were resisted in the experimental groups that received fractions of MMME in toluene, 1-butanol, and water at 50 and

100 mg/kg body weight) prior to the toxins. The butanol fraction (BF-MMME) was most effective in hepatoprotection, as supported by microsomal drug-metabolizing enzymes, prothrombin time, and DPPH studies (Balne et al. 2013).

3.4.11 Antihyperlipidaemic Activity

Administration of the leaf extract of *M. minuta* in male rate gerbils was found to reduce serum cholesterol and triglycerides by 31% and 63%, respectively, and liver cholesterol and triglycerides by 71% and 27%, respectively, compared to controls. In addition, the accumulation of cholesterol and triglycerides in the liver and aorta decreased, and excretions of cholesterol and triglycerides increased in faeces. The atheromatous plaques of the thoracic and abdominal aorta dissolved significantly (R.S. Gupta et al. 2000).

3.4.12 Antifertility Activity

The methanolic extract of *M. minuta* was checked for its antifertility activity in female Swiss albino mice. The levels of ascorbic acid and total cholesterol were elevated in the ovaries of the mice treated with *M. minuta*. The activities of the enzymes glucose-6-phosphate dehydrogenase (G6PDH) and Δ5-3-β-hydroxysteroid dehydrogenase were reduced significantly. The enzyme, Δ5-3-β-hydroxysteroid dehydrogenase participates in the biosynthesis of pregnenolone from progesterone, 17α-hydroxypregnenolone from 17α-hydroxyprogesterone, and androstenedione from dehydroepiandrosterone in the adrenal gland. The authors suggested that these observations indicate the antifertility activity of *M. minuta* in mice by inhibiting gonadal steroidogenesis (M. Gupta et al. 2002).

3.5 CONCLUSION

The use of ancient knowledge of *M. minuta* to treat disorders like cough, headache, diabetes, and many more is still in practice in rural India and Bangladesh. Biochemical analysis has identified the presence of a considerable number of steroids, triterpenoids, alkaloids, phenolic compounds, flavonoids, and catechin, which are responsible for properties such as antimicrobial, antioxidant, and analgesic activities. Further studies with the identified compounds will help to reveal their role and mode of action in various disorders. Detailed studies can open new avenues towards the development of new candidate drugs. Herbal drugs are vital in treating several disease conditions with little or no side effects. *M. minuta* has tremendous beneficial effects and is applied in different systems of traditional medication for the treatment of various diseases. *M. minuta*, a common aquatic pteridophyte with many important phytoconstituents, is used to treat insomnia and diabetes. The benefit of *M. minuta* is its wide geographical distribution which might offer an opportunity to develop formulations from various parts of the plant. This chapter puts together all the documented information on *M. minuta* available so far and hopes to reach future researchers for proper identification and utilization of these resources.

REFERENCES

Arokiyaraj, S., Bharanidharan, R., Agastian, P., and Shin, H. 2018. Chemical composition, antioxidant activity and antibacterial mechanism of action from *Marsilea minuta* leaf hexane: methanol extract. *Chemistry Central Journal* 12(1):105.

Balne, D., Pallerla, P., Vanapatla, S.R., and Bobbala, R.K. 2013. Hepatoprotective effect of whole plant extract fractions of *Marsilea minuta* Linn. *Asian Journal of Pharmaceutical and Clinical Research* 6(3):1000–1107.

Bhattamisra, S.K., Khanna, V.K., Agrawal, A.K., Singh, P.N., and Singh, S.K. 2008. Antidepressant activity of standardised extract of *Marsilea minuta* Linn. *Journal of Ethnopharmacology* 117(1):51–57.

Bhattamisra, S.K., Singh, P.N., and Singh, S.K. 2012. Effect of standardized extract of *Marsilea minuta* on learning and memory performance in rat amnesic models. *Pharmaceutical Biology* 50(6):766–772.

Bhattamisra, S.K., Singh, P.N., Singh, S.K., and Kumar, V. 2007. Anxiolytic activity of *Marsilea minuta* Linn in rodents. *Journal of Herbal Medicine and Toxicology* 1(1):15–20.

Chakraborty, R., De B., Devanna, N., and Sen S. 2013. Antitussive, expectorant activity of *Marsilea minuta* L., an Indian vegetable. *Journal of Advanced Pharmaceutical and Technology* 4(1):61–64.

Chatterjee, A., Dutta, C.P., Choudhury, B., Dey, P.K., Dey, C.D., Chatterjee, C., and Mukherjee, S.R. 1963. The chemistry and pharmacology of marsiline: a sedative and anticonvulsant principle isolated from *Marsilea minuta* Linn and *Marsilea rajasthanensis* Gupta. *Journal of Experimental Medical Sciences* 7:53–67.

Chowdhury, J., Majumdar, S., Roy, S., and Chakraborty, U. 2017. Antioxidant activity and phytochemical screening of two edible wetland pteridophytes *Diplazium esculentum* (Retz) Sw and *Marsilea minuta* L. – a comparative study. *World Journal of Pharmaceutical and Medical Research* 3(9):195–203.

de Britto, A.J., Gracelin, D.H.S., and Rathna Kumar, P.B.J. 2013. Qualitative and quantitative analysis of phytochemicals in *Marsilea minuta* linn. *International Journal of Pharma and Bio Sciences* 4(1):800–805.

Gopalakrishnan, K. and Udayakumar, R. 2014. GC-MS analysis of phytocompounds of leaf and stem of *Marsilea quadrifolia* (L.). *International Journal of Biochemistry Research & Review* 4(6):517–526.

Grover, J.K., Yadav, S., and Vats, V. 2002. Medicinal plants of India with antidiabetic potential. *Journal of Ethnopharmacology* 81(1):81–100.

Gupta, M., Mazumder, U.K., Datta, I., Battacharya, S., Mukherjee, S., and Manikandan, L. 2002. Studies of antifertility activity of *Marsilea minuta* Linn. *Indian Journal of Pharmaceutical Sciences* 64:176–178.

Gupta, R.S., Kumar, P., Sharma, A., Bharadwaj, T.N., and Dixit, V.P. 2000. Hypocholesterolemic activity of *Marsilea minuta* in gerbils. *Fitoterapia* 71(2):113–117.

Madhu, S., Kannabirran, V., Royal Frank, P., Sathish Reddy, M., and Gnanasekar, N. 2012. Evaluation of antidiabetic activity of *Marsilea minuta* Linn against alloxan induced diabetes in Albino rats. *IRJP* 3(8):223.

Madhu, S., Velmurugan, R., Gunasekaran, J., Chandrika Devi, D., Deepa, N., and Sundhararajan, R. 2015. Evaluation of antipyretic and analgesic activity of *Marsilea minuta* Linn. *International Research Journal of Pharmacy* 6(1):34–37.

Mahesh, B. and Satish, S. 2008. Antimicrobial activity of some important medicinal plant against plant and human pathogens. *World Journal of Agricultural Sciences* 4:839–843.

Mallikharjuna, P.B., Rajanna, L.N., Seetharam, Y.N., and Sharanabasappa, G.K. 2007. Phytochemical studies of *Strychnos potatorum* L.f. A medicinal plant. *E-journal of Chemistry* 4(4):510–518.

Marndi, S. and Kumar, S. 2017. *Marsilea minuta* L.: an aquatic medicinal fern of Odisha. *Journal of Biodiversity and Conservation* 1(2S):800–805.

Nagalingum, N., Schneider, H., and Pryer, K. 2006. Comparative morphology of reproductive structures in heterosporous water ferns and a reevaluation of the sporocarp. *International Journal of Plant Sciences* 167(4):805–815.

Okwu, D.E. 2005. Phytochemical, vitamins and mineral contents of two Nigeria medicinal plants. *International Journal of Molecular Medicine and Advance Sciences* 1:375–381.

Parihar, P. and Parihar, L. 2006. Some pteridophytes of medicinal importance from Rajasthan. *Indian Journal of Natural Products and Resources* 5:297–301.

Prashant, T., Kumar, B., Kaur, M., Kaur, G., and Kaur, H. 2011. Phytochemical screening and extraction: a review. *Internationale Pharmaceutica Sciencia* 1(1):98–106.

Rahmatullah, M., Kabir, A.A., Rahman, M.M., Hossan, M.S., Khatun, Z., and Khatun, M. 2010. Ethnomedicinal practices among a minority group of Christians residing in Mirzapur village of Dinajpur district, Bangladesh. *Advances in Natural and Applied Sciences* 4:45–51.

Revathi, M. and Catharin S.S. 2014. Phytochemical studies on leaves and Petiole extracts of *Marsilea minuta* L. *International Journal of Science and Research* 3:1872–1874.

Ripa, F.A., Nahar, L., Haque, M., and Islam, M.M. 2009. Antibacterial, cytotoxic and antioxidant activity of crude extract of *Marsilea quadrifolia*. *European Journal of Scientific Research* 33(1):123–129.

Sajini, R.J., Prema, S., and Chitra K. 2019. Phytoconstituents, pharmacological activities of *Marsilea minuta* L. (Marsileaceae) – an overview. *International Journal of Pharmaceutical Sciences and Research* 10(4):1582–1587.

Sarker, A.Q., Mondol, P.C., Alam, M.J., Parvez, J.S., and Alam, M.F. 2011. Comparative study on antitumor activity of three pteridophytes ethanol extracts *Journal of Agricultural Technology* 7(6):1661–1671.

Sen, S.K. and Behera L.M. 2008. Ethnomedicinal plants used by tribals of Bargarh district to cure diarrhoea and dysentery. *Indian Journal of Traditional Knowledge* 7(3):425–428.

Sen, S., Chakraborty, R., De, B., and Devanna, N. 2011. An ethnobotanical survey of medicinal plants used by ethnic people in West and South district of Tripura, India. *Journal of Forestry Research* 22:417–26.

Soni, P. and Singh, L. 2012. *Marsilea quadrifolia* Linn. – a valuable culinary and remedial fern in Jaduguda, Jharkhand, India. *Life Science* 2:99–106.

Tiwari, O.P., Bhattamisra, S.K., Singh, P.N., and Kumar, V. 2009. Adaptogenic anti-stress activity of standardised extract of *Marselia minuta* L. *Pharmacology Online* 1:290–290.

Upreti, K., Jalal, J.S., Tewari, L.M., Joshi, G.C., Pangtey, Y.P., and Tewari, G. 2009. Ethnomedicinal uses of pteridophytes of Kumaun Himalaya, Uttarakhand, India. *Journal of American Science* 5(4):167–70.

Vincent, P.C., Irudayaraj, V., and Johnson, M. 2012. Anti-bacterial efficacy of macroscopic, microscopic parts of sporophyte and in vitro cultured gametophyte of a fern *Cyclosorus interruptus* (Willd.) H. ito (Thelypteridaceae – Pteridophyta). *Journal of Chemical and Pharmaceutical Research* 4:1167–1172.

4 Traditional Uses, Phytochemistry, and Pharmacological Properties of *Hedychium coronarium*

Basudeba Kar, Jyotirmayee Lenka, Debasis Sahoo, Manaswini Dash, Bhaskar Chandra Sahoo, and Suprava Sahoo*
Centre for Biotechnology, School of Pharmaceutical Sciences, Siksha 'O' Anusandhan Deemed to be University, Bhubaneswar, India
* Corresponding author (Basudeba Kar): basudebakar@soa.ac.in

4.1 INTRODUCTION

Since time immemorial, medicinal plants have been in great demand for preparing curative drugs as they contain various constituents with therapeutic qualities. In 2017, it was roughly estimated that the global market for phyto drugs would be worth several hundred billion dollars (Ahn 2017). In developing countries, 80% of the world's population directly or indirectly rely on medicinal herbs to cure illness and ailments, according to the World Health Organization (Farnsworth et al. 1985; Smith-Hall et al. 2012; Abate et al. 2022). Knowing the therapeutic capabilities of these medicinal plants, pharmaceutical sectors are focusing more on these herbs for drug development (Atanasov et al. 2015).

Hedychium coronarium J. Koenig belongs to the family Zingiberaceae which is a monocotyledon perennial herb. It is generally known as white garland-lily, white ginger lily, or butterfly lily, and has oblong to oblong-lanceolate pointed leaves, and large bracts including heavily scented white flowers with yellow bases having erect pseudo-stems. During the year, August–October is the efflorescence time of the plant (Pachurekar and Dixit 2017; A.K. Arya et al. 2022). *H. coronarium* plants are indigenous to tropical Asia, particularly in the subcontinent of India. It is also found typically in the forest understorey in Australia, South-Eastern USA, Central America, and oceanic islets with warm climates (Ray et al. 2017). It is commonly called 'Flor de Mariposa' or white butterfly and is Cuba's national flower. However, it is considered an invasive weed in Brazil (de Souza and Correia 2007). This plant is used for various medications and is found in Indian and Chinese pharmacopoeia. For centuries in ancient India it has been consumed as a febrifuge, tonic, and excitant and used as folklore remedies in the treatment of tonsillitis, infected nostrils, and tumours (Bhandary et al. 1995; Jain and Prakash 1995). In Chinese pharmacopeia, it is mentioned in the treatment of lancinating pain, headache, inflammation, contusion, and extreme rheumatic pain (Manish 2013). In a study, *H. coronarium* was reported for its anti-cancerous, antioxidant, antihypertensive, diuretic, leishmanicidal, and antimalarial activities. In addition, it is also utilized in treating piles bleeding, irregular menstruation, and urinary tract stones (Bisht et al. 2012).

Essential oils are the key requirements of aromatherapy, a manifestation of alternative medicine having palliating effects which can induce relaxation and other stress relief (Lee et al. 2012).

DOI: 10.1201/9781003256830-4

The essential oils are produced by aromatic plants that contain secondary metabolites having many bioactivities including antioxidant, antimicrobial, and many more. These oils are used in many industries such as perfumery, pharmaceuticals, and flavourings because of their bio-safety, cost-effectiveness, and easy availability (Burt 2004; Bakkali et al. 2008; Abate et al. 2022; Gonfa et al. 2022). Nowadays, essential oils have become a subject of interest for researchers due to the high antioxidant quality of their bioactive components as an alternative to commercial products like butylated hydroxyanisole (BHA) and butylated hydroxytoluene (BHT) (Mimica-Dukic et al. 2004). *H. coronarium* essential oils obtained from rhizomes, leaves, flowers, and other parts have various therapeutic uses and are also used in aromatherapy (Lock et al. 2005; Murugappa and Kunapili 2006; Suresh et al. 2010; Ho 2011; Kiem et al. 2011; Donipati and Sreeramulu 2015; Nwe et al. 2019; Tse et al. 2019; Guzman et al. 2020; Manssor et al. 2020). Different geographical regions and environmental factors like climate, relief, and soil differ in their quantitative components (Djouahri et al. 2015).

Several studies on *H. coronarium* unveiled phytochemical variability due to different geographical locations (Lechat-Vahirua et al. 1993; Ali et al. 2002; Gurib-Fakim et al. 2002; dos Santos et al. 2010). In rhizome essential oil, the major terpenoid constituents are 1,8-cineole and β-pinene (Ray et al. 2017). The leaf oil of *H. coronarium* contains β-pinene (20.0%), linalool (15.8%), 1,8-cineole (10.7%), α-pinene (10.1%), and α-terpineol (8.6%) as major constituents whereas in root oil, mainly β-pinene (23.6%), α-humulene (17.1%), β-caryophyllene (13%), α-pinene (6.9%), and elemol (6.9%) are found (Chen et al. 2013; Singh et al. 2013). The volatile oils contain mainly sesquiterpenes and diterpenes in high quantities, which have significant roles in their bioactivities (Suresh et al. 2010). The genus *Hedychium* is marked by the labdane skeleton, and extracts of *H. coronarium* majorly contain coronarin D, a labdane diterpenoid that can show an anti-cancerous effect. A study observed that apoptosis and osteoclastogenesis suppression can be induced by blocking the inducible and constitutive nuclear factor kappa B pathway (Anand et al. 2008). The elemental composition of the rhizomes constitutes potassium, sulphur, phosphorous, and calcium and in the leaves, potassium, sulphur, chloride, and calcium analysed through the energy dispersive X-ray fluorescence (ED-XRF) method. The Association of Official Analytical Chemists (AOAC 2000) method found additional nutrient values in *H. coronarium* rhizomes, i.e. crude fat (4.27%), energy value (152.55 kcal/100 g), carbohydrates (25.17%), ash (7.34%), moisture (13.08%), protein (3.36%), and crude fibre (14.81%). However, the nutrient values in leaves showed moisture (13.89%), ash (8.21%), protein (7.46%), carbohydrates (12.93%), energy value (118.46 kcal/100 g), crude fibre (35.42%), and crude fat (4.10%). In addition to this, other rhizome and leave extracts indicated the presence of glycosides, flavonoids, organic acids, saponins, phenolic compounds, reducing sugars, terpenoids, alkaloids, carbohydrates, steroids, starch, and tannins (Nwe et al. 2018).

Since *H. coronarium* is one of the valuable medicinal, nutraceutical, commercial, and ornamental plants, it requires more attention for industrial exploration. Many important chemical compounds can be isolated from this plant to meet pharmaceutical demands. In this chapter, starting from traditional uses, the advanced laboratory investigations of this plant are highlighted. In addition, the roles of the phytochemicals of this plant in these practices are correlated and summarized. This may help scientific communities to gain more avenues to carry out *in vitro* and *in vivo* studies. Also, the pharmaceutical industries can use phytoconstituents for drug discovery against different deadly diseases.

4.2 TRADITIONAL USES

India has been an ethnobotanical country for centuries, using more than 7,000 plant species for various indigenous medicine and agriculture industries (Ahn 2017). *H. coronarium* of the ginger family is traditionally used as herbal medicine throughout the world (Sirirugsa 1999). In India, the herb is used in intrinsic cuisines, as a potent vegetable, and in cultural and religious beliefs (Sarangthem et al. 2013) (Table 4.1). This herb is mentioned in both Indian and Chinese pharmacopoeias for its

TABLE 4.1
Folk Medicinal Uses of *H. coronarium* in India

Sl. no.	Region	Local name	Plant part	Uses	Reference
1.	East Nimar region (Madhya Pradesh)	Ban haldi	Rhizome	Paste of rhizome is used externally for snakebite Decoction of rhizome is used for liver trouble	Ray et al. 2011
2.	Siddis of Uttara Kannada district, Karnataka	Kichchiligadda	Rhizome	On a daily basis, half a cup of cow's milk is mixed with a pinch of dried rhizome powder for treating diabetes For treating diphtheria, a fusion of the rhizome is mixed with sugar (sweetening purpose) and taken orally 3–4 times a day	Bhandary et al. 1995
3.	Valley districts of Manipur	Loklei	Rhizome and flower	Used to treat vomiting Flowers are used for religious rituals	Devi et al. 2014
4.	Baiga, Gond, Khariya, and Panka tribes of Amarakantak region of Madhya Pradesh	Gulbakawvali	Flower and rhizome	The fresh flower is crushed and two drops of the solution are applied externally to cure conjunctivitis	Kumar et al. 2004
5.	Tribal people of Apatani in the Eastern Himayan region of India	Dolon champa	Rhizome	Body ache	Kala 2005
6.	Kodagu district Karnataka	Kage suli	Rhizome and leaf	Sexual frailty: paste of rhizome is mixed with honey and taken orally Leaf and rhizome paste is applied externally for preventing microbial infections (antiseptic)	Lingaraju et al. 2013
7.	Manipur	Takhellei	Flowers, leaves, and rhizome	Flower is used for decoration and also offered to gods The decoction of leaves of *H. coronarium* is used for gargling in throat infection Rhizome starch is used as a febrifuge, tonic, and is used for rheumatic swelling Rhizomes along with young shoots are used as a vegetable	Sharma et al. 2011
8.	Didayi tribe of Malkangiri district of Odisha	Kulphul	Rhizome	Headache	Pattanaik et al. 2008
9.	Assam	Pakhila phul	Flowers, rhizome, and stem	Rhizome and flower extract prepared by boiling the rhizome in water Rhizome essential oil is used to treat fetid nostrils, tonic, febrifuge, and antirheumatic swellings, antihelminthic, mild tranquillizer, and tonic	Basak et al. 2010

(*continued*)

TABLE 4.1 (Continued)
Folk Medicinal Uses of *H. coronarium* in India

Sl. no.	Region	Local name	Plant part	Uses	Reference
10.	Miji, Mishmi, Nyishing, Nocte, Singhpo, Sherdukpen, Tangsa, and Khampati tribes of Arunachal Pradesh	Tora	Rhizome	Rhizome extract is used as expectorant, and for sprain and joint pain	Basak et al. 2010
11.	Tangkhul Tribe of Ukhrul District, Manipur	Hedichium/Yaimu	Rhizome and flower	Diabetes, urinary tract stones, digestion, nausea, vomiting, hiccups, asthma, internal injuries, antiseptic agent, and poor blood circulation	Leishangthem and Sharma 2014
12.	Gujjar and Bakerwals tribes of Poonch and Rajouri districts of Jammu & Kashmir	Jungli-Haldi	Rhizome	The rhizome is used for abdominal pain—10 g of sun-dried rhizome powder is mixed with cooked vegetables and can be taken at any time in a day	Shah et al. 2015
13.	Korwa, Gond, Nageshiya, Birhor, Pando, and Baiga tribes of Jashpur District of Chhattisgarh	Gulbakauli	Flower juice	Glaucoma Eye disease	Yadaw and Shrivastava 2020
14.	Ziro valley in Arunachal Pradesh	Dolon champa	Rhizome	Body ache	Kala 2005
15.	Andro village in Imphal, Manipur	Takhellei angouba	Rhizome	Decoction of rhizomes are used against fever	Singh et al. 2014
16.	Manipur	Lok-lei	Rhizome	Rhizomes are used for preparing eromba	Jain et al. 1995
17.	Tamenglong district, Manipur, Northeast India	Longnang, Kambui phuanmei	Young inflorescence, rhizomes, seeds, shoots	The shoots and young flower are eaten as vegetables Women wear it behind the ears for decoration purpose Rhizomes being used as analgesic The juice of matured seeds is used as shampoo	Daimei and Kumar 2014

medicinal applications. *H. coronarium* is not only an orally preserved beneficial species but also provides a broad spectrum of income sources. Commercially, it is used to make paper out of aerial stems containing 43–48% cellulose (Bisht et al. 2012). It is widely cultivated for its showy inflorescence and perfumery, used in the perfume industries (de Souza and Correia 2007). This plant is used in India and many regions of the world (Table 4.2). People of Hawaii generally call it wild shampoo ginger, and the mature seeds are used for hair and skin maintenance (Bhandary et al. 1995; Parida et al. 2013). Edible plant parts like flowers, stems, and roots are used as vegetables in Southeast Asia (Bisht et al. 2012). The strong and unique smell of the flower essential oil is due to the presence of terpenes: β-transocimenone, linalool, 1,8-cineole, α-farnesene, and α-terpineol (Báez et al. 2011).

TABLE 4.2
Folk Medicinal Uses of *H. coronarium* Outside India

Sl. no.	Plant part	Uses	Region	Reference
1.	Rhizome, leaf	Microbial infections Inflammatory complaints Parasitic infections Cancer	The Republic of Suriname (South America)	Mans et al. 2019
2.	Rhizome	Headache Skin diseases Inflammation Rheumatic pain	Vietnam	Kiem et al. 2011
3.	Rhizome	Diabetes Contusion inflammation Rheumatism pain Headache	China	Y.C. Kong, personal communication, 1977; Jain et al. 1995
4.	Rhizome	Stomach ache Febrifuge Schistosomiasis Fever	Indigenous ethnic people at Lore Lindu National Park, Central Sulawesi, Indonesia	Pitopang et al. 2019
5.	Roots and leaves	Leaves are used for fever Extracts of the roots dipped in hot water are used to treat internal pains	Dominica	Hodge and Taylor. 1957
6.	Leaves	Leaves are used for internal pains	Malaysia	Goh et al. 1984
7.	Rhizome, flower, stems	Applied externally for stiff or sore joints The rhizome is generally used as a stimulant and carminative Decoction of stem is gargled for treating tonsilitis The flower is used as vegetable	Thailand	Anderson 1986; Sirirugsa 1999
8.	Roots	Applied externally in the chest and arms to relieve pains To treat arthritis	Columbia	Laferriere 1994
9.	Leaves, roots, and seeds	To treat rheumatism, to induce diuresis, and to treat hypertension	Brazil	Ribeiro et al. 1986
10.	Leaves	To treat fever	Guaymi Indians in Western Panama	Joly et al. 1987
11.	Leaves	For treating jaundice	Bangladesh	Hasan et al. 2015
12.	Aerial part	Rheumatism and soothing	Peru	Valadeau et al. 2010
14.	Rhizome	Snakebites	Nicaragua	Coe and Anderson 2005

Leaves are eaten with betel nuts for treating abdominal pain. Boiled leaves of *H. coronarium* are applied to relieve stress and sore joints in Thailand (Anderson 1986; Sirirugsa 1999). In folk medicinal usage, various plant extracts and essential oils were used for treating diphtheria, headache, fever, rheumatism, diabetes, and cancer (Bhandary et al. 1995; Matsuda et al. 2002; Oh et al. 2006).

4.3 PHYTOCHEMISTRY

H. coronarium has many pharmacological activities, and is used in perfumery, in flavourings, and for commercial purposes. The components responsible for its activities were extracted from rhizomes and leaves and are mainly secondary metabolites like terpenoids, flavonoids, steroids, tannins, glucosides, carbohydrates, reducing sugars, phenolic compounds, organic acids, and alkaloids (Hill 1952; Panigrahy et al. 2017). In previous studies, labdane-type diterpenes and farnesane-type sesquiterpenes were identified from the rhizome part of the plant (Nakatani et al. 1994; Matsuda et al. 2002; Chimnoi et al. 2009).

The important components include coronarin A, coronarin B, coronarin D, labda-8(17),11,13-trien-15(16)-olide, coronarin D methyl ether, isocoronarin D, hedychicoronarian, 7β-hydroxylcalcaratarin A and E, peroxycoronarian D, 7β-hydroxyl-6-oxo-labda-8,12-diene-15,16-dial extracted from the rhizome part of *H. coronarium* (Itokawa et al. 1988; Taveira et al. 2005; Chimnoi et al. 2009; Lu et al. 2009; Kiem et al. 2011; Verma and Bansal 2013) (Table 4.3 and Figure 4.1). Phytochemical

TABLE 4.3
Phytochemical Constituents of Extracts of *H. coronarium*

Sl. no.	Compound name	Plant part	Reference
	(A) Diterpenes		
1.	(*E*)-15,16-Bisnorlabda-8(17),11-dien-13-one	Rhizome	Itokawa et al. 1980
2.	4-Hydroxy-3-methoxy ethyl cinnamate	Rhizome	Suresh et al. 2010
3.	(*E*)-Labda-8(17),12-diene-15,16-dial	Rhizome	Itokawa 1988
4.	Coronarin A	Rhizome	Kiem et al. 2011
5.	Coronarin B	Rhizome	Itokawa et al. 1988
6	Coronarin C	Rhizome	Taveira et al. 2005
7.	Coronarin D	Rhizome	Matsuda et al. 2002; Chimnoi et al. 2009
8.	Coronarin E	Rhizome	Matsuda et al. 2002
9.	Coronarin F	Rhizome	Itokawa et al. 1988
10.	Coronarin D methyl ether	Rhizome	Matsuda et al. 2002; Taveira et al. 2005
11.	Isocoronarin D	Rhizome	Nakatani et al. 1994; Taveira et al. 2005
12.	Labda-8(17),11,13-trien-15(16)-olide	Rhizome	Nakatani et al. 1994; Kiem et al. 2011
13.	Coronarin F (36)	Rhizome	Itokawa et al. 1988
14.	Coronarin G (15-methoxylabda-8(17),13-dien-16,15-olide)	Rhizome	Kiem et al. 2011
15.	Coronarin-H	Rhizome	Kiem et al. 2011
16.	Coronarin I (3β,7β,14-trihydroxy-15,16-epoxylabda	Rhizome	Kiem et al. 2011
17.	Cryptomeridiol	Rhizome	Kiem et al. 2011
18.	Hedychilactones A	Rhizome	Matsuda et al. 2002
19.	Hedychilactones B	Rhizome	Matsuda et al. 2002
20.	Hedychilactones C	Rhizome	Matsuda et al. 2002
21.	Hedyforrestin C	Rhizome	Kiem et al. 2011
22.	Pacovatin A	Rhizome	Suresh et al. 2010
23.	Cryptomeridiol	Rhizome	Suresh et al. 2010

TABLE 4.3 (Continued)
Phytochemical Constituents of Extracts of *H. coronarium*

Sl. no.	Compound name	Plant part	Reference
24.	4-Hydroxy-3-methoxy ethyl cinnamate	Rhizome	Suresh et al. 2010
25.	4-Hydroxy-3-methoxycinnamaldehyde	Rhizome	Suresh et al. 2010
26.	6-Oxo-7,11,13-labdatrien-17-al-16,15-olide	Rhizome	Suresh et al. 2010
27.	6-Oxo-7,11,13-labdatrien-16,15-olide	Rhizome	Suresh et al. 2010
28.	7,17-Dihydroxy-6-oxo-7,11,13-labdatrien-16,15-olide	Rhizome	Suresh et al. 2010
29.	Peroxycoronarin D	Rhizome	Chen et al. 2013
30.	Hedychicoronarin	Rhizome	Chen et al. 2013
31.	Methoxycoronarin D	Rhizome	Endringer et al. 2014
32.	Benzoyl eugenol	Rhizome	Taveira et al. 2005; Endringer et al. 2014
33.	Docosyl-(*E*)-ferulates	Rhizome	Jayaprakasam et al. 2006
34.	Eicosyl	Rhizome	Jayaprakasam et al. 2006
35.	Hedychenone	Rhizome	Matsuda et al. 2002; Suresh et al. 2010
36.	7-Hydroxy hedychenone	Rhizome	Matsuda et al. 2002
37.	16-Hydroxylabda-8(17),11,13-trien-15,16-olide	Rhizome	Kiem et al. 2011
38.	16-Methoxylabda-8(17),11*E*,13-trien-15,16-olide (hedycoronens B)	Rhizome	Kiem et al. 2011
39.	15-Methoxylabda-8(17), 11E, 13-trien-16, 15-olide (hedycoronens A)	Rhizome	Kiem et al. 2011
	(B) Sterols		
40.	Daucosterol	Rhizome	Kiem et al. 2011
41.	β-Sitosterol	Rhizome	Kiem et al. 2011
42.	Stigmasterol	Rhizome	Kiem et al. 2011
	(C) Flavonoids		
43.	5-Hydroxy-3,7,4'-trimethoxyflavone	Rhizome	Matsuda et al. 2002
44.	Chrysin	Rhizome and leaves	Redondo and Barbosa 2018
45.	Teptochrysin	Rhizome and leaves	Redondo and Barbosa 2018

investigation resulted in the isolation of 53 labdane-type diterpenes from the rhizome part of this plant (Chan and Wong 2015). Itokawa et al. (1988) reported coronarin A, B, C, and D from the rhizome methanolic extract, which belongs to cytotoxic labdane-type diterpenoids (Table 4.3 and Figure 4.1). Chen et al. (2013) and Singh et al. (2013) identified the major constituents from rhizome essential oil as β-pinene (20.0%), linalool (15.8%), 1,8-cineole (10.7%), α-pinene (10.1%), and α-terpineol (8.6%) while the leaf oils consist mainly of β-pinene (23.6%), α-humulene (17.1%), β-caryophyllene (13.0%), elemol (6.9%), and α-pinene (6.9%) (Table 4.4). Dash and Sheikh (2015) carried out the phytochemical screening of the rhizome methanolic extract of *H. coronarium* and found the presence of carbohydrates, flavonoids, saponins, steroids, and alkaloids. Mehta et al. (2014) performed semi-preparative high-performance liquid chromatography with a photodiode array detector for determining flavonoid, polyalcohol, and water content from the methanolic extracts of the flower part of the plant. Furthermore, Redondo and Barbosa (2018) used a colorimetric method using Folin-Ciocalteau's reagent to check the total phenolic content (TPC) in the sample of leaves and rhizome ethanolic extracts. The results showed the TPC in the leaves and rhizome extracts to be 1.95 and 1.48 (mg GAE/g sample), respectively. In a study by Kiem et al. (2011), the presence of sterols like daucosterol, stigmasterol, and β-sitosterol flavonoids in the

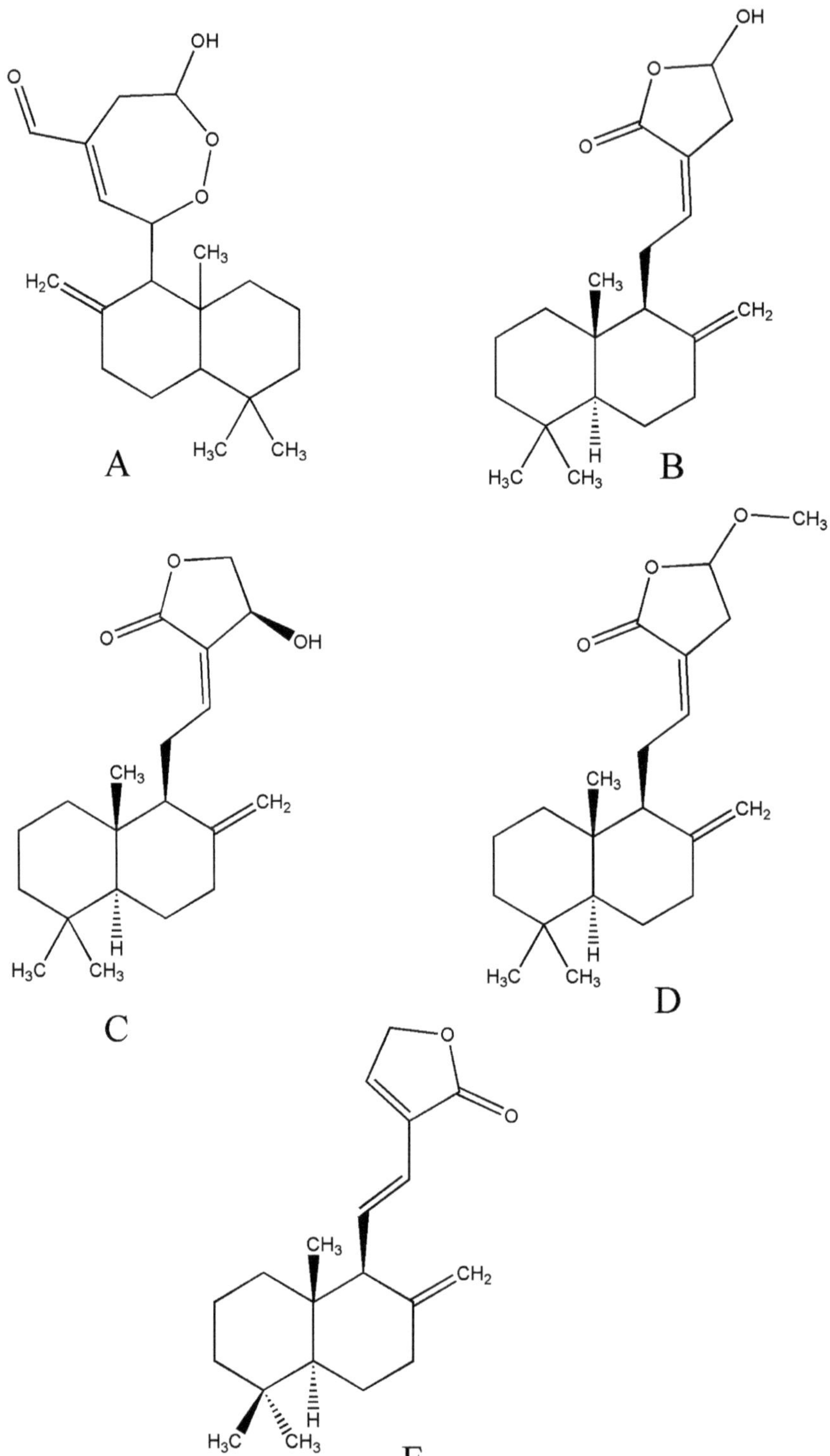

FIGURE 4.1 Major diterpene constituents of *H. coronarium*. (A) Coronarin B. (B) Coronarin D. (C) Isocoronarin D. (D) Coronarin D methyl ether. (E) Labda-8(17),11,13-trien-15(16)-olide.

TABLE 4.4
Different Volatile Components Found in the Essential Oil of *H. coronarium*

Sl. no.	Compound	Plant part	Reference
1.	α-Thujene	Rhizome	Ray et al. 2018
2.	α-Pinene	Rhizome	Joshi et al. 2008; Van et al. 2014; Shanmugam et al. 2015; Ray et al. 2018
3.	Camphene	Rhizome	Joshi et al. 2008; Ray et al. 2018
4.	Sabinene	Rhizome	Ray et al. 2018
5.	Myrcene	Rhizome	Ray et al. 2018
6.	*p*-Cymene	Rhizome	Ray et al. 2018
7.	Limonene	Rhizome	Ray et al. 2018
8.	1,8-Cineole	Rhizome, leaf, and flower	Joshi et al. 2008; Shanmugam et al. 2015; Miranda et al. 2015; Ray et al. 2018
9.	γ-Terpinene	Rhizome	Joshi et al. 2008; Ray et al. 2018
10.	Linalool	Rhizome	Ray et al. 2018
11.	Camphor	Rhizome	Ray et al. 2018
12.	Borneol	Stem and rhizome	Joshi et al. 2008; Shanmugam et al. 2015; Ray et al. 2018
13.	Terpinen-4-ol	Rhizome	Joshi et al. 2008; Ray et al. 2018
14.	α-Terpineol	Rhizome	Van et al. 2014; Shanmugam et al. 2015; Ray et al. 2018
15.	Eugenol	Rhizome	Ray et al. 2018
16.	α-Humulene	Root	Joshi et al. 2008; Van et al. 2014; Ray et al. 2018
17.	β-Caryophyllene	Root	Joshi et al. 2008; Van et al. 2014
18.	β-Ocimene	Flower	Shanmugam et al. 2015
19.	β-Cedrene	Rhizome	Joshi et al. 2008
20.	Ar-curcumene	Rhizome	Joshi et al. 2008; Ray et al. 2018
21.	Dehydroaromadendrane	Rhizome	Joshi et al. 2008
22.	10-Epi-γ-eudesmol	Rhizome	Joshi et al. 2008
23.	β-Eudesmol	Rhizome	Joshi et al. 2008
24.	(*E*)-β-Farnesene	Rhizome	Joshi et al. 2008
25.	Geraniol	Rhizome	Joshi et al. 2008
26.	β-Myrcen	Rhizome	Joshi et al. 2008
27.	γ-Muurolene	Rhizome	Joshi et al. 2008
28.	(*E*)-Nerolidol	Rhizome	Joshi et al. 2008; Ray et al. 2018
29.	α-Phellandrene	Rhizome	Joshi et al. 2008
30.	Spathulenol	Rhizome	Joshi et al. 2008; Ray et al. 2018
31.	Terpinolene	Rhizome	Joshi et al. 2008
32.	α-Terpinene	Rhizome	Joshi et al. 2008
33.	Germacrene D	Rhizome	Ray et al. 2018
34.	β-Bisabolene	Rhizome	Ray et al. 2018
35.	γ-Cadinene	Rhizome	Ray et al. 2018
36.	Elemol	Rhizome	Ray et al. 2018
37.	Caryophyllene oxide	Rhizome	Ray et al. 2018
38.	10-Epi-γ-eudesmol	Rhizome	Ray et al. 2018
39.	β-Eudesmol	Rhizome	Ray et al. 2018
40.	Coronarin E	Rhizome	Ray et al. 2018
41.	Carvacrol	Rhizome	Joshi et al. 2008); Ray et al. 2018)

rhizome of *H. coronarium* was mentioned. Moreover, Matsuda et al. (2002) reported the flavonoid compounds such as 5-hydroxy-3,7,4'-trimethoxyflavone obtained from the methanolic extract of *H. coronarium* rhizome.

Many previous studies on the phytochemical analysis of *H. coronarium* showed differences in essential oil composition collected from various eco-regions (Lechat-Vahirua et al. 1993; Ali et al. 2002; Gurib-Fakim et al. 2002; dos Santos et al. 2010). Northern and Southern regions of India showed considerable variation (Sabulal et al. 2007; Prakash 2010). A total of 44 components were identified in essential oil, including the monoterpenes and the sesquiterpenes. Among the monoterpenes, the hydrocarbon type was present abundantly as compared to oxygenated ones (Ray et al. 2018).

The primary constituents available in the study of Ray et al. (2018) were β-pinene (12.74–42.74%), 1,8-cineole (7.37–42.26%,) linalool (1.9–45.11%), coronarin E (1.01–39.57%), α-pinene (3.80–16.60%), *p*-cymene (1.20–13.38%), γ-terpinene (1.53–5.82%), and 10-epi-γ-eudesmol (1.11–4.86%) (Table 4.4). Coronarin E was identified as the major constituent for the first time from the rhizome oil by Ray et al. (2018). However, 1,8-cineole was reported as the major constituent in rhizome essential oil in different studies (Joy et al. 2007; Sabulal et al. 2007; Rodrigues et al. 2013). The gas chromatography–mass spectrometry analysis of *H. coronarium* leaf essential oil showed major constituents like 1,8-cineole (31.5%), β-pinene (13.7%), terpinen-4-ol (6%), and α-pinene (3.8%) collected from the areas of Choco, Colombia (Caballero-Gallardo et al. 2014) (Figure 4.2). The three major constituents reported in various works are mentioned in Table 4.5.

The gas chromatography flame ionization detector (GC-FID) of rhizome essential oil of *H. coronarium* analysed 44 phytochemicals in which the predominant compounds were

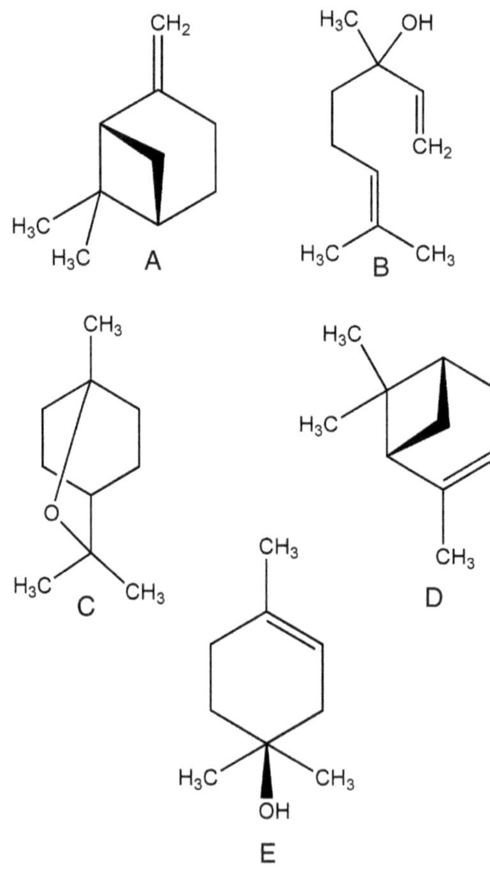

FIGURE 4.2 Major terpenoid constituents of *H. coronarium*. (A) β-Pinene. (B) Linalool. (C) 1,8-Cineole. (D) α-Pinene. E) α-Terpineol.

TABLE 4.5
Major Three Terpenoid Constituents of *H. coronarium* by Different Researchers

Sl. no.	First major constituent	Second major constituent	Third major constituent	Reference
1.	1,8-Cineole	β-Pinene	α-Pinene	Ray et al. 2018
2.	1,8-Cineole	β-Pinene	α-Terpineol	Joy et al. 2007
3.	1,8-Cineole	α-Terpineol	β-Pinene	Rodrigues et al. 2013
4.	1,8-Cineole	β-Pinene	α-Terpineol	dos Santos et al. 2010
5.	1,8-Cineole	β-Pinene	α-Pinene	Costa et al. 2019
6.	1,8-Cineole	β-Pinene	α-Terpineol	Lima et al. 2021
7.	1,8-Cineole	β-Pinene	α-Pinene	Calva et al. 2020
8.	β-Pinene	1,8-Cineole	α-Pinene	Shanmugam et al. 2015
9.	α-Pinene	β-Pinene	1,8-Cineole	Ali et al. 2002
10.	Linalool	Limonene	*Trans*-meta-mentha 2,8-dien	Prakash 2010

1,8-cineole (11.8–40.59%), β-pinene (11.07–42.74%), linalool (1.56–45.11%), coronarin E (1.01–39.57%), α-pinene (3.80–16.60%), *p*-cymene (1.05–8.89%), γ-terpinene (1.73–5.82%), and 10-epi-γ-eudesmol (1.11–4.86%) (Ray et al. 2018). The GC-FID analysis of flower essential oil showed major components as β-*trans*-ocimenone (28.1%), linalool (18.5%), 1,8-cineole (11.4%), α-terpineol (7.1%), 10-epi-γ-eudesmol (6.1%), sabinene (4.6%), and terpinen-4-ol (3.2%) (Lu et al. 2009).

The two-dimensional gas chromatography and time-of-flight mass spectrometry (GCxGC-TOF-MS) is an advanced chromatographic technique implemented for deeper characterization of terpenic composition found in the essential oil of *H. coronarium*. In a study by Ray et al. (2017), a total of 101 terpenic constituents were reported, out of which 35 alcohols, 34 hydrocarbons, 12 ketones, 7 aldehydes, 7 oxides, 3 ethers, and 3 esters were present. The principal components identified by GCxGC-TOF-MS were 1,8-cineole (37.6%), *p*-cymene (11.7%), *p*-mentha-1-en-8-ol (9.4%), terpinen-4-ol (5.47%), β-pinene (5.0%), camphene (4.2%), 4-carene (3.6%), α-pinene (2.3%), linalool (2.1%), and limonene (2.1%) (Ray et al. 2017).

4.4 PHARMACOLOGICAL ACTIVITIES

H. coronarium exhibits several bioactivities such as anthelmintic (Lima et al. 2021), antibacterial (Aziz et al. 2009), anti-inflammatory and analgesic (Shrotriya et al. 2007), antiplatelet (Guzman et al. 2020), antiallergic (Morikawa et al. 2002), antitumour (Oh et al. 2006), and cytotoxic effects (Itokawa et al. 1988), and anti-urolithiatic activity (Bahuguna and Kumar, 2014) (Figure 4.3). Pharmacological assessment of antifungal and antibacterial activities affirmed the presence of antimicrobial activities tested with rhizome essential oil, flower extract, and decoction of leaves (Joy et al. 2007; Aziz et al. 2009; Ho et al. 2011; Pandya et al. 2014).

This plant is popular for its anti-inflammatory activity due to the presence of 7-β-hydroxycalcaratarin A, calcaratarin A, coronarin A, and (*E*)-labda-(17),12-diene-15,16-dial extracted from the rhizome part of the plant (Chen et al. 2013). Coronarin A, B, C, D, E, and F were important compounds found in various parts of the plant (Verma and Bansal 2013). Coronarin D, a labdane diterpene, is a potential chemotherapeutic component used for inducing apoptosis (Anand et al. 2008), whereas coronarin A has the potential to inhibit the proliferation of human umbilical vein endothelial cells and act as cytotoxic to tumour cells (Aggarwal 2004). The anti-inflammatory activity possessed by the rhizome part of *H. coronarium* is exhibited by inhibiting 5-lipoxygenase (Braga et al. 2000).

Rhizome extracts of *H. coronarium* are rich in phenolics and have a high antioxidant quality, making it a potent bioactive resource for treating diseases due to oxidative stress (Panigrahy et al.

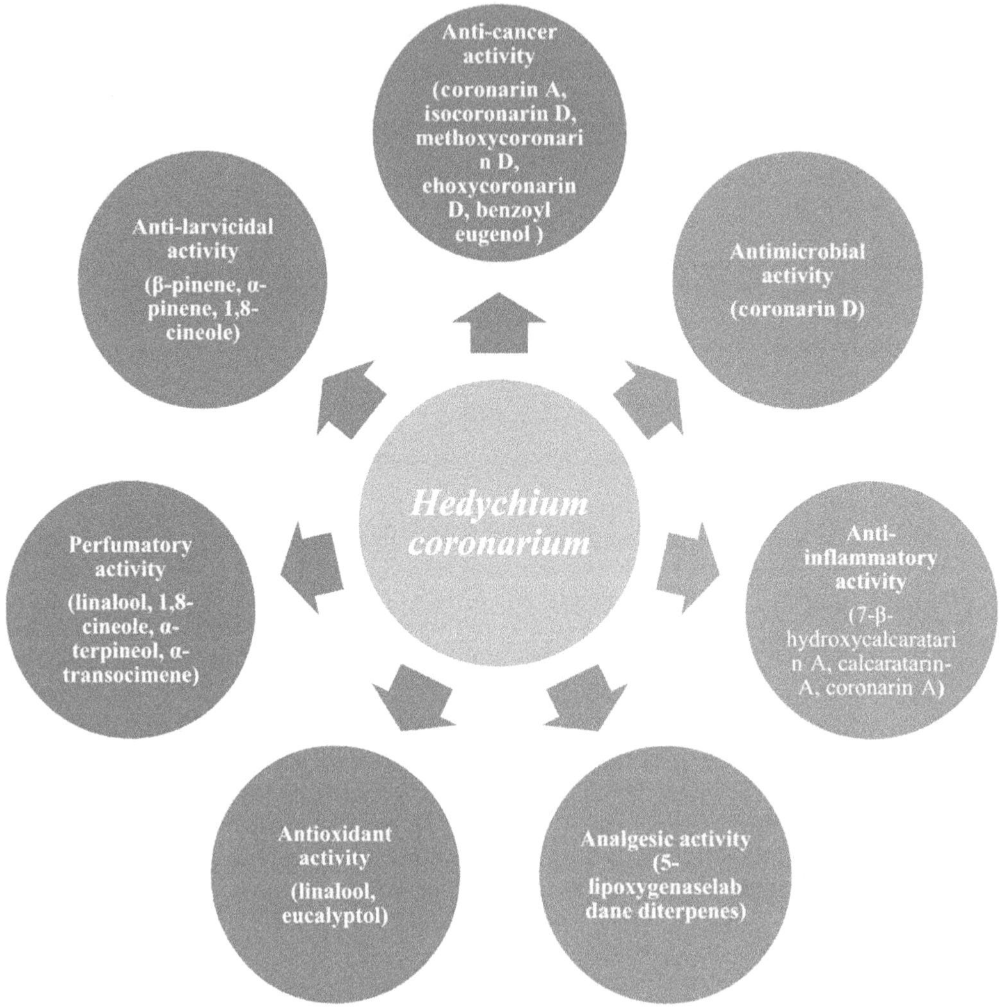

FIGURE 4.3 Pharmacological activity in *H. coronarium* due to its phytoconstituents.

2017). A study reported that the methanolic extract of *H. coronarium* rhizome has a potential anti-cancerous activity that suppresses MCF-7 cells of human breast cancer (Donipati and Sreeramulu 2015). Moreover, the ethanolic extract of the rhizome showed an antiproliferative effect on HeLa cervical cancer cells (Ray et al. 2019). In a diabetic rodent model, rats were given various extracts of plant leaves and were found to have lower glycaemic amounts (Tse et al. 2019). Furthermore, the rhizome-extracted oil showed remarkable activity against earthworms and tapeworms. The leaf and rhizome essential oils also exhibited mosquito larvicidal activities (Reuk-ngam et al. 2014).

4.4.1 Anti-Urolithiatic Activity

Kidney stones or urolithiasis is the presence of calculi in the urinary tract. Physiochemical events like nucleation, growth, supersaturation, aggregation, and retention within the kidneys result in urolithiasis. In the traditional system of medicine, the roots of *H. coronarium* are helpful in treating urolithiasis (Ahmed et al. 2016). This activity was evaluated by using alcoholic and aqueous extracts of the plant roots, which are potentially active for treating kidney stones. An *in vitro* model was used to study this activity of 70% v/v alcohol and chloroform extracts from the roots of *H. coronarium* (Bahuguna and Kumar 2014). Calcium oxalate was taken to induce kidney stones, and cystone was used as standard medicine. The phytochemical testing discloses the presence of saponin glycosides, fats, and volatile oil in the roots, whereas the aqueous extracts contain volatile oil and saponin glycosides. Both the alcoholic and aqueous extracts of *H. coronarium* roots contain saponin glycosides, but alcoholic extracts of roots were considered effective due to more potent pharma activity than aqueous extracts. In another *in vitro* study, Tailor and Goyal (2015) evaluated the effectiveness of ethanolic and aqueous extracts from roots and rhizomes of *H. coronarium* in removing experimentally prepared kidney stones like calcium oxalate. Based on the results of the study, this plant's ethanolic extracts of roots and rhizomes produced the highest rates of stone dissolution when compared to cystone at a concentration of 10 mg. Hence, alcoholic extracts can dissolve kidney stones more efficiently than any other extract (Bahuguna and Kumar 2014; Tailor and Goyal 2015).

4.4.2 Antimicrobial Activity

Humans share a symbiotic relationship with probiotics for good health as they can protect us from autoimmune diseases, boost the immune system, detoxify, and may even fight off stress (Turnbaugh et al. 2007). But sometimes there can be an imbalance because of the pathogenic microorganisms which can be treated by antimicrobial drugs (Khezerlou et al. 2018; Watkins 2018). *H. coronarium* showed a broad spectrum of antimicrobial activities determined by minimum inhibitory concentration (MIC) and the disc-diffusion method (Boyanova et al. 2005).

In a study by Ray et al. (2018), the antimicrobial activity was tested against three Gram-negative bacteria (*Escherichia coli*, *Enterococcus faecalis*, and *Pseudomonas aeruginosa*), two Gram-positive bacteria (*Bacillus subtilis* and *Staphylococcus aureus*), three fungal strains (*Aspergillus flavus*, *Aspergillus niger*, and *Fusarium oxysporum*), and *Candida albicans* (yeast). Chloramphenicol, fluconazole, and amphotericin B were used as positive controls for bacterial, fungal and yeast strains, respectively. The inhibition zone diameter ranged from 7 to 24 mm, of which the highest inhibition was found in *C. albicans* and the lowest was against the Gram-negative bacteria *E. faecalis*. The authors concluded that *H. coronarium* has better antifungal activity than antibacterial (Ray et al. 2018).

Antimicrobial activity of rhizome and leaf essential oils of *H. coronarium*, determined by the disc-diffusion method, showed inhibition against all bacterial and fungal strains used in one study (Gram-positive cocci, *S. aureus*, *Staphylococcus epidermis*, Gram-negative bacilli, *E. coli*, *P. aeruginosa*, and the yeasts, *C. albicans*, *C. glabrata*, *C. tropicalis*, *Malassezia pachydermatis* and *M. furfur*). The strongest activity of leaf and rhizome essential oils was found against *C. glabrata* followed by *M. furfur* and *C. albicans*; meanwhile, the weakest activity was seen in *E. coli* and *P. aeruginosa*. However, the methanolic and aqueous extracts of leaves and rhizome of this plant showed very little or no activity (Ho 2011).

Pandya et al. (2014) studied antifungal activity of crude extracts of flower against fungal strains *Alternaria*, *Fusarium*, and *Aapergillus flavus* in which the most efficient activity was shown by *Alternaria* having a minimum fungicidal concentration (MFC) of 100 μg/ml which is just 50% MFC of nystatin and 20% of greseofulvin used as standard drugs.

Coronarin D also showed antimicrobial activity isolated from the rhizome which was tested against Gram-positive bacteria (*S. epidermidis*, *S. aureus*, *E. faecalis*, *Bacillus cereus*), and Gram-negative bacteria (*E. coli*, *Salmonella typhimurium*, *P. aeruginosa*). Coronarin D was more effective against Gram-positive bacteria than Gram-negative bacteria. This is because coronarin D, a diterpene of the labdane type comprised of a decalin ring and an unsaturated lactone ring with one hydroxyl group, was considered a hydrophobic molecule that can easily enter and rupture the cell membrane of Gram-positive bacteria more effectively than Gram-negative bacteria. Assays with coronarin D demonstrated impressive activity against *B. cereus* at a MIC value of 6.25 µg/ml, and moderate activity against *S. aureus* and *S. epidermidis* at a MIC value of 12.5 µg/ml (Reuk-ngam et al. 2014).

4.4.3 Insecticidal Activity

Agriculture serves as a primary source of food supplies and is the backbone of the Indian economy. Pest insects can have adverse effects and are a major threat to crop production, but eradicating them can affect the ecological balance of the environment. In a study by Manssor et al. (2020), it was reported that the methanolic extract of *H. coronarium* leaves was effective against larvae of *Helicoverpa armigera* responsible for destroying 182 plant species. It was further seen that an effective increase in the concentration of plant extract is directly proportional to the mortality rate of the larvae. *H. coronarium* showed a 70% mortality rate with the highest concentration of 100 µg. In another study by S. Arya et al. (2022), the plant's insecticidal activity of rhizome chloroform oleoresin was observed by the leaf dip method against *Spodoptera litura*. *S. litura* is commonly known as the cotton cutworm. Its third instar and later stages cause maximum crop damage by inducing leaf defoliation. The larvae were observed after 24, 48, and 72 hours. After 72 hours, the maximum mortality rate was observed as 6.66%, 20%, 33.33%, and 60% at concentrations of 10, 25, 50, and 100 parts per million (ppm) of oleoresin, respectively. The highest insecticidal activity was observed at 100 ppm concentration and gradually decreased in a concentration-dependent manner. Tween 20 (1%) was taken as standard and the median lethal concentration (LC_{50}) values at 24, 48, and 72 hours after treatment were 0.07, 0.06, and 0.04%, respectively. Hence, *H. coronarium* could serve as a formidable bioinsecticide and may be utilized in the integrated pest management programme instead of chemical pesticides, which adversely affect the soil, environment, and healthy crops.

4.4.4 Anticancer Activity

Cancer is an uncontrollable growth of abnormal cells that destroys body tissue; when this spreads to other tissues it is referred to as metastasis. Cancer has become a leading cause of death worldwide. The most common cancers are blood, colon, breast, and rectum cancers. Many cancers can be cured if detected at an initial stage and treated efficiently and effectively. However, there is a chance of the cancer returning in the first two years or so after the treatment. To date, no medicines are able to eradicate cancer completely from the body. Therefore, many research activities are going on to discover new medicines including phytomedicines to eliminate cancer (Arruebo et al. 2011). In a literature study by Nwe et al. (2019), the antiproliferative potential of *H. coronarium* was reported. Aqueous and alcoholic extracts of rhizomes and leaves were taken to check the antiproliferative activity by using water-soluble tetrazolium salt 8 (WST-8) assays against cancer cell lines such as adenocarcinomic human alveolar basal epithelial (A549) cells, Michigan Cancer Foundation-7 (MCF-7) cells (breast cancer), and HeLa cells (cervix cancer). The alcoholic extract from leaves was effective against A549 and MCF7 cell lines with half-maximal inhibitory concentration (IC_{50}) values of 168.7793 µg/ml and 172.1678 µg/ml, respectively, whereas the aqueous extracts of leaves did not show any effect. The aqueous extract of the rhizome was not active against the cancer cells, but the alcoholic extract from the rhizome was effective against all three cancer cell lines with IC_{50} values of 139.9394 µg/ml, 160.0752 µg/ml, and 174.7494 µg/ml respectively. The alcoholic rhizome extract was found to be more effective against the three cancer cell lines than *H. coronarium* leaf extracts. Aziz et al. (2009) reported the potential cytotoxicity activity from various rhizome extracts such as methanol, ethyl acetate, and dichloromethane against brine shrimp nauplii. Among the samples, the methanolic extracts exhibited the highest toxicity.

4.4.5 Antioxidant Activity

Oxidative metabolism is the basis of human health, but the reaction products like reactive oxygen species can place oxidative stress on the human body. Antioxidants protect our cells against free radicals, which play a vital role in cancer, heart diseases, and others. In the human body, free radicals are generated when food breaks down or when pollution, radiation, or tobacco smoke is inhaled. Methanolic and ethanolic extracts of both leaves and rhizomes of *H. coronarium* have shown remarkable antioxidant activities. The ethanolic extract of whole plant parts exhibited 0.0% and 12.76% antioxidant activity at 10 μg/ml and 50 μg/ml, respectively (Lock et al. 2005). The antioxidant activity exhibited by methanolic rhizome extract was 81.4% (Chen et al. 2008) whereas the alcoholic leaf extract showed 90.1% reducing power (Chan et al. 2008).

Joshi et al. (2008) reported the antioxidant activity of *H. coronarium* rhizomes using different methods such as chelating metal properties, 2,2-diphenyl-1-picrylhydrazyl (DPPH) scavenging activity, and reducing power and found that the antioxidant activity increases with an increase in concentration. The results obtained indicate moderate chelating properties (58.11%), reducing power taken at an absorbance of 700 nm ranging from 0.242 to 0.605, and moderate scavenging activity (43.49%).

4.4.6 Analgesic Activity

Analgesics relieve pain by acting in the central and peripheral nervous systems without affecting consciousness. The analgesic agents can be narcotic and non-narcotic, negatively impacting human bodies (Dash 2016). To combat this, the bioactivity of *H. coronarium* can be used, having the least side effects on neuropharmacological problems. The analgesic activity of methanolic extracts of rhizomes was examined by tail immersion and acetic acid-induced writhing methods. The doses were given at 100, 200, and 400 mg/kg of body weight, and showed significant inhibition of the writhing response induced by acetic acid compared to the standard diclofenac sodium. The inhibition percentages were 26.15% at 100 mg/kg, 47.94% at 200 mg/kg, and 73.12% at 400 mg/kg whereas the standard drug showed 75.78% inhibition at 25 mg/kg body weight (Dash et al. 2011).

The neuropharmacological activity was also examined by using methanolic extract of rhizome in hole cross and open field tests in mice in which diazepam (3 mg/kg body weight) was taken as standard that showed suppression in motor activity (Kiem et al. 2011).

4.4.7 Antidiabetic Activity

Diabetes mellitus occurs when the body cannot take up glucose from the blood, resulting in hyperglycaemia which can be treated by taking insulin or by using antidiabetic drugs which can lower blood sugar levels. A study by Tavares et al. (2020) reported on the antidiabetic activity of aqueous extract of *H. coronarium* leaves. Animal models of two types were used: streptozotocin (STZ)-induced type 2 diabetes mellitus (T2DM) (Wistar rats; $n = 8$) and C57BKSdb/db mice ($n = 5$). After 28 days of observation, it was found that the levels of cholesterol, high-density lipoprotein, low-density lipoprotein, and triglyceride were improved in the STZ-induced T2DM model. It was also seen that the aldosterone levels were decreased significantly along with the lesion's degree of islet β-cells. In contrast, the insulin level was increased moderately by treatment with *H. coronarium* suggesting that aldosterone has an efficient role in this aspect (Tse et al. 2019).

4.4.8 Mosquito Larvicidal Activity

Malaria is a mosquito-borne disease caused by *Plasmodium* parasites. Malaria is likely to affect more people in tropical and subtropical developing countries and is regarded as a global threat.

Ho (2011) tested methanolic and aqueous extracts, including volatile oils of leaves and rhizome of *H. coronarium*, against *Aedes aegypti* larvae. Both the leaf and the rhizome oils showed mosquito larvicidal activity. The main components responsible for this activity are 1,8-cineole, α-pinene, and β-pinene in *H. coronarium*, which could be used as biodegradable mosquito larvicides of biotic origin.

4.4.9 Antiplatelet Activity

Platelets in the blood are essential for the blood clotting process. However, increased platelet aggregation can lead to many cardiovascular diseases and arterial thrombosis, which generally happens when the endothelium is damaged, activating the platelets to adhere to the exposed fibrous matrix. By suppressing platelet aggregation, the thrombolytic activity of blood platelets can be increased, which may reduce the risk of cardiovascular diseases (Patel and Watson 2018). The essential oil extracted from the rhizome and leaf of *H. coronarium* illustrated antiplatelet activity against the platelet aggregation raised by adenosine diphosphate (ADP). This mild platelet agonist causes platelet aggregation amplification and activates shape change (Murugappa and Kunapili 2006).

The leaf essential oil did not show any inhibition effect against platelet aggregation when it was activated by ADP (4 μM). However, it showed an inhibition percentage of 30.8% at 0.5 mg/ml and 71.7% at 1 mg/ml concentration when platelet aggregation was induced by collagen (Guzman et al. 2020).

However, the rhizome essential oil showed increased antiaggregant activity. This showed 18.9% at 0.25 mg/ml and 85.1% at 1 mg/ml inhibition activity platelet aggregation when subjected to ADP 4 μM as an agonist. But, when the was collagen used as an agonist, the inhibition resulted in 48.6% ± 0.9% at 0.25 mg/ml and 97.2% ± 2.1% at 1 mg/ml of essential oil (Guzman et al. 2020).

4.5 CONCLUSION AND FUTURE PERSPECTIVES

Specifically in developing countries, with many innovations in modernity, the health parameter has been deteriorating in parallel. To improve health parameters, synthetic drugs have become of keen interest for researchers. However, these synthetic drugs have high adverse effects on human health and the environment. As plant-derived products are safer and have fewer side effects, more focus should be placed on their robust bioactivities. The *H. coronarium* plant should be studied widely to generate global interest for researchers for intensive drug discovery. This chapter concludes that comprehensive information on the traditional uses, phytochemistry, and pharmacological activities of essential oils and extracts might increase demands in the agriculture and pharma industries. Further work needs to be carried out to identify more effective drug candidates to uplift scientific society and this plant.

REFERENCES

Abate, L., Tadesse, M.G., Bachheti, A., & Bachheti, R.K. (2022). Traditional and phytochemical bases of herbs, shrubs, climbers, and trees from Ethiopia for their anticancer response. *BioMed Research International* 2022:1589877.

Aggarwal, B.B. 2004. Nuclear factor-κB: the enemy within. *Cancer Cell* 6(3):203–208.

Ahmed, S., Hasan, M.M., & Mahmood, Z.A. 2016. *In vitro* urolithiasis models: an evaluation of prophylactic management against kidney stones. *Journal of Pharmacognosy and Phytochemistry* 5(3):28.

Ahn, K. 2017. The worldwide trend of using botanical drugs and strategies for developing global drugs. *BMB Reports* 50(3):111.

Ali, S., Sotheeswaran, S., Tuiwawa, M., & Smith, R.M. 2002. Comparison of the composition of the essential oils of *Alpinia* and *Hedychium* species—essential oils of Fijian plants, part 1. *Journal of Essential Oil Research* 14(6):409–411.

Anand, P., Mohankumar, C.J., Hema, P.S., Ichikawa, H., Aggarwal, B.B., Ajaikumar, B.K., & Nair, M.S. 2008. Coronarin D, a labdane diterpene, inhibits both constitutive and inducible nuclear factor-kappa B pathway activation, leading to potentiation of apoptosis, inhibition of invasion, and suppression of osteoclastogenesis. *Molecular Cancer Therapeutics* 7(10):3306–3317.

Anderson, E.F. 1986. Ethnobotany of hill tribes of northern Thailand. I. Medicinal plants of Akha. *Economic Botany* 40(1):38–53.

Arruebo, M., Vilaboa, N., Sáez-Gutierrez, B., Lambea, J., Tres, A., Valladares, M., & González-Fernández, Á. 2011. Assessment of the evolution of cancer treatment therapies. *Cancers* 3(3):3279–3330.

Arya, A.K., Durgapal, M., Bachheti, A., Joshi, K.K., Gonfa, Y.H., Bachheti, R.K., & Husen, A. (2022). Ethnomedicinal use, phytochemistry, and other potential application of aquatic and semiaquatic medicinal plants. *Evidence-Based Complementary and Alternative Medicine* 2022:4931556.

Arya, S., Kumar, R., Prakash, O., Latwal, M., Pandey, G., & Srivastava, S.K.R. 2022. Chemical composition, nematicidal, insecticidal and herbicidal activities of Hedychium coronarium J. Koenig rhizome oleoresin. *The Pharma Innovation Journal* 11(6):761–766.

Association of Official Analytical Chemists (AOAC). 2000. *Official Methods of Analysis of the Association of Official Analytical Chemists*, 17th ed. Washington, DC: AOAC.

Atanasov, A.G., Waltenberger, B., Pferschy-Wenzig, E.M., Linder, T., Wawrosch, C., Uhrin, P., Temml, V., Wang, L., Schwaiger, S., Heiss, E.H., & Rollinger, J.M. 2015. Discovery and resupply of pharmacologically active plant-derived natural products: a review. *Biotechnology Advances* 33(8):1582–1614.

Aziz, M.A., Habib, M.R., & Karim, M.R. 2009. Antibacterial and cytotoxic activities of *Hedychium coronarium* J. Koenig. *Research Journal of Agriculture and Biological Sciences* 5(6):969–972.

Báez, D., Pino, J.A., & Morales, D. 2011. Floral scent composition in *Hedychium coronarium* J. Koenig analyzed by SPME. *Journal of Essential Oil Research* 23(3):64–67.

Bahuguna, Y.M., & Kumar, N. 2014. Phytochemical and pharmacological evaluation of *Hedychium coronarium* J. Koening for antiurolithiatic activity. *World Journal of Pharmaceutical Sciences* 2(1):112–122.

Bakkali, F., Averbeck, S., Averbeck, D., & Idaomar, M. 2008. Biological effects of essential oils – a review. *Food and Chemical Toxicology* 46(2):446–475.

Basak, S., Sarma, G.C., & Rangan, L. 2010. Ethnomedical uses of Zingiberaceous plants of Northeast India. *Journal of Ethnopharmacology* 132(1):286–296.

Bhandary, M.J., Chandrashekar, K.R., & Kaveriappa, K.M. 1995. Medical ethnobotany of the Siddis of Uttara Kannada district, Karnataka, India. *Journal of Ethnopharmacology* 47(3):149–158.

Bisht, S., Bisht, N.S., & Bh, S. 2012. In vitro plant regeneration from seedling explants of *Hedychium coronarium* J. Koenig. *Journal of Medicinal Plants Research* 6(43):5546–5551.

Boyanova, L., Gergova, G., Nikolov, R., Derejian, S., Lazarova, E., Katsarov, N., Mitov, I., & Krastev Z. 2005. Activity of Bulgarian propolis against 94 Helicobacter pylori strains in vitro by agar-well diffusion, agar dilution and disc diffusion methods. *Journal of Medical Microbiology* 54(5):481–483.

Braga, F.C., Wagner, H., Lombardi, J.A., & de Oliveira, A.B. 2000. Screening Brazilian plant species for in vitro inhibition of 5-lipoxygenase. *Phytomedicine* 6(6):447–452.

Burt, S. 2004. Essential oils: their antibacterial properties and potential applications in foods—a review. *International Journal of Food Microbiology* 94(3):223–253.

Caballero-Gallardo, K., Pino-Benitez, N., Pajaro-Castro, N., Stashenko, E., & Olivero-Verbel, J. 2014. Plants cultivated in Choco, Colombia, as a source of repellents against *Tribolium castaneum* (Herbst). *Journal of Asia-Pacific Entomology* 17(4):753–759.

Calva, M.V., Viracucha, G.C., Papa, L.L., Barros, K.N., Pérez-Martínez, A., & Radice, M. 2020. Hedychium coronarium rhizomes essential oil. Main compounds and biological activity. An updated mini-review of the last 10 years. Presented at: MOL2NET'20, Conference on Molecular, Biomed., Comput. & Network Science and Engineering, 6th ed. Universidad Estatal Amazónica, Puyo (Ecuador).

Chan, E.W., Lim, Y.Y., Wong, L.F., Lianto, F.S., Wong, S.K., Lim, K.K., Joe, C.E., & Lim, T.Y. 2008. Antioxidant and tyrosinase inhibition properties of leaves and rhizomes of ginger species. *Food Chemistry* 109(3):477–483.

Chan, E.W., & Wong, S.K. 2015. Phytochemistry and pharmacology of ornamental gingers, *Hedychium coronarium* and *Alpinia purpurata*: a review. *Journal of Integrative Medicine* 13(6):368–379.

Chen, I.N., Chang, C.C., Ng, C.C., Wang, C.Y., Shyu, Y.T., & Chang, T.L. 2008. Antioxidant and antimicrobial activity of Zingiberaceae plants in Taiwan. *Plant Foods for Human Nutrition* 63(1):15–20.

Chen, J.J., Ting, C.W., Wu, Y.C., Hwang, T.L., Cheng, M.J., Sung, P.J., Wang, T.C., & Chen, J.F. 2013. New labdane-type diterpenoids and anti-inflammatory constituents from *Hedychium coronarium. International Journal of Molecular Sciences* 14(7):13063–13077.

Chimnoi, N., Sarasuk, C., Khunnawutmanotham, N., Intachote, P., Seangsai, S., Saimanee, B., Pisutjaroenpong, S., Mahidol, C., & Techasakul, S. 2009. Phytochemical reinvestigation of labdane-type diterpenes and their cytotoxicity from the rhizomes of *Hedychium coronarium. Phytochemistry Letters* 2(4):184–187.

Coe, F.G., & Anderson, G.J. 2005. Snakebite ethnopharmacopoeia of eastern Nicaragua. *Journal of Ethnopharmacology* 96(1–2):303–323.

Costa, R.O., Jose, C.M., Grombone-Guaratini, M.T., & Matos, D.M.S. 2019. Chemical characterization and phytotoxicity of the essential oil from the invasive *Hedychium coronarium* on seeds of Brazilian riparian trees. *Flora* 257:151411.

Dash, P.R. 2016. *Phytochemical Screening and Pharmacological Investigations on Hedychium coronarium.* Hamburg: Anchor Academic Publishing.

Dash, P.R., Nasrin, M., & Saha, M.R. 2011. Evaluation of analgesic and neuropharmacological activities of methanolic rhizome extract of *Hedychium coronarium. International Journal of Pharmaceutical Sciences and Research*, 2(4):979.

Dash, P.R., & Sheikh, Z. 2015. Preliminary studies on phytochemicals and cytotoxic activity of methanolic rhizome extract of *Hedychium coronarium. Journal of Pharmacognosy and Phytochemistry* 4(1):136–139.

de Souza, J.A., & Correia, M.C. 2007. Biological floral de *Hedychium coronarium* Koen. (Zingiberaceae). *Ornamental Horticulture* 13(1):21–30.

Devi, N.B., Singh, P.K., & Das, A.K. 2014. Ethnomedicinal utilization of Zingiberaceae in the valley districts of Manipur. *IOSR Journal of Environmental Science, Toxicology and Food Technology* 8(2):21–23.

Djouahri, A., Boualem, S., Boudarene, L., & Baaliouamer, A. 2015. Geographic's variation impact on chemical composition, antioxidant and anti-inflammatory activities of essential oils from wood and leaves of *Tetraclinis articulata* (Vahl) Masters. *Industrial crops and Products* 63:138–146.

Donipati, P., & Sreeramulu, S.H. 2015. *In vitro* anticancer activity of *Hedychium coronarium* against human breast cancer cell line MCF-7. *International Journal of Advanced Research* 3(10):1497–1501.

dos Santos, B.C., Barata, L.E., Marques, F.A., Baroni, A.C., Karnos, B.A., de Oliveira, P.R., & Guerrero, Jr P.G. 2010. Composition of leaf and rhizome essential oils of *Hedychium coronarium* Koen. from Brazil. *Journal of Essential Oil Research* 22(4):305–306.

Endringer, D.C., Taveira, F.S., Kondratyuk, T.P., Pezzuto, J.M., & Braga, F.C. 2014. Cancer chemoprevention activity of labdane diterpenes from rhizomes of *Hedychium coronarium. Revista Brasileira de Farmacognosia* 24:408–412.

Farnsworth, N.R., Akerele, O., Bingel, A.S., Soejarto, D.D., & Guo, Z. 1985. Medicinal plants in therapy. *Bulletin of the World Health Organization* 63(6):965.

Goh, S.H., Soepadmo, E., Chang, P., Banerjee, U., Chan, K.C., Deverre, J.R., Hadi, H., Loke, S.E., Nasrulhawq, A., Oo, S.L., & Taylor, C.E. 1984. Studies on Malaysian medicinal plants: preliminary results. In *Proceedings of the 5th Asian Symposium on Medicinal Plants and Spices* 5:473–483.

Gonfa, Y.H., Tessema, F.B., Gelagle, A.A., Getnet, S.D., Tadesse, M.G., Bachheti, A., & Bachheti, R.K. (2022). Chemical compositions of essential oil from aerial parts of *Cyclospermum leptophyllum* and its application as antibacterial activity against some food spoilage bacteria. *Journal of Chemistry* 2022:5426050.

Gurib-Fakim, A., Maudarbaccus, N., Leach, D., Doimo, L., & Wohlmuth, H. 2002. Essential oil composition of Zingiberaceae species from Mauritius. *Journal of Essential Oil Research* 14(4):271–273.

Guzman, L., Nerio, L.S., Venturini, W., Macias, J.P., Donoso, W., & Forero-Doria, O. 2020. Antiplatelet and antibacterial activities of essential oils obtained from rhizomes and leaves of *Hedychium coronarium* J. Koening. *Anais da Academia Brasileira de Ciências* 92(2):1–10.

Hasan, M.M., Rahman, M.A., Amin, R., Akter, M.H., Norin, N., Nargis, T., Das, P.R., Islam, M.T., Hossan, M.S., & Rahmatullah, M. 2015. Do folk medicinal plants have scientific validity behind their uses. Some notes on plants used in Jessore district, Bangladesh. *World Journal of Pharmaceutical Sciences* 4:169–179.

Hill, A.F. 1952. *Economic botany. A textbook of useful plants and plant products*, 2nd ed. New York: McGraw-Hill.

Ho, J.C. 2011. Antimicrobial, mosquito larvicidal and antioxidant properties of the leaf and rhizome of *Hedychium coronarium. Journal of the Chinese Chemical Society* 58(4):563–567.

Hodge, W.H., & Taylor, D. 1957. The ethnobotany of the island Caribs of Dominica. *Webbia* 12(2):513–644.

Itokawa, H., Morita, H., Katou, I., Takeya, K., Cavalheiro, A.J., de Oliveira, R.C., & Motidome, M. 1988. Cytotoxic diterpenes from the rhizomes of *Hedychium coronarium*. *Planta Medica* 54(4):311–315.

Itokawa, H., Morita, M., & Mihashi, S. 1980. Labdane and bisnorlabdane type diterpenes from *Alpinia speciosa* K. Schum. *Chemical and Pharmaceutical Bulletin* 28(11):3452–3454.

Jain, S.K., Fernandes, V.F., Lata, S., & Ayub, A. 1995. Indo-Amazonian ethnobotanic connections-Similar uses of some common plants. *Ethnobotany* 7:29–37.

Jain, S.K., & Prakash, V. 1995. Zingiberaceae in India: phytogeography. *Rheedea* 5(2):155.

Jayaprakasam, B., Vanisree, M., Zhang, Y., Dewitt, D.L., & Nair, M.G. 2006. Impact of alkyl esters of caffeic and ferulic acids on tumor cell proliferation, cyclooxygenase enzyme, and lipid peroxidation. *Journal of Agricultural and Food Chemistry* 54(15):5375–5381.

Joly, L.G., Guerra, S., Séptimo, R., Solís, P.N., Correa, M., Gupta, M., Levy, S., & Sandberg, F. 1987. Ethnobotanical inventory of medicinal plants used by the Guaymi Indians in Western Panama. Part I. *Journal of Ethnopharmacology* 20(2):145–171.

Joshi, S., Chanotiya, C.S., Agarwal, G., Prakash, O., Pant, A.K., & Mathela, C.S. 2008. Terpenoid compositions, and antioxidant and antimicrobial properties of the rhizome essential oils of different *Hedychium* species. *Chemistry & Biodiversity* 5(2):299–309.

Joy, B., Rajan, A., & Abraham, E. 2007. Antimicrobial activity and chemical composition of essential oil from *Hedychium coronarium*. *Phytotherapy Research* 21(5):439–443.

Kala, C.P. 2005. Ethnomedicinal botany of the Apatani in the Eastern Himalayan region of India. *Journal of Ethnobiology and Ethnomedicine* 1(1):1–8.

Khezerlou, A., Alizadeh-Sani, M., Azizi-Lalabadi, M., & Ehsani, A. 2018. Nanoparticles and their antimicrobial properties against pathogens including bacteria, fungi, parasites and viruses. *Microbial Pathogenesis* 123:505–526.

Kiem, P.V., Kim Thuy, N.T., Tuan Anh, H.L., Nhiem, N.X., Minh, C.V., Yen, P.H., Ban, N.K., Hang, D.T., Tai, B.H., Tuyen, N.V., Mathema, V.B., Koh, Y.S., & Kim, Y.H. 2011. Chemical constituents of the rhizomes of *Hedychium coronarium* and their inhibitory effect on the pro-inflammatory cytokines production LPS-stimulated in bone marrow-derived dendritic cells. *Bioorganic & Medicinal Chemistry Letters* 21:7460–7465.

Kumar, R., Suman, N.R., Dash, S.S. 2004. Traditional uses of plants by tribals of Amarakantak region, Madhya Pradesh. *Indian Journal of Traditional Knowledge* 3(4):383–390.

Laferriere, J.E. 1994. Medicinal plants of the lowland Inga people of Colombia. *International Journal of Pharmacognosy* 32(1):90–94.

Lechat-Vahirua, I., François, P., Menut, C., Lamaty, G., & Bessiere, J.M. 1993. Aromatic plants of French Polynesia. I. Constituents of the essential oils of rhizomes of three Zingiberaceae: *Zingiber zerumbet* Smith, *Hedychium coronarium* Koenig and *Etlingera cevuga* Smith. *Journal of Essential Oil Research* 5(1):55–59.

Lee, M.S., Choi, J., Posadzki, P., & Ernst, E. 2012. Aromatherapy for health care: an overview of systematic reviews. *Maturitas* 71(3):257–260.

Leishangthem, S., & Sharma, L.D. 2014. Study of some important medicinal plants found in Imphal-East district, Manipur India. *International Journal of Scientific and Research Publications* 4(9):1–5.

Lima, A.S., Junior, H.N., Costa-Junior, L.M., Monteiro, O.S., Maia, J.G., & da Rocha, C.Q. 2021. Anthelmintic effect of essential rhizome oil from *Hedychium coronarium* Koenig (Zingiberaceae) introduced in Northeastern Brazil. *Acta Tropica* 218:105912.

Lingaraju, D.P., Sudarshana, M.S., & Rajashekar, N. 2013. Ethnopharmacological survey of traditional medicinal plants in tribal areas of Kodagu district, Karnataka, India. *Journal of Pharmacy Research* 6(2):284–297.

Lock, O., Castillo, P., Doroteo, V., & Rojas, R. 2005. Antioxidant activity *in vitro* of selected Peruvian medicinal plants. *Acta Horticulturae* 675:103–106.

Lu, Y., Zhong, C.X., Wang, L., Lu, C., Li, X.L., & Wang, P.J. 2009. Anti-inflammation activity and chemical composition of flower essential oil from *Hedychium coronarium*. *African Journal of Biotechnology* 8(20):5373–5377.

Manish, M. 2013. Current status of endangered medicinal plant *Hedychium coronarium* and causes of population decline in the natural forests of Anuppur and Dindori districts of Madhya Pradesh, India. *International Research Journal of Biological Sciences* 2(3):1–6.

Mans, D.R., Djotaroeno, M., Friperson, P., & Pawirodihardjo, J. 2019. Phytochemical and pharmacological support for the traditional uses of Zingiberacea species in Suriname – a review of the literature. *Pharmacognosy Journal* 11(6S):1511–1525.

Manssor, N., Matthew, R., & Lopez, A.F. 2020. Insecticidal activity of Hydrocotyle leucocephala and Hedychium coronarium plant extracts against Helicoverpa armigera. *International Journal of Life Science and Pharma Research* 10(SP-11):84–90.

Matsuda, H., Morikawa, T., Sakamoto, Y., Toguchida, I., & Yoshikawa, M. 2002. Labdane-type diterpenes with inhibitory effects on increase in vascular permeability and nitric oxide production from *Hedychium coronarium. Bioorganic & Medicinal Chemistry* 10(8):2527–2534.

Mehta, J.P., Pandya, C.V., Parmar, P.H., Vadia, S.H., & Golakiya, B.A. 2014. Determination of flavonoids, phenolic acid and polyalcohol in *Butea monosperma* and *Hedychium coronarium* by semi-preparative HPLC photo diode array (PDA) detector. *Arabian Journal of Chemistry* 7(6):1110–1115.

Mimica-Dukic, N., Bozin, B., Sokovic, M., & Simin, N. 2004. Antimicrobial and antioxidant activities of *Melissa officinalis* L. (Lamiaceae) essential oil. *Journal of Agricultural and Food Chemistry* 52(9):2485–2489.

Miranda, C.A.S.F., Cardoso, M.D.G., Carvalho, M.L.M., Figueiredo, A.C.S., & Andrade, J.D. 2015. Chemical characterisation and allelopathic potential of essential oils from leaves and rhizomes of white ginger. *Revista Ciência Agronômica* 46:555–562.

Morikawa, T., Matsuda, H., Sakamoto, Y., Ueda, K., & Yoshikawa, M. 2002. New farnesane-type sesquiterpenes, hedychiols A and B 8, 9-diacetate, and inhibitors of degranulation in RBL-2H3 cells from the rhizome of *Hedychium coronarium. Chemical and Pharmaceutical Bulletin* 50(8):1045–1049.

Murugappa, S., & Kunapuli, S.P. 2006. The role of ADP receptors in platelet function. *Frontiers in Bioscience* 11:1977–1986.

Nakatani, N., Kikuzaki, H., Yamaji, H., Yoshio, K., Kitora, C., Okada, K., & Padolina, W.G. 1994. Labdane diterpenes from rhizomes of *Hedychium coronarium. Phytochemistry* 37(5):1383–1388.

Nwe, P.P., Min, M., & Aye, H.H. 2018. Screening of phytochemicals and antioxidant activity of the rhizomes and leaves of *Hedychium coronarium* J. Koenig (Ngwe-Pan). *Journal of the Myanmar Academy of Arts and Science* 17(1A):447.

Nwe, P.P., Min, M., & Khine, M.M. 2019. Investigation of some bioactivities of Hedychium coronarium J. Koenig (Ngwe-pan) rhizome and leaves. *Third Myanmar Korea Conference Research Journal* 3(5):1766–1772.

Oh, S., Jeong, I.H., Shin, W.S., Wang, Q., & Lee, S. 2006. Synthesis and biological activity of (+)-hedychilactone A and its analogs from (+)-sclareolide. *Bioorganic & Medicinal Chemistry Letters* 16(6):1656–1659.

Pachurekar, P., & Dixit, A.K. 2017. A review on pharmacognostical, phytochemical and ethnomedicinal properties of *Hedychium coronarium* J. Koenig an endangered medicine. *International Journal of Chinese Medicine* 1(2):49–61.

Pandya, C.V., Jadeja, A.J., & Golakiya, B.A. 2014. Antifungal activity of crude extracts of *Hedychium coronarium. International Journal of Research in Phytochemical and Pharmacological Sciences* 4:4–6.

Panigrahy, S.K., Kumar, A., & Bhatt, R. 2017. Antioxidant potentials of successive solvent extracts from the unexplored *Hedychium coronarium* rhizome. *Journal of Food Science and Technology* 54(10):3297–3306.

Parida, R., Mohanty, S., & Nayak, S. 2013. *In vitro* propagation of *Hedychium coronarium* Koen. through axillary bud proliferation. *Plant Biosystems – An International Journal Dealing with all Aspects of Plant Biology* 147(4):905–912.

Patel, K., & Watson, R.R. 2018. Chocolate and its component's effect on cardiovascular disease. In: *Lifestyle in Heart Health and Disease* (pp. 255–266). London: Academic Press.

Pattanaik, C., Reddy, C.S., & Murthy, M.S. 2008. An ethnobotanical survey of medicinal plants used by the Didayi tribe of Malkangiri district of Orissa, India. *Fitoterapia* 79(1):67–71.

Pitopang, R., Hamzah, B., Zubair, M.S., Amar, A.L., Fathurahman, F., Basri, Z., & Poulsen, A.D. 2019. Diversity of Zingiberaceae and traditional uses by three indigenous groups at Lore Lindu National Park, Central Sulawesi, Indonesia. *Journal of Physics: Conference Series* 1242(1):012039.

Prakash, O. 2010. Chemical composition and antibacterial activity of rhizome oils from *Hedychium coronarium* Koenig and *Hedychium spicatum* Buch-Ham. *Journal of Essential Oil-Bearing Plants* 13(2):250–259.

Ray, A., Dash, B., Sahoo, A., Nasim, N., Panda, P.C., Patnaik, J., Ghosh, B., Nayak, S., & Kar, B. 2017. Assessment of the terpenic composition of *Hedychium coronarium* oil from Eastern India. *Industrial Crops and Products* 97:49–55.

Ray, A., Jena, S., Dash, B., Kar, B., Halder, T., Chatterjee, T., Ghosh, B., Panda, P.C., Nayak, S., & Mahapatra, N. 2018. Chemical diversity, antioxidant and antimicrobial activities of the essential oils from Indian populations of *Hedychium coronarium* Koen. *Industrial Crops and Products* 112:353–362.

Ray, A., Jena, S., Dash, B., Sahoo, A., Kar, B., Patnaik, J., Panda, P.C., Nayak, S., & Mahapatra, N. 2019. *Hedychium coronarium* extract arrests cell cycle progression, induces apoptosis, and impairs migration and invasion in HeLa cervical cancer cells. *Cancer Management and Research* 11:483.

Ray, S., Sheikh, M., & Mishra, S. 2011. Ethnomedicinal plants used by tribals of East Nimar region, Madhya Pradesh. *Indian Journal of Traditional Knowledge* 10(2):367–371.

Redondo, M.S., & Barbosa G.B. 2018. Antioxidant activity and phenolic content of the ethanol extracts of *Hedychium coronarium* (Zingiberaceae) in Mindanao, Philippines. *Bulletin of Environment, Pharmacology and Life Sciences* 7:97–105.

Reuk-ngam, N., Chimnoi, N., Khunnawutmanotham, N., & Techasakul, S. 2014. Antimicrobial activity of coronarin D and its synergistic potential with antibiotics. *BioMed Research International* 2014:581985.

Ribeiro, R.D., De Melo, M.M., De Barros, F., Gomes, C., & Trolin, G. 1986. Acute antihypertensive effect in conscious rats produced by some medicinal plants used in the state of Sao Paulo. *Journal of Ethnopharmacology* 15(3):261–269.

Rodrigues, D.F., Arenas Velásquez, A.M., Cavaleiro, C., Salgueiro, L., Martins, G.Z., Magalhães, N.O., Martins, M.B., Cicarelli, R.M., & Moreira, R.R. 2013. Chemical composition and trypanocidal activity of the essential oils from *Hedychium coronarium* J. Koenig (Zingiberaceae). *International Scholarly Research Notices* 2013:639275.

Sabulal, B., George, V., Dan, M., & Pradeep, N.S. 2007. Chemical composition and antimicrobial activities of the essential oils from the rhizomes of four *Hedychium* species from South India. *Journal of Essential Oil Research* 19(1):93–97.

Sarangthem, N., Talukdar, N.C., & Thongam, B. 2013. Collection and evaluation of *Hedychium* species of Manipur, Northeast India. *Genetic Resources and Crop Evolution* 60(1):13–21.

Shah, A., Bharati, K.A., Ahmad, J., & Sharma, M.P. 2015. New ethnomedicinal claims from Gujjar and Bakerwals tribes of Rajouri and Poonch districts of Jammu and Kashmir, India. *Journal of Ethnopharmacology* 166:119–128.

Shanmugam, P.V., Yadav, A., & Chanotiya, C.S. 2015. Enantiomer differentiation of key volatile constituents from leaves, stems, rhizome and flowers of cultivated *Hedychium coronarium* Koenig from India. *Journal of Essential Oil Research* 27(2):101–106.

Sharma, G.J., Chirangini, P., & Kishor, R. 2011. Gingers of Manipur: diversity and potentials as bioresources. *Genetic Resources and Crop Evolution* 58(5):753–767.

Shrotriya, S., Ali, M.S., Saha, A., Bachar, S.C., & Islam, M.S. 2007. Anti-inflammatory and analgesic effects of *Hedychium coronarium* Koen. *Pakistan Journal of Pharmaceutical Sciences* 20(1):47–51.

Singh, B., Adhikari, D., Barik, S.K., & Chettri, A. 2013. *Pterocymbium tinctorium* (Merrill, 1901) (Magnoliophyta: Malvales: Sterculiaceae: Sterculioideae): new record from mainland India and extension of geographic distribution. *Check List* 9(3):622–655.

Singh, B., Borthakur, S.K., & Phukan, S.J. 2014. A survey of ethnomedicinal plants utilized by the indigenous people of Garo Hills with special reference to the Nokrek Biosphere Reserve (Meghalaya), India. *Journal of Herbs, Spices & Medicinal Plants* 20(1):1–30.

Sirirugsa, P. 1999. Thai Zingiberaceae: species diversity and their uses. *Pure and Applied Chemistry* 70:1–8.

Smith-Hall, C., Larsen, H.O., & Pouliot, M. 2012. People, plants and health: a conceptual framework for assessing changes in medicinal plant consumption. *Journal of Ethnobiology and Ethnomedicine* 8:43.

Suresh, G., Reddy, P.P., Babu, K.S., Shaik, T.B., & Kalivendi, S.V. 2010. Two new cytotoxic labdane diterpenes from the rhizomes of *Hedychium coronarium*. *Bioorganic & Medicinal Chemistry Letters* 20(24):7544–7548.

Tailor, C.S., & Goyal, A. 2015. Isolation of phyto constituents and *in vitro* antiurolithiatic by titrimetric method, antioxidant activity by 1, 1-diphenyl-2-picryl hydrazyl scavenging assay method of alcoholic roots and rhizomes extract of *Hedychium coronarium* J. Koenig plant species. *Asian Journal of Pharmaceutical and Clinical Research* 8(4):225–229.

Tavares, W.R., Barreto, M.D., & Seca, A.M. 2020. Uncharted source of medicinal products: the case of the *Hedychium* genus. *Medicines* 7(5):23.

Taveira, F.N., Oliveira, A.B., Souza Filho, J.D., & Braga, F.C. 2005 Epimers of labdane diterpenes from the rhizomes of *Hedychium coronarium*. J. Koenig. *Revista Brasileira de Farmacognosia* 15: 55–59.

Tse, L.S., Liao, P.L., Tsai, C.H., Li, C.H., Liao, J.W., Kang, J.J., & Cheng, Y.W. 2019. Glycemia lowering effect of an aqueous extract of *Hedychium coronarium* leaves in diabetic rodent models. *Nutrients* 11(3):629.

Turnbaugh, P.J., Ley, R.E, Hamady, M., Fraser-Liggett, C.M., Knight, R., & Gordon, J.I. 2007. The human microbiome project. *Nature* 449(7164):804–810.

Valadeau, C., Castillo, J.A., Sauvain, M., Lores, A.F., & Bourdy, G. 2010 The rainbow hurts my skin: medicinal concepts and plants uses among the Yanesha (Amuesha), an Amazonian Peruvian ethnic group. *Journal of Ethnopharmacology* 27(1):175–192.

Van Thanh, B., Dai, D.N., Thang, T.D., Binh, N.Q., Anh, L.D.N., & Ogunwande, I.A. 2014 Composition of essential oils of four *Hedychium* species from Vietnam. *Chemistry Central Journal* 8:54.

Verma, M., & Bansal, Y.K. 2013. Effect of additives on plant regeneration in *Hedychium coronarium* J. Koenig an endangered aromatic and medicinal herb. *International Journal of Pharmaceutical Sciences Review and Research* 23:105–110.

Watkins, K. 2018. Emerging infectious diseases: a review. *Current Emergency and Hospital Medicine Reports* 6(3):86–93.

Yadaw, P., & Shrivastava, S. 2020. A study of the use of some medicinal plants by tribes living in Jashpur district of Chhattisgarh state. *International Research Journal of Multidisciplinary Scope (IRJMS)* 1(4):45–51.

5 Pharmacognostical, Phytochemical, and Ethnomedicinal Review of *Enhydra fluctuans* Lour.

*Khoshnur Jannat, Md Nasir Ahmed, Tridib K. Paul, Chowdhury Alfi Afroze, Ommay Hany Rumi, Tohmina Afroze Bondhon, Anamul Hasan, Rownak Jahan, and Mohammed Rahmatullah**
Department of Biotechnology and Genetic Engineering, University of Development Alternative, Dhaka, Bangladesh
* Corresponding author (Mohammed Rahmatullah): rahamatm@hotmail.com

5.1 INTRODUCTION

Enhydra fluctuans Lour. (EF) is a tropical plant commonly known as enhydra, buffalo spinach, watercress in English, helencha in Bengali, harkuch in Hindi, godobos in Indonesian, and phak bung ruem in Thai. The plant is edible semiaquatic, and eaten as a vegetable. EF has a significant ethnomedicinal value for which it is widely used as a medicinal plant and vegetable (Ali et al. 2013; Arya et al. 2022). In Manipur state of Northeast India, it is known as komprek-tujombi, and tribal people cook the shoot for dietary purposes and eat the raw shoot as *singju* (Jain et al. 2011; Arya et al. 2022). A photograph of EF is shown in Figure 5.1.

5.2 ETHNOMEDICINAL USES

The plant is well known among the tribal people in North-East India, Bangladesh, and the Philippines. In the Garo tribe of Bangladesh, the Marakh sect of people use the plant as an antidiabetic medicine (Rahmatullah et al. 2012). Leaves and stems of the plant are fried and eaten by the tribal people of Oraon and Gor in Bangladesh (Azam et al. 2014). People from the Cachar district of Assam use the whole plant to treat constipation, and the stem is used for ulcers and gastric problems (Das et al. 2008). Among the Shan tribe of Assam, the plant is considered an edible food supplement (Pandey and Bora 1997). The aerial part of the plant is crushed finely and applied over pimples by the population of Dibrugarh district of North-East India.

Additionally, for treating diabetes, the Meitei-pangal community of Thoubal district of Manipur prescribes the plant's extract, which is obtained by boiling the plant which is cut into pieces at the nodes. Muslim herbalists of Thoubal district prescribe the liquid of boiled leaves mixed with sugar to treat kidney stones (Khan et al. 2010; Sarma et al. 2014). The Jaintia tribe of Meghalaya uses this plant leaf juice to treat skin diseases. The leaf juice is also prescribed as a laxative and orally for liver

DOI: 10.1201/9781003256830-5

FIGURE 5.1 Picture of Helencha (*Enhydra fluctuans*) taken from a pond side in Gazipur district, Bangladesh.

diseases by the tribe (Dolui et al. 2004; Jaiswal 2010). People from the Philippines use the plant to cure skin problems (Alom et al. 2021). The plant is also used as a remedy for snakebites in some wetland areas of India (Kumar et al. 2010). Indonesian Minangkabau tribal people use the plant for medicinal purposes, especially as an antidiabetic (Heylliward et al. 2012). In the Dhemaji district of Assam, Northeast India, a rural community uses plant leaves for antidiabetic effect. They make an extract of the plant leaves by crushing 125 g of leaves in 250 ml water and the extract is administered twice a day (Tarak et al. 2011). Tribal people of Manipur state, Northeast India prescribe fresh plant extract for diabetes (Jain et al. 2011).

5.3 PHYTOCHEMICALS PRESENT IN *ENHYDRA FLUCTUANS* ARE RESPONSIBLE FOR PHARMACOLOGICAL ACTIVATES

EF is rich in bioactive chemicals. Among all the phytochemicals flavonoids, isoflavone glycoside, essential oils, steroids, diterpenoids, sesquiterpene lactone, and melampolide are major phytochemicals (Mussel et al. 2000; Yadava and Singh 2007; Ghosh et al. 2013; Ahmed et al. 2017; Akhe et al. 2019; Saha et al. 2019; Barua et al. 2021; Koley et al. 2021). The reported constituents are listed in Table 5.1, and the structures of selected phytochemicals are shown in Figure 5.2.

The reported bioactivities of some phytochemicals are shown in Table 5.2.

5.4 PHARMACOLOGICAL ACTIVITIES

As the plant has various classes of phytochemicals, it has diverse pharmacological activities, which have been reported from time to time.

5.4.1 Antioxidant Activity

EF is well known for its antioxidant property. To show the antioxidant property, a 2,2-diphenyl-1-picrylhydrazyl (DPPH) radical scavenging activity test was performed by Kuri et al., where

TABLE 5.1
Reported Phytochemical Constituents of *E. fluctuans* Lour.

Plant part	Nature of phytoconstituents	Phytoconstituents	References
	Phytosterol	Cholesterol, sitosterol, stigmasterol, stigmasta-5,22,25-trien-3β-ol	Mussel et al. 2000
Stem, leaf	Germacranolide	Enhydrin, chloroenhydrin, fluctuanin, fluctuadin	Mussel et al. 2000; Barua et al. 2021
Whole plant	Kaurene derivatives, diterpenoid	Kauran-16-ol, kaur-16-en-19-oic acid, myricyl alcohol	Barua et al. 2021
Leaf	Essential oils	α-Pinene, (*E*)-β-ocimene, camphor, (*E*)-carophyllene, α-humulene, 1-octen-3-ol, linalool, longiverbenone, 1,2-dihydroperyllaldehydes (*cis*- and *trans*-4 isopropenylcyclohexanec arboxaldehydes), perillaldehyde, myrcene	Mussel et al. 2000; Barua et al. 2021
Leaf	Sesquiterpene lactone	8-Desacyl enhydrin tiglate, 8-desacyl enhydrin-[4-hydroxymethacrylate], 8-desacyl enhydrin-[4-hydroxytiglate], 8-desacyl enhydrin-[2,3-epoxyisobutyrate], 8-beta-methacryloyloxy-9alpha-acetoxy-14oxoacanthospermolide	Krishnaswamy and Ramji 1995
Whole plant	Flavonoids	Baicalein 7-*O*-glucoside and baicalein 7-*O*-diglucoside	Saha et al. 2019
Whole plant	Polyphenol	(+)-Catechin hydrate, vanillic acid, *p*-coumaric acid, ellagic acid	Akhe et al. 2019
Whole plant	Isoflavone glycoside	4',5,6,7-Tetrahydroxy-8-methoxyisoflavone-7-*O*-beta-D-galactopyranosyl-(1→3)-*O*-beta-D-xylopyranosyl-(1→4)-*O*-alpha-l-rhamnopyranoside	Yadava and Singh 2007
Leaf	Melampolide	8-Deeopxyangeloyl-8-[2-hydroxy-3-chloro-isobutyroyl]-enhydrin	Barua et al. 2021
Leaf		Arabinogalactone	Ghosh et al. 2013
Helencha leaves (only tender ones with characteristic flavour)	Hydroxycinnamic acids	Campesteryl ferulate, 24-Methylcholestanol ferulat, 1,2-diferuloylgentiobiose	Koley et al. 2021
	Hydroxybenzoic acid	Paeoniflorin	Koley et al. 2021
	Dihydroflavonol	Dihydrokaempferol-3-*O*-rutinoside	Koley et al. 2021
	Flavanol	(−)-Epicatechin-(2a-7) (4a-8)-epicatechin 3-*O*-galactoside	Koley et al. 2021
	Flavanone	Neoeriocitrin	Koley et al. 2021
	Flavone	Luteolin-7-*O*-malonyl-glucoside, luteolin-7-*O*-rhamnoside, diosmetin-7-*O*-rutinoside	Koley et al. 2021
	Flavonols	Kaempferol-3-*O*-(600-malonyl-glucoside, Quercetin-3-*O*-rutinoside, quercetin-3-*O*-(OH-FA) sophoroside, quercetin-3-*O*-(SI) sophoroside 7-*O*-glucoside, isorhamnetin-3-*O*-rutinoside, spinacetin-3-*O*-(200-feruloylglucosyl) (1-[6)-[apiosyl(1-[2)]-glucoside	Koley et al. 2021
	Isoflavonoid	6"-*O*-Acetyldaidzin	Koley et al. 2021
	Methylated flavonol	Laricitrin-3-*O*-hexoside	Koley et al. 2021
	Tetramethoxyflavone	Tetramethylscutellarein (40,5,6,7-tetramethoxyflavone)	Koley et al. 2021
	Lignan	Sesaminol-2-*O*-triglucoside, secoisolariciresinol di-*O*-glucoside, lariciresinol-sesquilignan	Koley et al. 2021
	Other polyphenols	Demethyloleuropein, arbutin, ligstroside-aglycone	Koley et al. 2021
Whole plant		Gibberellin A9, gibberellin A13, 4-hydroxy farnesyl acetate	Barua et al. 2021

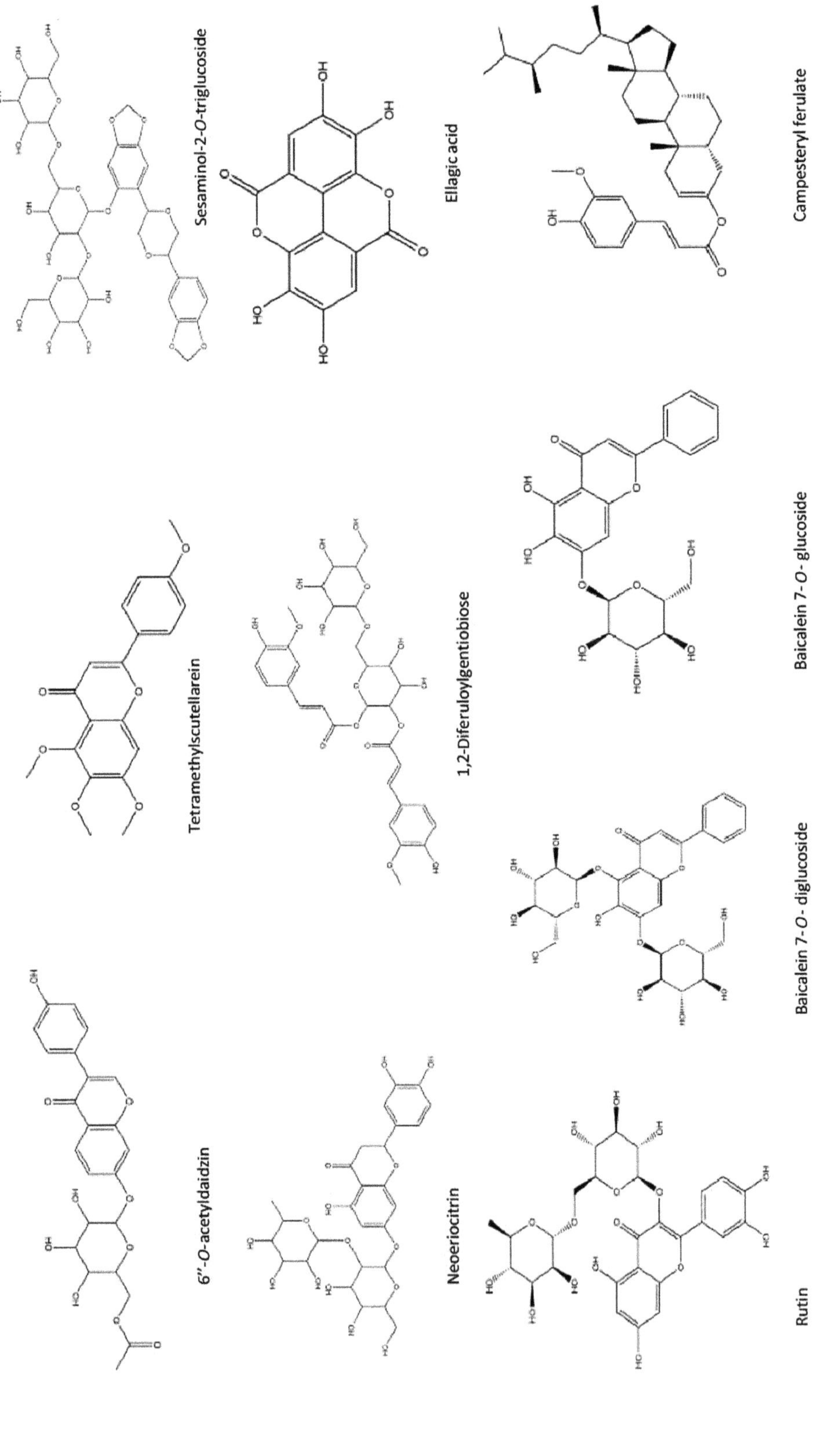

FIGURE 5.2 Structure of phytoconstituents present in Helencha (*E. fluctuans*).

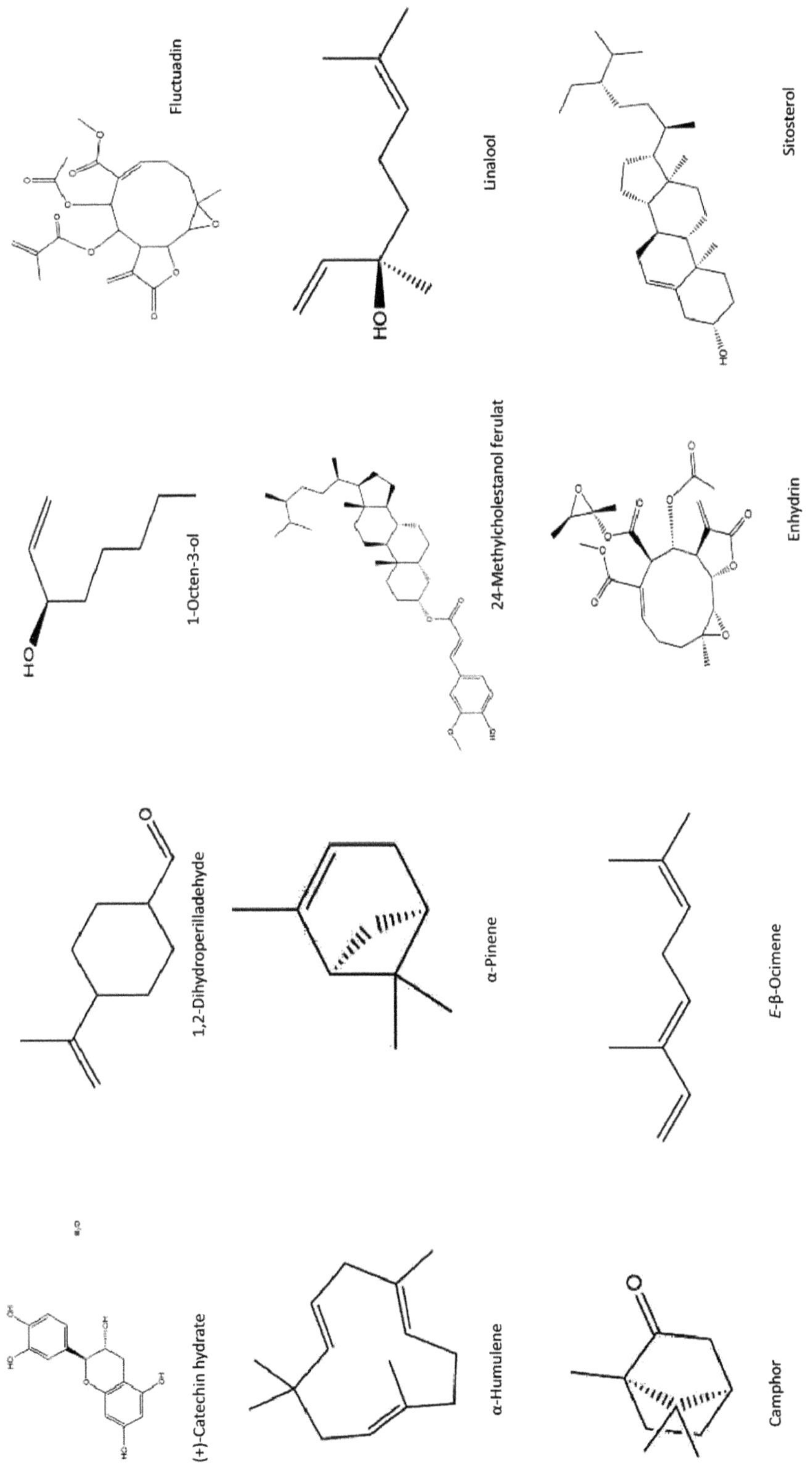

FIGURE 5.2 (Continued)

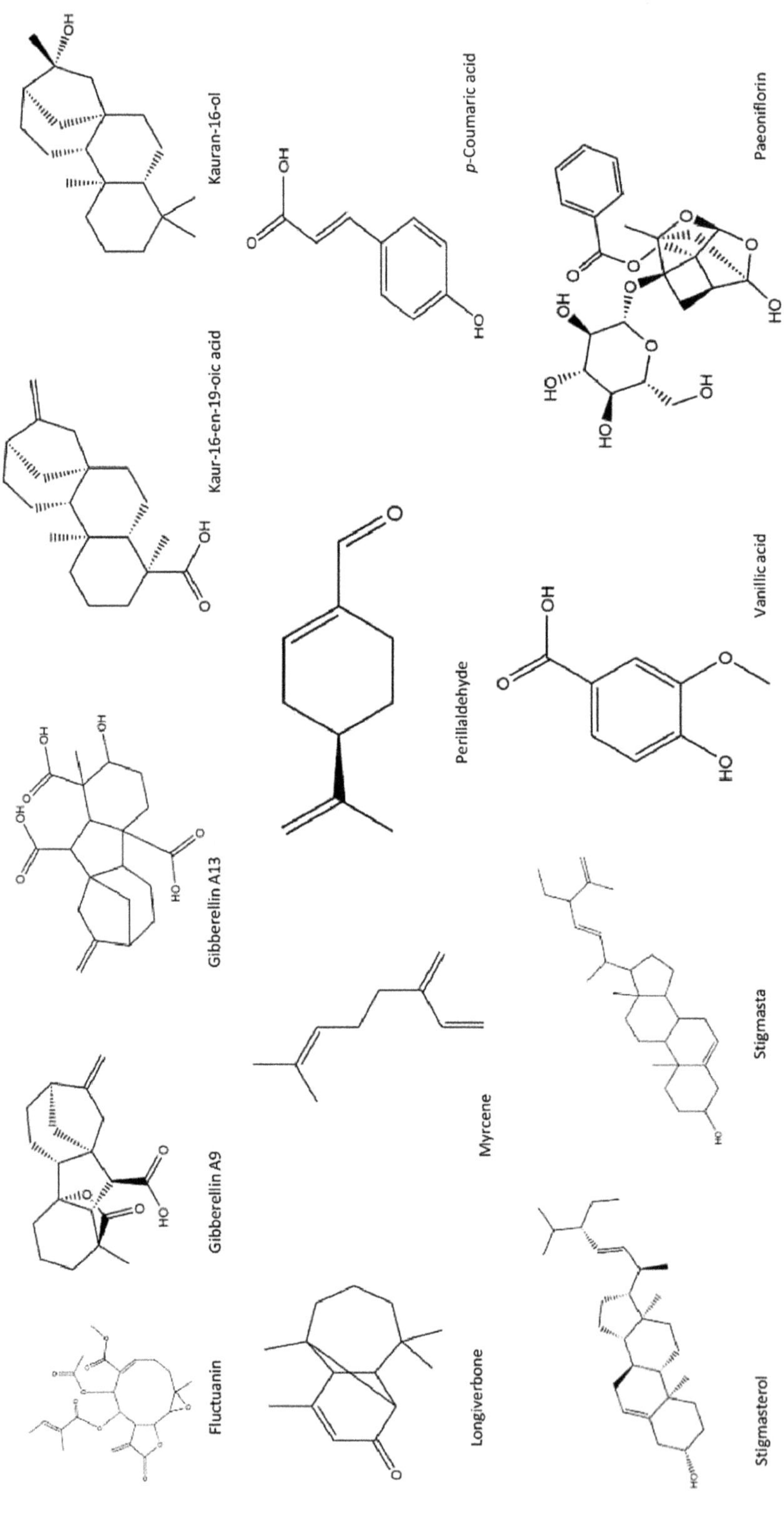

FIGURE 5.2 (Continued)

TABLE 5.2
Bioactivity of Selected Phytochemicals of *E. fluctuans* Lour.

Phytochemical	Bioactivity	References
Sitosterol	Antidiabetic, anti-inflammatory, analgesic, anxiolytic, sedative effect, immunomodulatory, anticancer, hepatoprotective, antibacterial	Fraile et al. 2012; Dighe et al. 2016; Ododo et al. 2016; Sharmila and Sindhu 2016; Babu et al. 2020
Stigmasterol	Anti-inflammatory	Zeb et al. 2017
Diterpenoids	Antimicrobial, anti-inflammatory, antitumour, antiviral	Eksi et al. 2020
Enhydrin	American trypanosomiasis, leishmaniasis, antihyperglycaemic	Frank et al. 2013; Serra-Barcellona et al. 2017; Ulloa et al. 2017
α-Pinene	Anti-inflammatory, chondroprotective, antimicrobial	Rufino et al. 2014; Ložienė et al. 2018
Camphor	Insecticidal, antimicrobial, antiviral, anticoccidial, anti-nociceptive, anticancer, antitussive activities, skin penetration enhancer, analgesic	Chen et al. 2013; da Sliva et al. 2015
Longiverbenone	Antibacterial, cytotoxic, antiproliferative, antioxidant	Rahman et al. 2008; Yagi et al. 2016
Linalool	Antimicrobial, anti-inflammatory, anticancer, antioxidant, anti-hyperlipidaemic, antinociceptive, analgesic, anxiolytic, anti-depressive, neuroprotective	Kamatou et al. 2008; Pereira et al. 2018
Perillaldehyde	Intestinal anti-inflammatory, vasodilation, antidepressant, anticancer, antioxidant, antimicrobial	Igarashi et al. 2013; Uemura et al. 2018
Ellagic acid	Anti-mutagenic, antimicrobial, antioxidant properties, inhibitors of the human immunodeficiency virus (HIV), an inhibitor of aldolase reductase (thus protects from blindness, kidney damage, paralysis, heart attacks, and loss of limbs associated with the two types of diabetes), anti-inflammatory	Sepúlveda et al. 2011
p-Coumaric acid	Antioxidant, anti-inflammatory, analgesic, anti-antimicrobial, anti-necrotic, anti-cholestatic, anti-amoebic	Aldaba-Muruato et al. 2021
Luteolin-7-*O*-rhamnoside	Antiproliferative, antioxidative, anti-inflammatory	Park et al. 2019
Quercetin 3-rutinoside (rutin)	Antibacterial, antiprotozoal, antitumour, anti-inflammatory, antiallergic, antiviral, cytoprotective, vasoactive, hypolipidaemic, antiplatelet, antispasmodic, antihypertensive, prevention of neuroinflammation, promotion of neural crest cell survival, sedative, anticonvulsant, analgesic, anti-Alzheimer, antidiabetic, anti-ulcer, anticataract and ophthalmic effect, anticancer, hepatoprotective	Ganeshpurkar et al. 2017; Patel et al. 2019

they used a methanolic extract of the whole plant. The test showed significant antioxidant activity compared to ascorbic acid (Kuri et al. 2014). In another study to determine the antioxidant properties, ethanolic extract, petroleum ether extract, and chloroform extract of the aerial parts of EF were used. In the DPPH scavenging test, at a concentration of 500 g/ml, the ethanolic extract was able to scavenge 61.67% of DPPH free radical while the other two extracts gave a lower percentage than the ethanolic extract. In the nitric oxide scavenging activity and assay for superoxide ion scavenging activity, ethanolic extract gave the highest result (Swain et al. 2012a). The antioxidant activity of EF was measured in a different system of assays (DPPH, superoxide radical scavenging assay, hydroxyl radical scavenging assay, and lipid peroxidation assay). In all the assays, the aqueous extract of EF showed significant results (Dasgupta et al. 2007). In a DPPH radical scavenging study, aqueous extract of EF leaves was found most potent among the other fractions of extract (Kamal et al. 2019). Half-maximal inhibitory concentration (IC_{50}) values of 619.55, and 115.67 μg/ml were found in a DPPH radical scavenging study, studied with methanolic and ethanolic extract, respectively. The result was compared to the standard ascorbic acid with an IC_{50} value of 17.81 μg/ml (Haoya et al. 2018). Aqueous extract of EF demonstrated lower antioxidant properties (IC_{50} 40.63 μg/ml) than ethanol extract (IC_{50} 22.53 μg/ml) in a DPPH radical scavenging test. Both the results were higher than the result for vitamin C whose IC_{50} value was reported at 15.55 μg/ml (Htun et al. 2018). Aqueous extract of EF reportedly has a prophylactic effect. The study was performed against arsenic-induced ($NaAsO_2$ 5 mg/kg) hepatotoxicity in male Swiss albino mice. Reactive oxygen species production, lipid peroxidase, and NADPH oxidase production were altered when treated with an aqueous extract of EF (Dua et al. 2016). The antioxidant property of the methanolic extract of EF was examined by determining total antioxidant activity (TAA), ferric reducing antioxidant power (FRAP), DPPH radical scavenging assay, 2,2'-azino-bis(3-ethylbenzothiazoline-6-sulphonic acid) (ABTS) radical scavenging assay, and superoxide scavenging activity assay. For all these assays the extract showed potent antioxidant activity with IC_{50} values of TAA= 5.99 ± 0.63 (mg AAE/g of extract), FRAP = 28.23 ± 2.80 (mM Fe^{2+}/mg of extract), DPPH = 113.84 ± 2.73 μg/ml, ABTS = 98.84 ± 8.85 μg/ml, and superoxide = 146.91 ± 13.14 μg/ml (Adarsha et al. 2021).

The antioxidant activity of extracts from different fractions, such as methanol, chloroform, ethyl acetate, and *n*-butanol, were evaluated using different *in vitro* antioxidant assays. In the DPPH assay, the ethyl acetate fraction exhibited the highest scavenging activity with an IC_{50} value of 23.4 μg/ml. Ethyl acetate again showed the highest inhibition of nitric oxide (IC_{50} = 26.1 μg/ml) in the nitric oxide scavenging assay. Maximum superoxide anion scavenging activity was demonstrated by ethyl acetate (IC_{50} = 56.7 μg/ml). Superoxide anions are known to damage biomacromolecules directly or indirectly. In hydroxyl radical scavenging activity, *n*-butanol showed the highest activity with IC_{50} = 86.7 μg/ml (Sannigrahi et al. 2010b). In both FRAP and DPPH radical scavenging activity assays, methanol extract of EF leaves exhibited significant antioxidant properties (Panigrahi and Bhatnagar 2015). A 24 kDa arabinogalactan (carbohydrate polymer) was found in EF leaves, which had antioxidative properties (Ghosh et al. 2013). In another DPPH radical scavenging assay, methanolic extract of aerial parts of EF exhibited antioxidant properties with an IC_{50} value of 113.27 μg/ml (Lopa et al. 2021).

5.4.2 Anti-inflammatory and Analgesic Activity

During screening for anti-inflammatory activity, two flavonoids (baicalein 7-*O*-glucoside and baicalein 7-*O*-diglucoside) were isolated from the ethyl acetate fraction of EF. The test was performed by carrageenan, histamine-induced acute inflammation, and Freund's complete adjuvant-induced chronic inflammation in rats. Total flavonoids of EF showed potent anti-inflammatory activity. At doses of 200 mg/kg and 400 mg/kg, the total flavonoids showed 27.05% and 55.49% analgesic activity against acetic acid-induced writhing when the extract was administered orally in Swiss albino mice; total flavonoids of EF further increased the pain threshold in mice in the hot plate method (Sannigrahi et al. 2011).

The ethyl acetate fraction of the leaf extract of EF was used to obtain flavonoids, which were tested to examine the anti-inflammatory activity through cyclooxygenase (COX) and lipoxygenase (LOX) assays. The flavonoid fraction inhibited the COX-2 enzyme with an IC_{50} of 80 μg/ml. The fraction also inhibited the 5-LOX enzyme with an IC_{50} of 92 μg/ml (Satyajit and Pradhan 2012). Other studies also found some analgesic properties of the plant extract. Methanolic extract of the plant (200 and 500 mg/kg) showed promising analgesic activity when administered orally in the acetic acid-induced writhing and tail-flick method in Swiss albino mice (Rahman et al. 2002). Analgesic activity (hot plate method) and anti-inflammatory activity (formalin-induced paw oedema) were studied in male Wistar albino rats. Ethanolic extract of EF leaves and stems was administered to the rats at a dose of 300 mg/kg. The dose of the extract significantly reduced paw inflammation and pain (Roy et al. 2021).

5.4.3 Antidiabetic Activity

EF has been reported to have antidiabetic properties. A study was carried out by administering the aqueous-alcoholic extract (200 mg/kg) of the plant to streptozotocin-induced adult Long–Evan rats. Twenty-one days of treatment with the EF extract significantly reduced the blood glucose level of the experimental rats (Hasan et al. 2019). Dutta et al. studied the hypoglycaemic effect of EF by checking the inhibition of α-amylase and α-glucosidase enzymes. Both methanol and acetone extract of EF leaves demonstrated higher α-glucosidase inhibitory activity (with IC_{50} of 6.55 ± 0.087 mg/ml and 6.00 ± 0.047 mg/ml, respectively) than α-amylase inhibitory activity (with IC_{50} of 3.42 ± 0.243 mg/ml and 5.39 ± 0.051 mg/ml, respectively), which significantly revealed that EF has antidiabetic properties (Dutta et al. 2016). A study by Delfita et al. showed that ethanolic extract of EF aerial parts could reduce blood glucose levels in the oral glucose tolerance test. The study was designed by taking 25 diabetic rats who were orally administered three doses of extract: 200, 500, and 1000 mg/kg. Only the 500 mg/kg dose showed significant antihyperglycaemic activity (Delfita et al. 2021a). Another study by Delfita et al. confirmed the antidiabetic property of the plant by administering the *n*-hexane fraction of aerial parts of EF to alloxan-induced diabetic Wistar rats (Delfita et al. 2021b). To check the antidiabetic effect of the plant, a study was done to examine the protective effect against insulitis. Insulitis is an inflammation of pancreatic β cells where proinflammatory cytokines such as tumour necrosis factor-α (TNF-α), interleukin-1β (IL-1β), and interferon-γ (IFN-γ) are found in an increased amount in the islet of Langerhans. Insulitis is considered as directly involved in the pathogenicity of diabetes. In this study, a total of 30 male rats were taken, where 25 were alloxan-induced diabetic rats and five were kept normal as control. The *n*-hexane fraction of the EF aerial part was administered at doses of 57.03, 114.06, and 171.09 mg/kg body weight, and the experiment was carried out for 21 days. The best protective effect against insulitis was found at a dose of 57.03 mg/kg, which showed a reduced blood glucose level, an increase in insulin level, and increased pancreatic β-cell numbers (Lundberg et al. 2017; Böni-Schnetzler et al. 2019; Zhao et al. 2020; Delfita et al. 2022).

5.4.4 Antimicrobial Activity

Methanol and acetone extracts of EF leaves were studied against some Gram-positive and Gram-negative bacteria and fungi. Methanolic extract of EF leaves moderately inhibited the growth of *Bacillus cereus*, *Escherichia coli*, *Shigella shiga*, and *Shigella sonnei* with a 6–14 mm zone of inhibition. An antifungal activity study carried out with the disc diffusion assay method using nystatin disc (100 µg/disc), methanolic extract of EF showed antifungal activity against *Aspergillus niger* and *Fusarium* spp. (Amin et al. 2012). Methanolic extract of EF leaf was examined against *E. coli*, *Pseudomonas aeruginosa*, *Staphylococcus aureus*, and *Micrococcus luteus* and significant zone inhibition was observed (Bhakta et al. 2008). Aqueous extract of EF leaf was examined against various Gram-positive and Gram-negative bacteria such as *Staph. aureus*, *Bacillus licheniformis*, *Bacillus brevis*, *Bacillus subtilis*, *Staphylococcus epidermidis*, *P. aeruginosa*, *E. coli*, *Vibrio cholerae*, *Shigella flexneri*, and a fungal species, *Candida krusei*. The extract showed antimicrobial properties with an inhibition zone of more than 20 mm (Thatoi et al. 2008). In the disc diffusion method, methanolic extract of EF displayed a 12–24 mm zone of inhibition against two Gram-positive bacteria, *B. cereus* and *Listeria monocytogenes*, and four Gram-negative bacteria, *Acinetobacter baumannii* and *P. aeruginosa*, *E. coli* ATCC 25922, and *L. monocytogenes* MTCC 657. The aqueous extract demonstrated a 13–21 mm zone of inhibition against Gram-negative bacteria except for *P. aeruginosa*, and a 10–17 mm zone of inhibition against Gram-positive bacteria (Sikdar et al.

2020). Acetone and ethanolic extract of EF displayed promising inhibitory effects against different bacterial species such as *Staph. aureus*, *Staphylococcus saparophyticus*, *E. coli*, *Salmonella typhi*, and *Shigella dysenteriae* when inoculated in an amount of 16 mg/ml. The diameter of the zone of inhibition was 17–21 mm for ethanolic extract and 17–22 mm for acetone extract (Ahmed et al. 2016). Polyphenolic components of EF leaf extract were tested against several clinical-pathological bacteria. The extract was found to be most effective against *S. flexneri* with a minimum inhibitory concentration (MIC = 0.25 μg/ml) and *Salmonella paratyphi* A (MIC = 0.5 μg/ml). The extract was also found to be effective against *P. aeruginosa* ATCC 27853 (MIC = 4 μg/ml) and *P. aeruginosa* clinical strain (MIC = 16 μg/ml). The growth of biofilm inhibition was also tested and found effective against *P. aeruginosa* (Koley et al. 2021). Another study was done by Sarma et al. where they took methicillin-resistant *Staphylococcus aureus* (MRSA) (Gram positive), and extended-spectrum beta-lactamase (ESBL) and metallo beta-lactamase-producing *E. coli*, and *Klebsiella pneumoniae* (Gram negative). All three species were resistant to certain commonly used antibacterial drugs such as ampicillin, oxacillin, cefotaxime, and ceftazidime (Sarma et al. 2016). Toluene and methanol extract of leaves and stems were studied, where toluene extract showed a promising antibacterial effect (range of zone of inhibition 7 ± 0.5 mm to 20 ± 0.5 mm) on the three bacterial species. On the other hand, the effect of methanol extract was not consistent (Sarma et al. 2016). Among the extract fraction of EF leaves (aqueous, carbon tetrachloride, chloroform, and petroleum ether), the carbon tetrachloride fraction showed antibacterial activity on *Bacillus megaterium*, *Staph. aureus*, *E. coli*, and *P. aeruginosa* with the zone of inhibitions at 7 μg/disc, 8 μg/disc, 8 μg/disc, and 7 μg/disc, respectively (Kamal et al. 2019).

Another antibacterial study was performed on *Staph. aureus*, *E. coli*, *P. aeruginosa*, and *Salmonella enteritidis* species where the ethanolic extract (400 μg/disc) of the whole plant only inhibited *E. coli* growth with a zone of inhibition of 12 mm (Haoya et al. 2018). Ethanol extract of EF was tested on *B. subtilis*, *Staph. aureus*, *P. aeruginosa*, *Bacillus pumilus*, *E. coli*, and *C. albicans* species for its antimicrobial property. In the agar-well diffusion method, the extract gave a more potent result against *B. pumilus* with a zone of inhibition of 35 mm, whereas for other microbes, the range was 19–35 mm (Htun et al. 2018). Antibacterial activity of methanolic extract of EF was determined by the disc diffusion method, which was performed on two Gram-positive bacteria, *B. cereus* ATCC 10876 and *Staph. aureus* ATCC 29740, and two Gram-negative bacteria, *E. coli* ATCC 25922 and *Salmonella* sp. ATCC 14028. The extract only worked on the two Gram-positive bacteria *B. cereus* and *Staph. aureus*, with an inhibition zone of 9.02 and 7.12 mm, respectively. Antifungal activity was tested by poisoned food technique on *A. niger* ATCC 16404, and *C. albicans* ATCC 10231. The results were calculated by percentage inhibition by recording the zone of inhibition. The extract exhibited 25.0% inhibition of *A. niger* growth and 30.5% inhibition of *C. albicans* (Rana et al. 2014).

Ethanol extract of EF leaf and rhizome was examined for their antibacterial activity. A disc diffusion method was done on different bacteria: *S. dysenteriae*, *S. typhi*, *S. paratyphi*, *B. cereus*, *B. subtilis*, *E. coli*, *P. aeruginosa*, *Staph. aureus*, *V. cholerae*, and *B. megaterium*. The range of zone of inhibition was found to be 7–11 mm, among which the extract showed the highest zone of inhibition against *B. cereus* at 11 mm. The extract did not work against *S. paratyphi* and *B.megaterium* (Reddy et al. 2014). Alcohol, chloroform, petroleum ether, and water extract of whole plant of EF were checked for their antibacterial property against nine species of bacteria, namely *B. subtilis*, *B. cereus*, *B. megaterium*, *Staph. aureus*, *S. dysenteria*, *S. sonnei*, *S. typhi*, *E. coli*, and INABA-ET (*Vibrio*). In the disc diffusion method, the zone of inhibition was calculated. The chloroform extract did not work against any of the bacteria. Alcoholic extract displayed a zone of inhibition (8–14 mm) against six types of bacteria. It did not work against *S. typhi*, *E. coli*, and INABA-ET (*Vibrio*). Petroleum ether extract of EF did not show any inhibition against the two species of *Shigella* but worked on other species (zone of inhibition 7–11 mm). Water extract exhibited antibacterial activity against three species of *Bacillus* (zone of inhibition 11–13 mm) and did not work on the other six species (Ahmed et al. 2005).

5.4.5 Cytotoxic Activity

Amin et al. (2012) studied the cytotoxic effect with the brine shrimp lethality test. They found that acetone extract of leaves of EF was active against brine shrimp nauplii (*Artemia salina*). Another study found cytotoxic activity in the carbon-tetrachloride fraction of EF leaves extract with a median lethal concentration (LC_{50}) value of 0.84 μg/ml. The study was done with a brine shrimp lethality test against *A. salina* (Kamal et al. 2019). Ethanol and methanol extracts of EF leaves showed cytotoxic activity in the concentration range of 400 μg/ml to 0.78 μg/ml. The test was done on brine shrimp nauplii and the LC_{50} value of methanol extract was 3.29 μg/ml and the LC_{50} value of ethanol extract was 27.446 μg/ml, whereas the LC_{50} value of a control, vincristine sulphate, was 0.632 μg/ml (Haoya et al. 2018). An *A. salina* (brine shrimp) lethality test was performed to check the cytotoxic effect of ethanol and aqueous extracts of EF, which showed LD_{50} values of 989 μg/ml and 224.66 μg/ml, respectively (Htun et al. 2018). Chloroform extract of fresh leaves of EF showed the highest cytotoxic activity of 93% among all other extracts (hexane, methanol, acetone) in another brine shrimp lethality test (Panigrahi and Bhatnagar 2015).

5.4.6 Anticancer Activity

Human cell lines for lung cancer (A549), cervix cancer (HeLa), breast cancer (MCF7), normal human fibroblast (WI-38), liver cancer (HePG2), pancreatic cancer (PSN-1), and pancreatic ductal adenocarcinoma (PANC-1) were used to test the anticancer activities. Aqueous extract of EF did not show any antiproliferative activity against the cancer cell lines, but ethanolic extract showed significant antiproliferative activity in HePG2, HeLa, and PANC-1 cancer cell lines (Htun et al. 2018). To determine the anticancer activity of EF extract, Ehrlich's ascites carcinoma (EAC) was induced in 48 mice and 12 mice remained as control among 60 mice. Forty-eight mice (grouped II–V) were injected intraperitoneally with the EAC cell line. After 24 hours, methanol extract was administered to group II–IV at 200 and 400 mg/kg doses for nine consecutive days. Group V was given 5-fluorouracil, a standard anticancer drug, at a dose of 20 mg/kg. On the 10th day, six mice from each group were sacrificed for the anticancer study and found to have a significant tumour cell volume decrease along with a restoration of antioxidant enzyme activity (Sannigrahi et al. 2010a). In producing anticancer drugs, cancer cell selectivity is a major problem. Tumour necrosis factor-related apoptosis-induced ligand (TRAIL) selectively kills various cancer cells without killing the normal cells. TRAIL resistance can cause a decrease in its therapeutic use. Germacranolides, isolated from methanol extract of EF aerial parts, were found to have potent TRAIL-resistance overcoming activity when tested against TRAIL-resistant human gastric adenocarcinoma (AGS) cells (Ahmed et al. 2017).

5.4.7 Other Pharmacological Activities

5.4.7.1 Thrombolytic

EF has been reported to possess thrombolytic activity. The extract of the plant might be responsible for plasminogen activation that produces clot lysis activity (Kamal et al. 2019). A concentration-dependent result of clot lysis was found in an *in vitro* experiment using crude methanolic extract (2, 4, 6, 8, 10 mg/ml) of leaves of EF. The leaf extract showed a significant clot lysis property of 46.91% (Kamal et al. 2019). The significant result was exhibited by 10 mg/ml extract (31% thrombolytic activity). A positive control, streptokinase, was also checked for its thrombolytic activity in comparison with the extract (Kuri et al. 2014).

5.4.7.2 Cytoprotective

Aqueous extract of EF exhibited a cytoprotective effect when checked against lead-induced cytotoxicity on hepatocytes. Both *in vitro* and *in vivo* studies showed that the aqueous extract can restore cell viability (Dua et al. 2016). Another study looked at the cytoprotective effect of aqueous extract of EF. In this study, $NaAsO_2$ (10 μM) was used to induce toxicity in murine hepatocytes as an *in vitro* study. The treatment with the extract of 400 μg/ml restored cell viability. An *in vivo* study was done to exhibit the prophylactic effect of the extract (400 mg/kg) against $NaAsO_2$ (10 mg/kg) toxicity (Dua et al. 2016).

5.4.7.3 Central Nervous System Depressant

Three fractions of methanolic extract (benzene, chloroform, and ethyl acetate at concentrations of 100 mg/kg) were studied in Swiss albino mice to check the central nervous system depressant activity of EF. The fractions were significantly effective, demonstrated by motility depressant, sedative, anticonvulsant, and antistress activity (Roy et al. 2011).

5.4.7.4 Hepatoprotective

For the hepatoprotective study, petroleum ether, chloroform, and ethanol extracts of 200 mg/kg were studied on CCl_4-induced hepatotoxic rats. Among the three extracts, ethanolic extract showed a significant improvement when compared to the standard drug, silymarin at 25 mg/kg (Swain et al. 2012b).

5.4.7.5 Phagocytic

Patil et al. (2008) studied the phagocytic activity of aqueous extract of EF leaves. Different concentrations of extract (25, 50, and 100 μg/ml) were studied. The result showed that this extract possesses stimulated chemotactic, phagocytic, and intracellular killing potency of human neutrophils when tested using *in vitro* phagocytosis methods.

5.4.7.6 Antidiarrhoeal

The antidiarrhoeal activity of aqueous and methanolic extracts of EF was exhibited when used in castor oil-induced diarrhoeal mice. Both extracts showed satisfactory results at the dose of 250 mg/kg. An *in vitro* antimicrobial experiment was done to check the inhibition of diarrhoea-causing microbes. Both extracts inhibited the growth of *S. dysenteriae*, *S. flexnari*, and *S. boydii* (Uddin et al. 2005).

5.4.8 CONCLUSION

This chapter has discussed the ethnomedicinal uses and pharmacological and phytochemical efficacies of the tropical herb EF. The plant possesses various pharmacological properties such as antidiabetic, anti-inflammatory, anticancer, antiviral, antibacterial, insecticidal, analgesic, antidiarrhoeal, antioxidant, neuroprotective, antiplatelet, anti-ulcer, phagocytic, and cytoprotective activities due to its diverse phytoconstituents (Table 5.2). The plant also has wide ethnomedicinal application in treating intestinal disorders, diabetes, constipation, kidney stones, skin diseases, and liver diseases that have been corroborated by a secondary published database of pharmacological activities. It is expected that the EF plant will spur further scientific research and hopefully lead to the discovery of new lead compounds and prophylactics and therapeutics.

REFERENCES

Adarsha, R., S. Bratati, R. Alokesh, M.A. Kumar & R. Sudipta. 2021. Antioxidant and antibacterial activities of phytochemicals in methanolic extract of five underutilized leafy vegetables. *Research Journal of Biotechnology* 16(8):1–10.

Ahmed, A.M.A., M.S. Rahman & M.N. Anwar. 2005. Antibacterial activities of extracts obtained from five important medicinal plants of Bangladesh. *European Journal of Scientific Research* 11(3):444–448.

Ahmed, F., U.K. Karmakar, M.A. Arai, N. Ishikawa, S.K. Sadhu & M. Ishibashi. 2017. Germacranolides from *Enhydra fluctuans* with TRAIL-resistance abrogating activity. *Natural Product Communication* 12(8):1169–1170.

Ahmed, T., M. Kamruzzaman, M.M. Islam, M. Hasanuzzaman, A. Ahmed & D.K. Paul. 2016. *Journal of Medicinal Plants Studies* 4(3):241–247.

Akhe, F.K., M.M. Khatun, S. Ray, S.T. Arafat, S.J. Uddin, H. Hossain, M.N. Siddiqui & M.G. Sarower. 2019. Profiling of bioactive compounds in watercress (*Enhydra fluctuans* Lour.) by high performance liquid chromatography. *PharmacologyOnLine* 1:65–75.

Aldaba-Muruato, L.R., J. Ventura-Juárez, A.M. Perez-Hernandez, A. Hernández-Morales, M.H. Muñoz-Ortega, S.L. Martínez-Hernández, B. Alvarado-Sánchez & J.R. Macías-Pérez. 2021. Therapeutic perspectives of *p*-coumaric acid: anti-necrotic, anti-cholestatic and anti-amoebic activities. *World Academy of Sciences Journal* 3(5):47.

Ali, M.R., M.M. Billah, M.M. Hassan, S.M.R. Dewan & M. Al-Emran. 2013. *Enhyrda fluctuans* Lour: a review. *Research Journal of Pharmaceutical and Technology* 6(9):927–929.

Alom, S., P. Sadhu & R. Bezbaruah. 2021. A comprehensive review on ethnobotanical and pharmacological uses of the tropical herb *Enhydra fluctuans* Lour. *Current Trends in Pharmaceutical Research* 8(1):119–137.

Amin, M.R., R. Mondol, M.R. Habib & M.T. Hossain. 2012. Antimicrobial and cytotoxic activity of three bitter plants-*Enhydra flactuans*, *Andrographis paniculata*, and *Clerodendrum viscosum*. *Advanced Pharmaceutical Bulletin* 2(2):207–211.

Arya, A.K., M. Durgapal, A. Bachheti, K.K. Joshi, Y.H. Gonfa, R.K. Bachheti & A. Husen 2022. Ethnomedicinal use, phytochemistry, and other potential application of aquatic and semiaquatic medicinal plants. *Evidence-Based Complementary and Alternative Medicine* 2022:4931556.

Azam, M.N.K., M.N. Ahmed, M.M. Rahman & Rahmatullah M. 2014. Ethnomedicines used by the Oraon and Gor tribes of Sylhet district, Bangladesh. *American-Eurasian Journal of Sustainable Agriculture* 7(5):391–402.

Babu, S. & S. Jayaraman. 2020. An update on β-sitosterol: a potential herbal nutraceutical for diabetic management. *Biomedicine & Pharmacotherapy* 131:110702.

Barua, A., M.S. Alam, M. Junaid, Y. Akter, S.S. Afrose, T. Sharmin, R. Akter & S.M.Z. Hosen. 2021. Phytochemistry, traditional uses and pharmacological properties of *Enhydra fluctuans* Lour: a comprehensive review. *Current Pharmaceutical Biotechnology* 22(8):1061–1068.

Bhakta, J., P. Majumdar & Y. Munekage. 2008. Antimicrobial efficacies of methanol extract of *Asteracantha longifolia*, *Ipomoea aquatica* and *Enhydra fluctuans* against *Escherichia coli*, *Pseudomonas aeruginosa*, *Staphylococcus aureus* and *Micrococcus luteus*. *The Internet Journal of Alternative Medicine* 7(2):125–130.

Böni-Schnetzler, M. & D.T. Meier. 2019. Islet inflammation in type 2 diabetes. *Seminar in Immunopathology* 41(4):501–513.

Chen, W., I. Vermaak & A. Viljoen. 2013. Camphor – a fumigant during the black death and a coveted fragrant wood in ancient Egypt and Babylon – a review. *Molecules* 18(5):5434–5454.

Das, A.K., B.K. Dutta & G.D. Sharma. 2008. Medicinal plants used by different tribes of Cachar District, Assam. *Indian Journal of Traditional Knowledge* 7(3):446–454.

Dasgupta, N. & B. De. 2007. Antioxidant activity of some leafy vegetables of India: a comparative study. *Food Chemistry* 101(2):471–474.

da Silva, E.T., A.S. Araújo, A.M. Moraes, L.A. Souza, M.C.S. Lourenço, M.V.N. Souza, J.L. Wardell & S.M.S.V. Wardell. 2015. Synthesis and biological activities of camphor hydrazone and imine derivatives. *Scientia Pharmaceutica* 84(3):467–483.

Delfita, R., D. Dahelmi, D. Tjong & S. Suhatri. 2021a. Effect of *Enhydra fluctuans* on kidney function in alloxan-induced diabetic rats. *Open-Access Macedonian Journal of Medical Sciences* 9(A):1187–1194.

Delfita, R., D. Dahelmi, D. Tjong & S. Suhatri. 2022. Protective effect of *Enhydra fluctuans* DC. aerial against insulitis in alloxan-induced diabetic rats. *Jouranal of Research in Pharmacy* 26(1):180–188.

Delfita, R., D.H. Tjong, D. Dahelmi & S. Suhatri. 2021b. Hypoglycemic effects of *Enhydra fluctuans* aerial extract on alloxan-induced diabetic rats. *Journal of Physics: Conference Series* 1940:012058.

Dighe, S., B.S. Kuchekar & S.B. Wankhede. 2016. Analgesic and anti-inflammatory activity of β-sitosterol isolated from leaves of Oxalis corniculate. *International Journal of Pharmacological Research* 6(3):109–113.

Dolui, A.K., H.K. Sharma, T.B. Marein & T.C. Lalhriatpuii. 2004. Folk herbal remedies from Meghalaya. *Indian Journal of Traditional Knowledge* 3(4):358–364.

Dua, T.K., S. Dewanjee & R. Khanra. 2016. Prophylactic role of *Enhydra fluctuans* against arsenic-induced hepatotoxicity via anti-apoptotic and antioxidant mechanisms. *Redox Report* 21(4):147–154.

Dua, T.K., S. Dewanjee, R. Khanra, S. Joardar, S. Barma, S. Das, M. Zia-Ul-Huq & V. De Feo. 2016. Cytoprotective and antioxidant effects of an edible herb, *Enhydra fluctuans* Lour. (Asteraceae), against experimentally induced lead acetate intoxication. *PLoS One* 11(2):e0148757.

Dutta, J. & M.C. Kalita. 2016. *In vitro* hypoglycemic evaluation of seven culinary plants of northeast India against type 2 diabetes. *Asian Journal of Pharmaceutical and Clinical Research* 9(2):209–212.

Eksi, G., S. Kurbanoglu & S.A. Erdem. 2020. Analysis of diterpenes and diterpenoids. Recent advancement in natural product analysis. *ScienceDirect* 2020:313–345.

Fraile, L., E. Crisci, L. Córdoba, M.A. Navarro, J. Osada & M. Montoya. 2012. Immunomodulatory properties of beta-sitosterol in pig immune responses. *International Immunopharmacology* 13(3):316–321.

Frank, F.M., J. Ulloa, S.I. Cazorla, G. Maravilla, E.L. Malchiodi, A. Grau, V. Martino, C. Catalán & L.V. Muschiettti. 2013. Trypanocidal activity of *Smallanthus sonchifolius*: Identification of active sesquiterpene lactones by bioassay-guided fractionation. *Evidence-based Complementary and Alternative Medicine* 2013:627898.

Ganeshpurkar, A. & A.K. Saluja. 2017. The pharmacological potential of rutin. *Saudi Pharmaceutical Journal* 25(2):149–164.

Ghosh, D., S. Ray, K. Ghosh, V. Micard, U.R. Chatterjee, P.K. Ghoshal & B. Ray. 2013. Antioxidative carbohydrate polymer from *Enhydra fluctuans* and its interaction with bovine serum albumin. *Biomacromolecules* 14(6):1761–1768.

Haoya, N.A., M.S. Islam, S. Sultana, S.J. Mahia, F.I. Sukorno & M. Biswas. 2018. Phytochemical screening and evaluation of antioxidant, antibacterial, and cytotoxic activities of methanolic and ethanolic extracts of *Enhydra fluctuans* Lour. r (helencha) (family: Asteraceae) stems and leaves. *International Journal of Plant & Soil Science* 23(6):1–14.

Hasan, M.N., F. Sabrin, B. Rokeya, M.S.H. Khan, M.U. Ahmed, A. Matondo, M.M. Billah & S. Akter. 2019. Glucose and lipid-lowering effects of *Enhydra fluctuans* extract in cadmium treated normal and type-2 diabetic model rats. *BMC Complementary and Alternative Medicine* 19(1):278.

Heylliward, G., A.N. Bakhtiar, N. Nurainas, B. Saad, & G. Arjet. 2012. *Eksplorasi Pengetahuan Lokal Etnomedisin dan Tumbuhan Obat Indonesia Berbasis Komunitas*. Padang, Sumatera Barat: Universitas Andalas.

Htun, M.T.M., M.T. Aung, N.N. Than & D.H. Ngwe. 2018. Investigation of some bioactivities of *Peperomia pellucida* L. (thit-yay-gyi) and *Enhydra fluctuans* L. (kana-phaw). *Journal of the Myanmar Academy of Arts and Sciences* 16(1B):193–208.

Igarashi, M. & Y. Miyazaki. 2013. A review of bioactivities of *Perilla*: progress in research on the functions of *Perilla* as medicine and food. *Evidence-based Complementary and Alternative Medicine* 2013:925342.

Jain, A., M. Sundriyal, S. Roshnibala, R. Kotoky, P.B. Kanjilal, H.B. Singh & R.C. Sundriyal. 2011. Dietary use and conservation concern of edible wetland plants at indo-burma hotspot: a case study from northeast India. *Journal of Ethnobiology and Ethnomedicine* 7:29.

Jaiswal, V. 2010. Culture and ethnobotany of Jaintia tribal community of Meghalaya North East India – a mini review. *Indian Journal of Traditional Knowledge* 9(1):38–44.

Kamal, S., S.R. Rony, S. Sharmin, F.R. Laboni & M.H. Sohrab. 2019. Phytochemical and pharmacological potentials of *Enhydra fluctuans* available in Bangladesh. *Journal of Pharmaceutical Research International* 29(4):1–11.

Kamatou, G.P.P. & A.M. Viljoen. 2008. Linalool – a review of a biologically active compound of commercial importance. *Natural Product Communications* 3(7):1183–1192.

Khan, H.M. & P.S. Yadava. 2010. Antidiabetic plants used in Thoubal District of Manipur. *Indian Journal of Traditional Knowledge* 9(3):510–514.

Koley, K.T., Z. Khan, D. Oulkar, T. Banerjee, A. Sing, S.G. Karkute & S. Banerjee. 2021. Coupling the high-resolution LC-MS characterization of the phenolic compounds with the antimicrobial and antibiofilm properties of helencha (*Enhydra fluctuans* Lour.). *Journal of Food Science and Technology* 58:4755–4765.

Kumar, S. & S. Narain. 2010. Herbal remedies of wetlands macrophytes in India. *Internal Journal of Pharmacy and Biological Sciences* 1(2):1–12.

Kuri, S., M.M. Billah, S.M.M. Rana, Z. Naim, M.M. Islam, M. Hasanuzzaman, M.R. Ali & R. Banik. 2014. Phytochemical and in vitro biological investigations of methanolic extracts of Enhydra Fluctuans Lour. *Asian Pacific Journal of Tropical Biomedicine* 4(4):299–305.

Krishnaswamy, N.R. & Ramji, N. 1995. Sesquiterpene lactones from *Enhydra fluctuans*. *Phytochemistry*, 38(2):433–435.

Lopa, S.S., M.Y. Al-Amin, M.K. Hasan, M.S. Ahammed, K.M.M. Islam, A.H.M.K. Alam, T. Tanaka & M.G. Sadik. 2021. Phytochemical analysis and cholinesterase inhibitory and antioxidant activities of *Enhydra fluctuans* relevant in the management of Alzheimer's disease. *International Journal of Food Science* 2021:8862025.

Ložienė, K., J. Švedienė, A. Paškevičius, V. Raudonienė, O. Sytar & A. Kosyan. 2018. Influence of plant origin natural α-pinene with different enantiomeric composition on bacteria, yeasts and fungi. *Fitoterapia* 127:20–24.

Lundberg, M., P. Seiron, S. Ingvast, O. Korsgren & O. Skog. 2017. Insulitis in human diabetes: a histological evaluation of donor pancreases. *Diabetologia* 60(2):346–353.

Mussel, A., A. Bighelli, T.M. Hoi, M.T.P. Thao, R.H. Thai & J. Casanova. 2000. Dihydroperillaldehyde from *Enhydra fluctuans* Lour. essential oil. *Flavor and Fragrance Journal* 15(5):299–302.

Ododo, M.M., M.K. Choudhury & A.H. Dekebo. 2016. Structure elucidation of β-sitosterol with antibacterial activity from the root bark of *Malva parviflora*. *SpringerPlus* 5(1):1210.

Pandey, A.K. & H.R. Bora. 1997. Edible plants of Shan Tribe of Assam. *Ancient Science of Life* 16(4):258–276.

Panigrahi, R. & S. Bhatnagar. 2015. Cytotoxic, phytochemical and antioxidant potential of marshy herb *Enhydra fluctuans* Lour. *International Journal of Ethnobiology & Ethnomedicine* 1(1):1–6.

Park, C.M. & S. Young-Sun. 2019. Luteolin and luteolin-7-*O*-glucoside protect against acute liver injury through regulation of inflammatory mediators and antioxidative enzymes in GalN/LPS-induced hepatitic ICR mice. *Nutrition Research and Practice* 13(6):473–479.

Patel, K. & D.K. Patel. 2019. The beneficial role of rutin, a naturally occurring flavonoid in health promotion and disease prevention: a systematic review and update. *Bioactive Food as Dietary Interventions for Arthritis and Related Inflammatory Diseases* 2019:457–479.

Patil, K., P. Majumder, & R.R. Wadekar. 2008. Effect of *Enhydra fluctuans* Lour. leaf extract on phagocytosis by human neutrophils. *Journal of Natural Remedies* 8(1):76–81.

Pereira, I., P. Severino, A.C. Santos, A.M. Sliva & E.B. Souto. 2018. Linalool bioactive properties and potential applicability in drug delivery system. *Colloids and Surfaces B: Biointerfaces* 171:566–578.

Rahman, M.S. & M.N. Anawr. 2008. Antibacterial and cytotoxic activity of longiverbenone isolated from the rhizome of *Cyperus scariosus*. *Bangladesh Journal of Microbiology* 25(1):82–84.

Rahman, M.T., N. Begum, M. Alimuzzaman & M.O. Khan. 2002. Analgesic activity of *Enhydra fluctuans*. *Fitoterapia* 73(7–8):707–709.

Rahmatullah, M., N.K. Azam, Z. Khatun, S. Seraj, F. Islam, A. Rahman, S. Jahan & M.S. Aziz. 2012. Medicinal plants used for the treatment of diabetes by the Marakh sect of the Garo tribe living in Mymensingh district, Bangladesh. *African Journal of Traditional, Complementary and Alternative Medicine* 9(3):380–385.

Rana, S.M.M., M.M. Billah, M.S. Hossain, A.K.M. Saifuddin, S.K.M.A. Isam, S. Banik, Z. Naim & G.S. Raju. 2014. Susceptibility to microorganisms to selected medicinal plants in Bangladesh. *Asian Pacific Journal of Tropical Biomedicine* 4(11):911–917.

Reddy, J.M.P., I.M. Shareef, S.M. Gopinath, K.S. Dayananda, A. Mandal, B. Sreekanth & K.M. Purushotham. 2014. Antibacterial activity of some indigenous medicinal plants. *Global Journal of Research on Medicinal Plants and Indigenous Medicine* 3(3):75–79.

Roy, K.K., M.K.I. Mollah, M.M. Reja, D. Shil & R.K. Maji. 2021. Evaluation of analgesic and anti-inflammatory activity of ethanolic extract of *Enhydra fluctuans* on male Wistar rats. *Journal of Applied Pharmaceutical Research* 9(1):37–42.

Roy, S.K., K. Mazumder, & A. Islam. 2011. Pharmacological evaluation of *Enhydra fluctuans* aerial parts for central nervous system depressant activity. *PharmacologyOnLine* 1:632–643.

Rufino, A.T., M. Ribeiro, F. Judas, L. Salgueiro, M.C. Lopes, C. Caveleiro & A.F. Mendes. 2014. Anti-inflammatory and chondroprotective activity of (+)-α-pinene: structural and enantiomeric selectivity. *Journal of Natural Products* 77(2):264–269.

Saha, S. & S. Paul. 2019. A review on phytochemical constituents and pharmacological properties of *Enhydra fluctuans* Lour. *Journal of Pharmacognosy and Phytochemistry* 8(2):887–893.

Sannigrahi, S., U.K. Mazumdar, A. Mondal, D. Pal, S.L. Mishra & S. Roy. 2010a. Flavonoids of *Enhydra fluctuans* exhibit anticancer activity against Ehrlich's ascites carcinoma in mice. *Natural Product Communication* 5(8):1239–1242.

Sannigrahi, S., U.K. Mazumder, D. Pal, M.L. Mishra & S. Maity. 2011. Flavonoids of *Enhydra fluctuans* exhibits analgesic and anti-inflammatory activity in different animal models. *Pakistan Journal of Pharmaceutical Science* 24(3):369–375.

Sannigrahi, S., U.K. Mazumder, D. Pal, S. Parida & S. Jain. 2010b. Antioxidant potential of crude extract and different fractions of *Enhydra fluctuans* Lour. *Iranian Journal of Pharmaceutical Research* 9 (1):75–82.

Sarma, U., V.V. Borah, K.K. Saikia & N. Hazarika. 2014. *Enhydra fluctuans*: a review on its pharmacological importance as a medicinal plant and prevalence and use in north-east India. *International Journal of Pharmacy and Pharmaceutical Sciences* 6(2):48–50.

Sarma, U., V.V. Borah, K.K. Saika, & N.K. Hazarika. 2016. Screening of *Enhydra fluctuans* for phytochemical composition and broad-spectrum antibacterial activity against clinical bacterial isolates. *Journal of Herbs, Spices, and Medicinal Plants* 22(4):300–308.

Satyajit & D. Pradhan. 2012. Natural flavonoids isolated from the leaves of *Enhydra fluctuans* inhibits cyclooxygenase-2 and 5-lipooxygenase inflammation in various models. *International Journal of Research in Pharmacology and Pharmacotherapeutics* 1(1):65–71.

Sepúlveda, L., A. Ascacio, R. Rodríguez-Herrera, A. Aguilera-Carbó & C.N. Aguilar. 2011. Ellagic acid: biological properties and biotechnological development for production processes. *African Journal of Biotechnology* 10(22):4518–4523.

Serra-Barcellona, C., N.C. Habib, S.M. Honoré, S.S. Sánchez & S.B. Genta. 2017. Enhydrin regulates postprandial hyperglycemia in diabetic rats by inhibition of α-glucosidase activity. *Plant Foods for Human Nutrition* 72(2):156–160.

Sharmila, R. & G. Sindhu. 2017. Evaluate the antigenotoxicity and anticancer role of β-sitosterol by determining oxidative DNA damage and the expression of phosphorylated mitogen-activated protein kinases, C-fos, C-jun, and endothelial growth factor receptors. *Pharmacognosy Magazine* 13(49):95–101.

Sikdar, S., S. Roy, T.N. Banu, M.A. Mondal & S. Mandal. 2020. Phytochemical analysis and antibacterial property assessment of helencha (*Enhydra fluctuans*; family: Asteraceae) extract. *GSC Biological and Pharmaceutical Sciences* 12(02):136–142.

Swain, P.K., S.C. Dinda, D.P. Nayak, B. Kar & V.J. Patro. 2012a. Antioxidant activity of *Enhydra fluctuans* Lour. aerial parts. *Journal of Phytotherapy and Pharmacology* 1(2):23–34.

Swain, P.K., V.J. Patro, S.C. & D.P. Nayak. 2012b. Hepatoprotective activity of *Enhydra fluctuans* lour. Aerial parts against ccl4 induced hepatotoxicity in rats. *International Journal of Research in Ayurveda and Pharmacy* 3(6):893–896.

Tarak, D., N.D. Namsa, S. Tangjang, S.C. Arya & B. Rajbonshi. 2011. An inventory of ethnobotanicals used as antidiabetic by a rural community of Dhemaji district of Assam, Northeast India. *Journal of Ethnopharmacology* 138(2011):345–350.

Thatoi, H.N., S.K. Panda, S.K. Rath & S.K. Dutta. 2008. Antimicrobial activity and ethnomedicinal uses of some medicinal plants from Similipal Biosphere Reserve, Orissa. *Asian Journal of Plant Sciences* 7(3):260–267.

Uddin S.J., M.M. Ferdous, R. Rouf, M.S. Alam, M.A.M. Sarkar, & J.A. Shilpi. 2005. Evaluation of antidiarrhoeal activity of *Enhydra fluctuans*. *Journal of Medical Sciences* 5(4):324–327.

Uemura, T., T. Yashiro, O. Rei, S. Naoki, N. Tadaaki, H. Masakazu, K. Shoko, N. Chiharu & A. Gen-ichiro. 2018. The intestinal anti-inflammatory activity of perillaldehyde. *Journal of Agricultural and Food Chemistry* 66(13):3443–3448.

Ulloa, J.L., R. Spina, A. Casasco, P.B. Petray, V. Martino, M.A. Sosa, F.M. Frank & L.V. Muschietti. 2017. Germacranolide-type sesquiterpene lactones from *Smallanthus sonchifolius* with promising activity against *Leishmania mexicana* and *Trypanosoma cruzi*. *Parasites & Vectors* 10(1):567.

Yadava, R.N. & S.K. Singh. 2007. Novel bioactive constituents from *Enhydra fluctuans* Lour. *Natural Product Research* 21(6):481–486.

Yagi, S., R. Babiker, T. Tzanova & H. Schohn. 2016. Chemical composition, antiproliferative, antioxidants, and antibacterial activities of essential oils from aromatic plants growing in Sudan. *Asian Pacific Journal of Tropical Medicine* 9(8):763–770.

Zeb, M.A., S.U. Khan, T.U. Rahman, M. Sajid & S. Seloni. 2017. Isolation and biological activity of beta-sitosterol and stigmasterol from the roots of *Indigofera heterantha*. *Pharmacy and Pharmacology International Journal* 5(5):204–207.

Zhao, B., F. Wu, X. Han, W. Zhou, Q. Shi & H. Wang. 2020. Protective effects of acarbose against insulitis in multiple low-dose streptozotocin-induced diabetic mice. *Life Sciences* 263:118490.

6 Medicinal Uses, Phytochemistry, and Pharmacological Properties of *Acorus calamus*

Muhammad Qamar,[1] Naveed Ahmad,[2] Tariq Ismail,[1] Tuba Esatbeyoglu,[3] Saeed Akhtar,[1,] and Muhammad S. Mubarak[4,*]*
[1] Institute of Food Science and Nutrition, Bahauddin Zakariya University, Multan, Pakistan
[2] Multan College of Food & Nutrition Sciences, Multan Medical and Dental College, Multan, Pakistan
[3] Institute of Food Science and Human Nutrition, Department of Food Development and Food Quality, Gottfried Wilhelm Leibniz University Hannover, Hannover, Germany
[4] Department of Chemistry, The University of Jordan, Amman, Jordan
[*] Corresponding authors (Muhammad. S. Mubarak): mmubarak@ju.edu.jo; (Saeed Akhtar) saeedakhtar@bzu.edu.pk

6.1 INTRODUCTION

Acorus calamus is a monocot, grass-like perennial, tall-growing plant primarily found in Europe, North America, and Asia in damp places like riverbanks, swamps, and ponds. This plant belongs to the Acoraceae family (Singh et al. 2011). The name comes from two Greek words, '*akoros*' meaning fragrant rhizomes, and '*calamus*', which means cane. Due to essential oil (EO), the leaves and rhizomes have a pleasant, sweet odour (Batra et al. 2012; Arya et al. 2022). The plant has a wide cultural reputation with numerous folk names, centuries-old history, and extensive use. It is also known as sweet flag, rat, muskrat, calamus, flag root or sway, sweet calomel, Vacha, sweet sedge, sweet cane, sweet myrtle, and pine root (Motley 1994). This plant persisted as an under-utilized herb with exceptional functional and therapeutic features and has been used in China and India's traditional/conventional medicine systems because of its therapeutic properties (Batra et al. 2012; Wu et al. 2009; Lee et al. 2011). The plant has also been extensively used to treat ailments like diarrhoea, colic ache, diabetes, fever, cough, inflammation, bronchitis, tumours, haemorrhoids, depression, general debility, and numbness, and as an antidote for several poisonings in Indonesia, America, and India (Vaidyaratnam 1994; Nadkarni 1998; Gilani et al. 2006; Si et al. 2010). Regarding the use of this plant in paediatrics, it has been described in the Siddha medicinal system. In southern India, the paste of *A. calamus* is applied in speech therapy to sharpen memory in children (Meena et al. 2010). It is also used to treat dyspepsia, abdominal discomfort, dysentery, and flatulence in Pakistan (Ahmad et al. 2006).

DOI: 10.1201/9781003256830-6

The use of *A. calamus* in traditional medicine systems was also accredited to its pharmacological aspects such as its antioxidant and anti-inflammatory potential, immunomodulation, antitumour, antidiabetic, antimicrobial, and analgesic behaviour (MacGaw et al. 2002; Mehrotra et al. 2003; Kim et al. 2009; Gaidhani et al. 2009; Wu et al. 2009; Jayaraman et al. 2010; Si et al. 2010; Chaubey et al. 2018; Loying et al. 2019; Pawar et al. 2020; Derymedvid et al. 2021; Dinev et al. 2021; Vakayil et al. 2021). The activities are associated with the presence of bioactive compounds in *A. calamus* EO including β-gyrjunene, acorenone, β-asarone, calamendiol, isoshyobunine, α-calacorene, α-selinene, camphene, shyobunone, and calamusenone (Tariq et al. 2010).

6.2 TRADITIONAL MEDICINAL CLAIMS

From a historical perspective, *A. calamus* has numerous medicinal uses. This plant has been used to provide relief from headaches in American civilizations. Chinese people used the plant to reduce inflammation and bowel problems. Regarding traditional systems of medicine, the plant remained an integral therapeutic agent. Major health claims associated with this plant are in treating bronchitis, appetite loss, colic, chest pain, dysentery, twinge, upset stomach, rheumatism, mental disorders, and vascular illness, and acting as a sedative (Kirtikar and Basu 1987). In Pakistan, people use *A. calamus* to treat abdominal discomfort, dysentery, dyspepsia, and flatulence (Ahmad et al. 2006).

The Ayurvedic medicinal system believes that *A. calamus* contains laxative, diuretic, emetic, stimulant, bitter tonic, expectorant, antispasmodic, aromatic, aphrodisiac, anthelmintic, and carminative potential. In addition, various health-related conditions such as chronic diarrhoea, bronchial catarrh, dysentery, colic, recurrent fever, asthma, otitis media, cough, and abdominal and glandular tumours have been treated with *A. calamus* (Anonymous 2001). Western herbal medicines use this plant to treat digestive problems including colic, gas, bloating, and poor digestion. The plant also helps reduce discomfort, bloating, and headaches. Small doses of *A. calamus* are claimed to reduce acidity, but higher intakes may induce it. Furthermore, the plant has been used to treat eczema, rheumatism, and liver and kidney disorders. It is taken with jaggery and honey when it is acidic, but for dyspepsia, it is taken with salt and water to induce vomiting (Bangasen 1984).

To treat vamana and dyspepsia, *A. calamus* is used as an emetic agent (Chunekar and Pandey 1998). The plant has additionally been used as a carminative and antispasmodic agent to relieve constipation and stomach discomfort; it adds bulk to faeces and increases urine output, especially when taken with milk and water in the swollen bladder (Bangasen 1984). Rhizome powder of this plant is used to treat cough, fever, stomach discomfort, epilepsy, and other ailments in paediatrics (Chellaiah et al. 2006; Ignacimuthu et al. 2006). In the Indian traditional system of Ayurveda, *A. calamus* rhizomes are used to treat brain-related disorders of psychoneurosis, hysteria, memory loss, epilepsy, and schizophrenia (Nadkarni 1998; Prajapati et al. 2003). Shah (2005) reported that this plant improves the brain's grasping, cognitive, and memory abilities when used with milk for a month.

The rhizome paste made by mixing powder made from rhizomes with honey enhanced the newborn's seminal power, eyesight, learning skills, and mental growth. Rhizome paste also stimulated the neurological system and hence alleviated depression. Jadhav (1994) reported that the paste made with ghee, gold, and water improved the infants' gripping and memory.

A. calamus has also been used as a skin ointment for treating skin-related disorders in powder, enemas, balms, tablets, and ghee (Kirtikar and Basu 1987; Anonymous 2001). Eczema and other related problems have been treated with *A. calamus*. The powder form was used to treat maggot-infested wounds and infections (Kirtikar and Basu 1987). The plant's fumes also minimized discomfort and swelling in pile masses when tested against haemorrhoids. When mixed with mustard seed paste, it was used to treat hydrocele (Shastri and Pandey 1997). Figure 6.1 includes pictures of the plant along with the phytochemicals present and its pharmacological activities.

FIGURE 6.1 Traditional medicinal claims and pharmacological activities of *A. calamus*.

6.3 PHYTOCHEMICALS

A qualitative screening analysis of *A. calamus* indicated the presence of flavonoids and saponins; these compounds account for the antimicrobial activity of the plant (Pokhrel and Chaulagain 2020). Other reports have shown the presence of triterpenoids and alkaloids in *n*-hexane and ethanol extracts, while the chloroform extract has only triterpenoids (Muchtaromah et al. 2018). On the other hand, the acetone extract of *A. calamus* contained phytosterols, alkaloids, tannins, and saponins, while ethanol and petroleum ether extracts contained alkaloids, saponins, and tannins. Moreover, research findings indicated that the acetone extract has the highest number of phytochemicals. In addition, asarone was the major constituent among all detected phytochemicals as revealed by gas chromatography–mass spectrometry (GC-MS) analysis of the extracts (Porwal et al. 2021).

Decades earlier, GC-MS analysis showed that European native triploid *A. calamus* var. *calamus* and Indian-native tetraploid *A. calamus* var. *angustatus* contain 184 and 93 compounds, respectively, with β-asarone as an integral constituent. Hydrocarbons dominated the overall composition, followed by alcohols, carbonyl compounds, phenols, oxide compounds, and furans (Mazza 1985). In another study, 243 volatile constituents were detected in an alcohol extract of *A. calamus* var. *calamus* using

GC-MS. Similarly, alcohols, carbonyls, and hydrocarbons were among the compounds analysed in the study carried out by Mazza (1984). Using the micellar electrokinetic capillary chromatography (MEKC) technique, β-asarone, a phenylpropanoid, was found in a diploid variety of *A. calamus* rhizome taken from the US wetland and triploid type of *A. calamus* rhizome obtained commercially from India. The Indian variety was noted to have more β-asarone contents (4.4% w/w), followed by the US-based variety with 0.2% w/w β-asarone (Hanson et al. 2005).

The simultaneous distillation–extraction (SDE) approach was used to obtain volatile organic compounds from rhizomes of Nepalese *A. calamus*, subsequently characterized by GC-MS. Characterization has shown the presence of 53 volatile organic compounds; 11 alcohols, 3 esters, 14 aldehydes, 19 hydrocarbons, 1 furan, 1 miscellaneous (*N*-containing), and 4 ketones. The total yield of these volatile compounds was 7493.59 mg/kg. In addition, major bioactive compounds included 46.78% β-asarone, 11.09% farnesol, 0.41% linalool, 14.15% (*E,Z*)-2,4-decadienal, 6.10% methyl eugenol, 0.11% (*E*)-caryophyllene, 0.06% α- and β-pinene, 0.7% ocimeme, 0.39% β-elememe, 0.03% camphor, and 0.26% aromadendrene (Gyawali and Kim 2009). Previously, β-asarone was identified and quantified using high-performance liquid chromatography–tandem mass spectrometry (HPLC-MS/MS) from plasma samples of three female albino Wistar rats orally fed with *A. calamus* rhizome extracts (Nandakumar et al. 2013). Similarly, 14 chemical constituents, which made up the 100% oil composition, were identified and quantified wherein β-asarone (68.96%), shyobunone (8.65%), *p*-methoxyphenyl butanone (4.16%), and 4-hydroxy-4-methyl-2-pentanone (8.73%), were among the predominant compounds, while isoelemicine (2.68%), *cis*-ocimene (1.44%), δ-cadinene (0.84%), and muurolene (1.92%) were the minor compounds (Joshi and Bashyal 2018).

Dhivya et al. (2019) employed HPLC equipped with a photodiode array detector to quantify *β*-asarone from *A. calamus* rhizome. The EO from the dry rhizome of *A. calamus* was extracted by Clevenger-type equipment with a hydrodistillation technique. Methanol offered maximum extraction of β-asarone as 980.840 μg/g, almost 98.08%, followed by ethyl acetate, *n*-hexane, acetone, and chloroform. A study conducted to evaluate the chemical composition using GC-MS analysis indicated that β-asarone was among the predominant compounds (82.42%). In contrast, the calarene, euasarone, and 2-methoxy-3-allylphenol were the least abundant; they were found in the range of 2.41%, 1.92%, and 1.91%, respectively (Loying et al. 2019).

Similarly, when analysed by GC-MS, the leaf methanol extract of *A. calamus* revealed 36 peaks, including 16 compounds, which were identified using previously published procedures. These compounds include, among others, *cis*-dimethylstilbene, 2,3-dimethoxy-5(2-propenyl)-2,5-cyclohexadiene-1, butyl 3,4,4,4-tetrafluro-3-trifluromethylbutanate, neophytadiene, phthalic acid butylhexyl ester, phthalic acid 6-ethyl-3-octylbutyl ester, and phthalic acid didecyl ester. It was also reported that these compounds have biological properties and must be considered for plant-based drugs and insecticide development (Lekshmi and Kensa 2020). Muthusamy et al. (2022) recently performed an HPLC analysis using *A. calamus* methanol extract to identify the concentration of α- and β-asarone. For *in vitro* plantlets, the concentrations of α- and β-asarone were 0.338 and 2.187 μg/g dry weight (DW), respectively, whereas, *in vivo* plantlets, the concentrations of α- and β-asarone were 0.093 and 2.060 ng/g DW. Similarly, a study aimed to identify bioactive metabolites responsible for the antimicrobial activity of *A. calamus* rhizome using gas GC-MS was carried out and reported five compounds named spiro [2,4,5,6,7,7a-hexahydro-2-oxo-4,4,7a-trimethylbenzofuran]-7,2'-(oxirane), 9-octadecenoic acid (*Z*)-, phenylmethyl ester, asarones, shyobunone, benzene, and 1,2,3-trimethoxy-5-(2-propenyl) (Prabha and Kumar 2021). In a preliminary phytochemical screening using thin layer chromatography, *A. calamus* methanol extract had phenolic acids, flavonoids, tannins, steroids, saponins, and terpenoids (Balaji and Shanthakumari 2020).

Alkaloids with a diverse carbon skeleton (14 compounds) including 4-pyrazin-2-yl-but-3-ene-1,2-diol, 2*R*-4-(5′-methyl-pyrazin)-2-yl-but-3-ene-1,2-diol, methyl 2-(4-hydroxyphenyl)acetyl glycinate, neotatarine, acortatarin A, perillascens, tatarine A, telitoxine, 2-oxo-1-pyrrolidineacetic acid, *N*-ethoxycarbonylmethylpyrrolidin-2-one, α-pyrrolidone acetic methyl ester, piperazine,

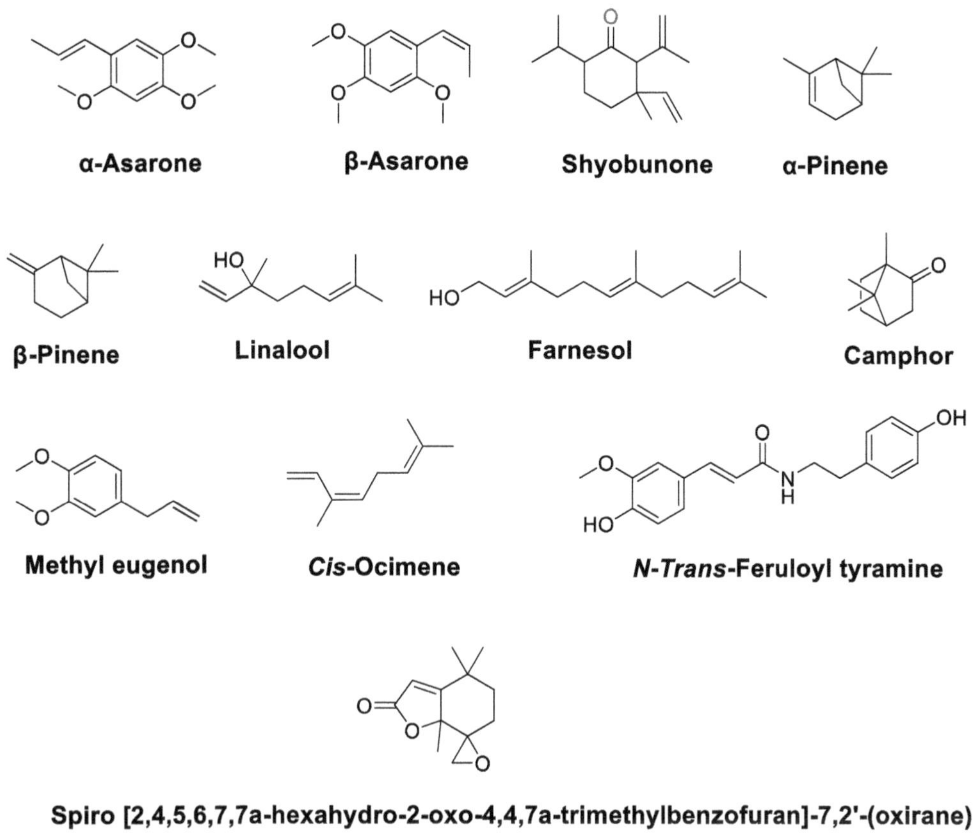

FIGURE 6.2 Chemical structures of compounds present in *A. calamus*.

3-(4-hydroxy-3-methoxyphenyl)-*N*-[2-(4-hydroxyphenyl)-2-methoxyethyl]acrylamide, and *N*-*trans*-feruloyl tyramine were identified from the ethanol extract of *A. calamus* rhizomes. The first four compounds were identified using nuclear magnetic resonance, while the other ten were identified using previously published procedures (Hao et al. 2021). Figure 6.2 shows the chemical structures of some compounds present in *A. calamus*

6.4 PHARMACOLOGICAL POTENTIAL

6.4.1 Anti-inflammation and Immunomodulation Effect

Research findings revealed the immunomodulatory and anti-inflammatory response of underutilized therapeutic herbs like *A. calamus*. The anti-inflammatory potential of *A. calamus* to cure chronic and acute inflammation was assessed using a rat model. The carrageenan-induced paw oedema, croton oil granuloma pouch inflammatory response, and cotton pellet granuloma formation were significantly reduced by orally administered extracts. Similarly, the rhizomes extract exhibited a profound reduction in chronic and acute inflammation in immunological models (Varde et al. 1988; Vohra et al. 1989). The immunomodulatory potential of *A. calamus* rhizome ethanolic extract was also verified in inhibiting mitogen (phytohemagglutinin) and purified protein derivative as an antigen proliferation which enhanced marginal blood mononuclear cells, interleukin (IL)-2, and nitric oxide production in humans (Mehrotra et al. 2003). Furthermore, *A. calamus* leaf extract, when tested on human keratinocyte HaCaT cells, revealed multiple mechanisms to inhibit pro-inflammatory cytokines (Kim et al. 2009). The acetone extract (25–75%) of *A. calamus* was tested in an albino rat, which reversed paw inflammation to normal in 30 minutes (Lad et al. 2010).

Although *A. calamus* was considered a traditional remedy in inflammatory responses, its biological function in human skin cells remained unresolved. EO of *A. calamus* rhizome has also revealed a promising anti-inflammatory potential. It was tested in an egg-albumin denaturation assay with a half-maximal inhibitory concentration (IC_{50}) of 172.41 μg/ml and in normal sodium diclofenac (IC_{50} was 21.0 μg/ml) (Loying et al. 2019). In addition, Tanaka et al. (2001) showed a preventive upshot of the methanol extract of *A. calamus* rhizome (12.5 μg/ml) in murine endothelial and myeloid leukaemia cells.

Chamlagai and Singh (2016) also reported the anti-inflammatory response of an aqueous extract from *A. calamus* rhizome. These researchers tested the extract in a red blood cell membrane stabilization assay in an *in vitro* study. They concluded that a 10 mg/ml concentration of the extract exhibited significant inhibition in haemolysis and stabilized the membrane. Aqueous, butanol, and *n*-hexane extracts of *A. calamus* were also tested against lipoxygenase (LOX) and cyclooxygenase (COX)-arbitrated eicosanoid production by arachidonic acid. The butanol extract repressed the lipoxygenase product 1 (LP1) and COX-mediated thromboxane B2 (TXB2) production. When underlying signalling pathways were investigated, it was presumed that butanol extract acted on protein kinase C (PKC) by inhibiting the phospholipase C (PLC) path in platelets (Ahmed et al. 2014).

The anti-inflammatory, capillary strengthening, and membrane protective potential of *A. calamus* extract as a conditional therapeutic agent were compared with quercetin and diclofenac sodium. A dose-dependent anti-exudative action of *A. calamus* leaf extract was established in the rat models, which used histamine, carrageenan, and zymosan-mediated oedema in the paw. Carrageenan and histamine were effective to a greater extent (79.5% and 50%, respectively), but hydroalcoholic extract exhibited the maximum activity in zymosan-mediated oedema (dealcoholized extract *A. calamus* leaves was 43.5% effective). Considering the activity, the extracts have a comparative potential to quercetin but are less effective than diclofenac sodium. The leaf extracts of *A. calamus* decreased vascular penetrability by 1.2–1.4 times, stabilized the membrane, and strengthened the capillary. The membrane-stabilizing activity of *A. calamus* leaf extracts was 31.3–40.7% in this model of spontaneous haemolysis of erythrocytes.

Regarding this action, the extracts reliably enhanced the action of drugs, i.e. diclofenac sodium and quercetin. The dealcoholized extracts of *A. calamus* leaves at a concentration of 1 ml/kg showed the maximum activity, but further research is still needed to use this as an alternative therapeutic agent (Kalko et al. 2021). Figure 6.1 displays the traditional medicinal claims and pharmacological activities of *A. calamus*.

6.4.2 Antioxidant Potential

The antioxidant potential of the *Acorus* spp. Has been widely assessed and tested against reactive oxygen species (ROS) and free radicals-associated syndromes. In this respect, the *A. calamus* ethyl acetate extract exhibited potent antioxidant activity as evaluated by the 1,1-diphenyl-2-picrylhydrazyl (DPPH) assay (Acuna et al. 2002). Similarly, 0.2 g/ml of the extract has also shown 86.43% DPPH radical scavenging potential (Govindarajan et al. 2003). The antioxidant potential of *A. calamus* also limits ROS production because of constant noise exposure. The noise-induced brain stress was also protected in rats by ethyl acetate and methanol extracts of *A. calamus*. These variations were appraised by quantifying the activities of catalase (CAT), glutathione peroxidase (GPx), superoxide dismutase (SOD), reduced glutathione (GSH), vitamins E and C, lipid peroxidation (LPO), and thiols. Stress reduction is mainly due to the presence of β-asarone. Similarly, α-asarone was also tested for antioxidant activity in noise-induced brain stress by administering intraperitoneally 30 minutes before exposure for 30 days. Results revealed that the stress reduction caused by α-asarone is similar to β-asarone's stress reduction mechanism (Manikandan and Devi 2005).

Similarly, *A. calamus* ethanol extract was tested for antioxidant effect in acetaminophen-induced toxicity in male albino rats. Administration of acetaminophen elevated haemoglobin and leukocyte count, differential leukocyte count, granulocytes, mean corpuscular volume, packed cell volume, body weight, platelet concentration, and uric acid. *A. calamus* extract significantly enhanced the functioning of renal SOD, CAT, and GPx, and decreased malondialdehyde concentration in acetaminophen-treated rats. The extract also inhibited acetaminophen-induced haemolysis. Histopathological examination revealed a protective effect of the extract against acetaminophen-induced necrosis and renal impairment (Palani et al. 2010). Moreover, rhizome and leaf extracts of *A. calamus*, are traditionally used in neuroprotection. In rats, *A. calamus* hydroalcoholic extract was neuroprotective in middle cerebral occlusion-induced ischaemia (Shukla et al. 2006). The constituents like α- and β-asarone significantly repressed glutamate-induced toxicity and *N*-methyl-D-aspartate (NMDA) in primary cortical cultures (Cho et al. 2002). The *in vitro* radio-protective potential of *A. calamus* extract also confirmed its protective role in limiting DNA damage, lipid peroxidation in supercoiled plasmid DNA (pBR322), and alkaline single-cell gel electrophoresis. The free radical scavenging potential of the extract protected DNA and membrane from γ-radiation in human peripheral blood leukocytes and murine cells (Sandeep and Nair 2010).

Research findings also showed that *A. calamus* acetonitrile, acetone, alcoholic, and aqueous extracts exhibit antioxidant potential in DPPH, ferric ion reducing antioxidant power assay (FRAP), and the 2,2′-azinobis (3-ethylbenzothiazoline-6-sulphonic acid) (ABTS) scavenging activity for free radicals assay. The acetone extract was the most potent, followed by the acetonitrile, methanol, and aqueous extracts (Bahukhandi et al. 2013). In addition, the aqueous extract exhibited a promising antioxidant activity as assessed by lipid peroxidation (LPO), DPPH, and ion chelation assays (Manju et al. 2013). Another *in vitro* study used DPPH, FRAP, hydroxyl radical, nitric oxide, reduction aptitude, and scavenging power of superoxide radical to assess the antioxidant potential of aqueous and ethanol extracts from the whole plant of *A. calamus*. The presence of flavonoids and phenolics in *A. calamus* extracts contributed notably to the antioxidant effects. The half-maximal inhibitory concentration (IC_{50}) for the ethanol extract in scavenging the activities of nitric oxide, superoxide radical, hydroxyl oxide, and DPPH were 118.8, 38.3, 109.9, and 54.8 μg/ml, respectively. The FRAP values and the permanent latency of mentioned results were found to augment depending on the dose (Barua et al. 2014). Hydroalcohol and ethanol were used for *A. calamus* root and rhizome extraction to assess the antioxidant activity using the DPPH assay, and were compared with silymarin and butylated hydroxyanisole (BHA). Both of these extracts (hydroalcohol and ethanol) showed free radical scavenging values of 56.71 ± 19.54 and 59.13 ± 18.95, respectively (Elayaraja et al. 2010; Sujitha et al. 2013). The *A. calamus* EO also showed prominent antioxidant activity when tested using DPPH and β-carotene/linoleic acid bleaching assays (Shukla et al. 2013). An IC_{50} value of 704 μg/ml for the methanol extract free radical scavenging activity was also obtained (Ahmed et al. 2009).

Findings indicated that the EO contains the highest amount of an active component, α-asarone (7.9%), which is strongly believed to exert antioxidant activity. Loying et al. (2019) reported significant free radical scavenging activity of *A. calamus* EO in the DPPH assay with an IC_{50} value of 1.68 μg/ml. EO from the leaves of *A. calamus* showed an IC_{50} value of 10.79 μg/ml in the DPPH radical scavenging assay (Chaubey et al. 2018). Results regarding the radical scavenging activity of extracts and EO are also in line with the results of Manikandan and Devi (2005), Devi and Ganjewala (2011), and Chaubey et al. (2018). Similarly, a notable antioxidant role was determined for the methanol extract of *A. calamus* using the nitric oxide and DPPH assays, which revealed IC_{50} values of 236.60 and 192.99 μg/ml, respectively (Chakraborty et al. 2020). Moreover, the IC_{50} of *A. calamus* EO, assessed by the DPPH assay, was 108.71 μg/ml (Khanal et al. 2020). The DPPH assay validated the methanol extract's antioxidant potential with a slightly higher IC_{50} value of 3.74 μg/ml, similar to the standard (ascorbic acid with IC_{50} = 3.56 μg/ml) (Pokhrel and Chaulagain 2020).

The total phenolic fraction and the antioxidant potential of *A. calamus* fluctuate significantly throughout the year and its growth stages. Phenolic contents were at a higher concentration (1.5–2.0

times) in the senescence stage compared to the vegetative state. Similarly, a higher (1.5–3.0 times) antioxidant activity was observed in the fruiting and flowering phase compared to the shoot bud bursting phase. Furthermore, the concentration of phenolics, tannins, and FRAP activity was maximal during October, with respective values of 9.11 mg GAE/g, 26.0 mg TAE/g, and 5.86 mM AAE/100 g. Still, the radical scavenging activity was the highest in July, as assessed through DPPH and ABTS assays with values of 6.66 mM and 27.51 mM AAE/100 g. The climate factors of temperature and rainfall have shown a positive relationship ($p <0.01$) with DPPH and ABTS activities ($r = 0.523$ and $r = 0.534$, respectively). The phenolic contents and antioxidant potential were significantly related ($p <0.01$) regarding DPPH ($r = 0.304$) and ABTS assays ($r = 0.307$). Interestingly, research findings showed that October is the most suitable period for attaining the maximum antioxidant activity from *A. calamus* extract (Bahukhandi et al. 2021).

On the other hand, the aqueous extracts of *A. calamus* were used to develop eco-friendly, cost-effective zinc oxide nanoparticles ($ZnONP_S$), extensively cast off in Kerala's conventional Ayurvedic medicinal system. Standard ascorbic acid at 1000 μg/ml exhibited 99% total antioxidant capacity, with $ZnONP_S$ to 66% (Puthur et al. 2020). Other plants like *Chlorella vulgaris*, *A. calamus*, *Lemna minuta*, and *Scenedesmus dimorphus* were tested for their antioxidant potential using the DPPH assay. The highest antioxidant activity (methanol extract) was shown in *S. dimorphus*, followed by *A. calamus*, *L. minuta*, and *C. vulgaris* (Dinev et al. 2021). The oxidative and endoplasmic reticulum stress induction by tunicamycin and L-glutamate was significantly suppressed by α-asarone, leading ultimately to cell death. The L-glutamate-induced production of ROS and tunicamycin-induced kinase phosphorylation (RNA-like endoplasmic reticulum kinase (PERK) was also curbed by *A. calamus* extracts and α-asarone, signifying their use as novel therapeutic agents for neuroprotection (e.g. Alzheimer's disease) (Mikami et al. 2021).

6.4.3 Neuroprotective Effect

To assess the memory-enhancing and learning potential of *A. calamus*, different concentrations (25, 50, and 100 mg/kg) of ethanol extract were orally and intraperitoneally administered to male rates using a Y maze and shuttle box model. The findings revealed an enhancement in spatial recognition data and acquisition-recalling (Naderi et al. 2010). Likewise, when intraperitoneally administered at a concentration of 0.5 ml/kg, an ethanol extract of *A. calamus* significantly inhibited conditional avoidance response, potentiated pentobarbitone-created sleep periods, and protected against pentylenetetrazole (PTZ)-induced convulsions (40–60%) in rats. The extract did not exhibit a slight random effect on male rats' fighting, motor activity, and aggressive response (Vohora et al. 1990).

Rota-rod performance and grid walking: similarly, an ethanol and water (1:1, v/v) extract of *A. calamus* rhizomes was investigated to assess its neuroprotective potential in rats by inducing middle cerebral artery occlusion (MCAO) ischaemia. Results revealed that the rota-rod performance was significantly impaired accompanying grid walking in rats after 72 hours of MCAO-induced ischaemia compared to sham-operated rats. Additionally, lipid peroxidation increased; however, the SOD and GSH levels in experimental rats decreased. When administered with the extract at 25 mg/kg concentration, experimental rats showed a substantial improvement in neurobehaviour response. The extract caused a significant reduction in malondialdehyde levels, while levels and activity of GSH and SOD in corpus striatum and cortex were enhanced. Moreover, *A. calamus*-treated rats have also shown a reduction in contralateral cortical infarct surface compared to control animals (19% vs 33%). Overall, the neurological functioning score was improved in the tested animals indicating a neuroprotective potential of *A. calamus* (Shukla et al. 2006).

Using a rat model, saponin-rich extracts of *A. calamus* (SRE-AC) were employed to ameliorate controlled cortical injury (CCI)-induced neuropathic pain. CCI induction significantly increased mechanical and thermal hyperalgesia, functional sciatic index with decreased nerve conduction velocity, and histopathological and biochemical alterations. However, oral administration of pregabalin and SRE-AC notably ameliorated the threshold for nociceptive pain, electrophysiological changes, and sciatic functional index in a dose-dependent manner in CCI-induced animals (Muthuraman and Singh 2012).

A few compounds from *A. Calamus* extract were isolated and tested for their neuroprotective and cell viability activity in H_2O_2-induced SH-Y5Y cell line (neuroblastoma). Results showed that compounds such as α-asaronol, β-asaronol, 3-(2,4,5-trimethoxyphenyl)propan-1-ol, and 1,2,4-trimethoxy-5-(3-methoxy propyl)benzene exhibit protection against neuronal oxidative stress in all tested concentrations from 0.01 to 100 μM with β-asaronol exerting the highest neuroprotective potential (Bai et al. 2020). In addition, the methanol extract of *A. calamus* rhizome was given to rats at three concentrations of 250, 500, and 750 mg/kg to evaluate its potential against 6-hydroxy dopamine-induced Parkinson's disease. A marked improvement in the motor balance and locomotor coordination in 6-hydroxydopamine (6-OHDA)-lesioned rats was observed. Extracellular dopamine in striatum and tyrosine hydroxylase expression in substantia nigra were also significantly enhanced. Moreover, the extract also increased the level of DJ-1 with decreasing α-synuclein levels as assessed through western blotting analysis. Hence, it was proposed that *A. calamus* extract may be a promising anti-Parkinson's disease agent (Rabadia et al. 2014).

6.4.4 Anticancer Activity

Research findings showed that the *A. calamus* methanol extract inhibits growth in the human gastric cell line (AGS). Furthermore, angiogenesis in human umbilical vein endothelial cells was also inhibited by extracts and EO of *A. calamus*. It was also revealed that treatment of AGS cells with EO and *A. calamus* extract caused GI cell cycle arrest and down-regulation of Oct4 and nucleostemin (NS). In EO, compounds like valencene, epiprezizaene, and isocyclocitral were identified by GC-MS (Haghighi et al. 2017). In addition, *A. calamus* ZnONPs competently repressed the capability of SK-MEL-3 cells (melanoma cell line) (Vakayil et al. 2021).

6.4.5 Antimicrobial Activity

Extracts of *A. calamus*, when tested against fungal strains of *Trichophyton rubrum*, *Microsporum gypseum*, and *Penicillium marneffei* showed IC_{50} values of 0.2, 0.2, and 0.4 mg/ml. However, when tested against yeast strains, namely *Candida albicans*, *Saccharomyces cerevisiae*, and *Cryptococcus neoformans*, a moderate activity with 0.1–1.0 mg/ml minimum inhibitory concentration (MIC) was observed. Electron microscope scans revealed that the extract caused a collapse of treated hyphae and conidia due to cell damage and loss of cellular fluid (Phongpaichit et al. 2005). Similarly, the rhizome and leaf extracts, when tested against fungal (*Aspergillus flavus*, *Aspergillus niger*, and *Microsporum canis*) and yeast (*Cryptococcus gastricus* and *C. albicans*) strains, MIC values of 2–4 and 4–8 mg/ml, respectively, were achieved. Additionally, when α- and β-asarones were tested for antifungal and anti-yeast potential, results revealed a strong activity compared to rhizome and leaf extracts. The study, therefore, suggests the presence of α- and β-asarones in these extracts, which correspond to antimicrobial activity. Moreover, β-asarone exhibited antimicrobial potential comparable to α-asarone (Devi and Ganjewala 2009).

Similarly, the ethyl acetate extract and β-asarone from *A. calamus* rhizome significantly inhibited the growth of *C. albicans* at 0.5 mg/ml with complete fungicidal at 8 mg/ml. The lowest concentration of β-asarone killed 99.99% of inoculum within 2 hours of exposure. Likewise, the sub-inhibitory concentration also inhibited biofilm development and morphogenesis. In addition, the compound was non-toxic to human red blood cells at MIC values. The growth inhibitory effect was indicated by treating the cells at growth inhibition concertation with β-asarone, which inhibited cell growth and ergosterol biosynthesis (Rajput and Karuppayil 2013). In addition, the peroxidase enzyme from *A. calamus*, purified through chromatography and tested by gel filtration using Superose 1210/300 GL column, exhibited antifungal activity (Ghosh 2006). Similarly, *A. calamus* rhizomes revealed a strong antimicrobial effect against *Staphylococcus aureus*, *Bacillus subtilis*, and *Pseudomonas aeruginosa* with a MIC of 0.25 (Sabitha et al. 2003). The susceptibility of *B. subtilis* and *Mycobacterium* spp. Against calamus oil was also reported (Radušienė et al. 2006). Another study revealed that the *A. calamus* ethyl acetate leaf and rhizome extracts inhibit Gram-positive and Gram-negative bacterial growth except in the *Escherichia coli* strain. Furthermore, the α- and β-asarones were also reported to have similar activity (Devi and Ganjewala 2009). The antimicrobial potential is likely attributed to α- and β-asarones (Bhuvaneswari and Balasundaram, 2006; Devi and Ganjewala 2009). When used below the cytotoxic levels of herpes simplex virus (HSV)-1 and HSV-2, the *A. calamus* rhizome alcohol extract significantly inhibited viral propagation in both HSV-1 and HSV-2 cells suggesting the antiviral potential of *A. calamus* and β-asarone (Badam 1995). Table 6.1 lists the biological activities of different *A. calamus* extracts.

The activity of *A. calamus* against *S. aureus* (Gram positive) and *E. coli* and *Salmonella typhi* (Gram negative) was determined using the agar-well diffusion method. Results showed that the methanol, chloroform, and *n*-hexane *A. calamus* rhizome extracts inhibit microbial growth in both types (Pokhrel and Chaulagain 2020). The most significant inhibition zone (7.26 and 3.28 mm) for *E. coli* and *S. typhi* was caused by the chloroform extract, followed by ethanol (5.90 and 3.07 mm) and *n*-hexane (5.33 and 2.95 mm). The MIC and minimum bactericidal concentration (MBC) of the three solvents were 0.39% and 0.78%, respectively, indicating a comparable antimicrobial potency (Muchtaromah et al. 2019). On the other hand, when verified against bacteria (*E. coli* and *S. aureus*) and fungus (*C. albicans*), the dried rhizome extract of *A. calamus* demonstrated antimicrobial and antifungal activity. The MIC values for *E. coli*, *S. aureus*, and *C. albicans* were 2%, 3%, and 3%, with a 9.80, 9.50, and 8.67 mm zone of inhibition, respectively (Susanah et al. 2018). Moreover, the EO extracted from *A. calamus* L. (wild) rhizome, when tested for antimicrobial potential against Gram-positive (*S. aureus*) and Gram-negative bacteria (*E. coli*, *Klebsiella pneumoniae*, and *Proteus vulgaris*) exhibited a profound antibacterial effect (Joshi and Bashyal 2018).

Similarly, research findings also indicated that rhizome extracts of *A. calamus* (ethyl acetate, *n*-butanol-, and *n*-hexane) significantly inhibit *E. coli*, *S. aureus*, and *C. albicans* in descending order, whereas the aqueous extract was unable to inhibit microbial growth. Furthermore, *A. calamus* ZnONPS-coated fabrics were reported to inhibit microbial propagation with maximum inhibition against *A. baumanii*, *P. aeruginosa*, and *K. oxytoca* (Vakayil et al. 2021). These fractions' total phenolic and flavonoid content were 1820.51, 329.71, 237.81, and 74.36 mg GAE 100/g, and 2978.34, 399.07, 142, 23, 14.80 mg QE 100/g, respectively. Thus, a positive correlation was observed between flavonoid and phenolic contents and antimicrobial potential (Rita et al. 2019).

The phytochemical content and antimicrobial potential of *Allium sativum*, *Curcuma anga*, and *A. calamus* were tested against *S. aureus* and *E. coli* in aqueous and ethanol fractions with three combinations (36:36:28 (C_1), 40:30:30 (C_2), and 35:40:45 (C_3). The C_3 from the aqueous extract exerted the highest inhibitory potential against *S. aureus* and *E. coli*. C1 and MBC revealed the highest MIC of ethanol extract at 0.39% concentration at 0.78%. In contrast, the highest MIC against *E. coli* by the aqueous extract was shown at 25% concentration of C_1 with 50% MBC. In comparison, 0.39% ethanol extract (C_3) revealed the highest inhibition at 0.39% with MBC at 0.78% (Muchtaromah et al. 2018). Similarly, rhizome EO at a concentration of 500 µg/ml exhibited an inhibition zone of 16 mm and 12 mm against *Aspergillus fumigates* and *S. cerevisiae*, respectively. Additionally, when tested against *B. subtilis* and *B. cereus* the inhibition zones were 12 and 8 mm, respectively (Loying et al. 2019). On the other hand, the methanol extract exerted potent antimicrobial and preventive properties (Pawar et al. 2020).

TABLE 6.1
Biological Activities of *A. calamus* Extracts

Origin	Plant part	Extraction solvent	Model	References
India	Rhizome	Ethanol	Immunomodulatory	Mehrotra et al. 2003
Korea	Leaf	Water	Anti-inflammatory	Kim et al. 2009
–	Rhizome	Acetone	Anti-inflammatory	Lad et al. 2010
India	Rhizome essential oil	–	Anti-inflammatory Antioxidant Antibacterial	Loying et al. 2019
Ukraine	Leaf	Hydroalcoholic	Anti-inflammatory	Kalko et al. 2021
USA	Rhizome	Ethyl acetate	Antioxidant	Acuna et al. 2002
India	β-Asarone (rhizome)	Ethyl acetate Methanol	Antioxidant	Manikandan and Devi 2005
India	Rhizome	Ethanol	Antioxidant	Palani et al. 2010
India	Aerial parts	Ethanol	Antioxidant	Palani et al. 2009
India	Rhizome	50% ethanol	Antioxidant	Sandeep and Nair 2010
India	Rhizome	Water	Antioxidant	Manju et al. 2013
Japan	Rhizome	50% ethanol	Antioxidant	Mikami et al. 2021
India	Rhizome	Acetone	Antioxidant	Bahukhandi et al. 2013
India	Rhizome	Ethanol	Antioxidant	Barua et al. 2014
–	Rhizome	Hydroalcoholic	Antioxidant	Elayaraj et al. 2010
–	Rhizome	Ethanol	Antioxidant	Sujitha et al. 2013
India	Rhizome essential oil	–	Antioxidant	Shukla et al. 2013
India	Leaf essential oil	–	Antioxidant	Chaubey et al. 2018
India	Rhizome	Methanol	Antioxidant	Chakraborty et al. 2020
Nepal	Rhizome essential oil	–	Antioxidant Antibacterial	Pokhrel and Chaulagain 2020
India	Rhizome	Methanol	Antioxidant	Bahukhandi et al. 2021
Indonesia	Rhizome	*n*-Hexane	Antibacterial	Muchtaromah et al. 2019
Indonesia	Rhizome	Ethanol	Antibacterial	Susanah et al. 2018
India	Rhizome essential oil	–	Antibacterial	Joshi and Bashyal 2018
Indonesia	Rhizome	Water	Antibacterial	Rita et al. 2019
India	Rhizome	Methanol	Antibacterial	Pawar et al. 2020
Lithuania	Leaf essential oil	–	Antibacterial	Radušienė et al. 2006
India	CeO_2-NPs using rhizome	Water	Antibacterial	Altaf et al. 2021
Thailand	Rhizome	Methanol	Antifungal	Phongpaichit et al. 2005
India	Rhizome Leaf	Ethyl acetate	Antifungal Antibacterial	Devi and Ganjewala 2009
India	Rhizome	Ethyl acetate	Antifungal	Rajput and Karuppayil 2013
Thailand	Crude ethanol extract	Ethanol	Antifungal	Dethoup et al. 2019
India	BFNPs using rhizome	Ethyl alcohol	Antifungal	Thakur et al. 2020

The ethanol extract of *A. calamus* was tested against eight plant pathogens (fungi), namely *Celletotrichum capsica*, *Sclerotium rolfsii*, *Alternaria brassicicola*, *Rhizoctonia solani*, *Phytophthora palmivora*, *Lasiodiplodia theobromae*, *Bipolaris oryzae*, and *Pyricularia oryzae*. The extract at the concentration level of 10 g/L was effective against all fungi except *R. solani* and *B. oryzae* as determined by the chromatographic method. In addition, β-asarone and galangin isolated from petroleum ether and chloroform extracts showed antifungal activity against both pathogens at MICs of 0.5 and 0.125–0.25 g/L (Dethoup et al. 2019). Furthermore, the antifungal activity of barium ferrite nanoparticles (BFNPs) was evaluated *in vitro* using *A. calamus* against *Alternaria alternats*,

Marssonina rosae, *Fusarium oxysporum*, and *Colletotrichum gloesporiodes*. Results indicated that BFNPs exhibit activity against various phytopathogenic fungi; the highest mycelial growth inhibition (76.67%) was observed at 600 mg/L against *F. oxysoporum* mycelium (Thakur et al. 2020). Recently, using *A. calamus* aqueous extract, the anti-biofilm activity of cerium oxide nanoparticles (CeO_2-NPs) was investigated against Gram-positive and Gram-negative bacteria. Results showed that treatment with CeO_2-NPs suppressed the test bacteria biofilms by over 75%. Moreover, microscopy confirmed that treatment with CeO_2-NPs inhibits bacterial growth and colonization on solid supports. The test bacteria's colonization and biofilm growth on the glass surface was also reduced. There was also a dose-dependent suppression of preformed biofilms; adding CeO_2-NPs to the culture media significantly reduced the production of the test bacteria's exopolysaccharides (Altaf et al. 2021).

6.4.6 Other Biological Activities

Published research showed that *A. calamus* extract prevents the development of $FeCl_3$-induced epileptogenesis by moderating antioxidant enzymes (Pradhan et al. 2007). The anti-mutagenic effect of *A. calamus* has been studied using *Salmonella typhimurium* strains; results indicated a reduction in revertant colonies against NaN_3-induced mutagenicity (Aqil et al. 2008). Similarly, the EO of *A. calamus* reduced the intestinal fundus spontaneous contractions and charcoal intestinal transit, ultimately affecting the rat gastrointestinal motor activity. Rats were given charcoal suspension, and after an hour, they were orally fed with *A. calamus* EO at the rate of 50, 100, and 200 mg/kg, and atropine at 2 mg/kg. Animals were sacrificed 30 minutes later, and the lengths of the small intestine and the path taken by the charcoal were measured. The EO strongly reduced spontaneous fundus contractions when tested at concentrations of 10 and 100 g/ml. It was shown that atropine and all tested doses of EO notably shortened the time charcoal travelled through the digestive tract (Nešić et al. 2018).

Moreover, *A. calamus* alcohol extract, its main ingredient α-asarone, and *Tinospora cordifolia* (standard drug) were examined against histomorphometric alterations in various hippocampal regions in restraint-induced stressed rats. Results revealed a decrease in the adverse effects caused by stress due to improvement in the hippocampus morphology. The use of *A. calamus* as an adjuvant therapy seems to be a promising niche for coping with stress as it plays a role in neuronal stress adaptation (Subamalani et al. 2019).

6.5 CONCLUSION AND FUTURE PERSPECTIVES

The overall profile of *A. calamus* shows richness in numerous active phytochemicals. The reviewed literature provides clear evidence suggesting the potential of leaves and rhizomes of *A. calamus* in neurological and metabolic modulation. The major bioactive constituents responsible for pharmacological outcomes are calamine, asarones (α and β), and eugenol. This plant's phytochemical profile and therapeutic potential may vary based on geographical alteration. For example, the US Food and Drug Administration banned *A. calamus*, rendering β-asarone a carcinogen after conducting animal trials in 1974. The study used calamus oil, which contains 80% β-asarone while Indian and European genotypes contain much lower β-asarone concentrations.

Regarding toxicity studies, the concentration of mycotoxin, heavy metals, and pesticide residue(s) must also be considered. Therefore, the mechanism of action and biological potential of compounds other than eugenol and asarones still require further elucidation to develop therapeutics. Multicentre clinical trials, with even systematized and planned studies are required to formulate a standardized *A. calamus* extract. Additionally, pharmacokinetic/dynamic studies of active phytochemicals within a physiological system to explore the pharmacological role of these compounds are also required.

REFERENCES

Acuna, U.M., D.E. Atha, J. Ma, M.H. Nee & E.J. Kennelly. 2002. Antioxidant capacities of ten edible North American plants. *Phytotherapy Research* 16(1):63–65.

Ahmad, S., A. Ali, S. Bibi, K.B. Marwat & G. Hassan. 2006. Ethnobotanical study on some medicinal plants of Ouch, district lower Dir, Pakistan. *Pakistan Journal of Plant Sciences* 12 (1):65–71.

Ahmed, F., J.N.N.S. Chandra, A. Urooj & K.S. Rangappa. 2009. In vitro antioxidant and anticholinesterase activity of *Acorus calamus* and *Nardostachys jatamansi* rhizomes. *Journal of Pharmacy Research* 2(5):830–883.

Ahmed, S., G.U.L. Saima, Z.U.H. Muhammad & M.S. Stanković. 2014. Pharmacological basis of the use of *Acorus calamus* L. in inflammatory diseases and underlying signal transduction pathways. *BoletínLatinoamericano y del Caribe de Plantas Medicinales y Aromáticas* 13(1):38–46.

Altaf, M., S. Manoharadas & M.T. Zeyad. 2021. Green synthesis of cerium oxide nanoparticles using *Acorus calamus* extract and their antibiofilm activity against bacterial pathogens. *Microscopy Research and Technique* 84(8):1638–1648.

Anonymous. 2001. *The Wealth of India: A Dictionary of Indian Raw Materials &Industrial Products*, 1st ed. India: National Institute of Science Communication.

Aqil, F., M. Zahin & I. Ahmed. 2008. Antimutagenic activity of methanolic extracts of four ayurvedic medicinal plants. *Indian Journal of Experimental Biology* 46:668–672.

Arya, A.K., M. Durgapal, A. Bachheti, K.K. Joshi, Y.H. Gonfa, R.K. Bachheti & A. Husen 2022. Ethnomedicinal use, phytochemistry, and other potential application of aquatic and semiaquatic medicinal plants. *Evidence-Based Complementary and Alternative Medicine* 2022:4931556.

Badam, L. 1995. In vitro studies on the effects of Acorus calamus extract and β-asarone on Herpes viruses. *Deerghayu International* 11:16–18.

Bahukhandi, A., S. Rawat, A.K. Jugran, I.D. Bhatt & R.S. Rawal. 2021. Seasonal variation in phenolics and antioxidant activity of *Acorus calamus* Linn.: an important medicinal plant of Himalaya. *National Academy Science Letters* 44(1):13–15.

Bahukhandi, A., S. Rawat, I.D. Bhatt & R.S. Rawal. 2013. Influence of solvent types and source of collection on total phenolic content and antioxidant activities of *Acorus calamus* L. *National Academy Science Letters* 36(1):93–99.

Bai, Y., Y. Sun, J. Xie, B. Li, Y. Bai, D. Zhang & X. Zheng. 2020. The asarone-derived phenylpropanoids from the rhizome of *Acorus calamus* var. *angustatus* Besser. *Phytochemistry* 170:112212.

Balaji, R.M. & D. Shanthakumari. 2020. Investigation on phytochemical and antimicrobial properties of *Acorus calamus* extract. *PalArch's Journal of Archaeology of Egypt/Egyptology* 17(9):6744–6755.

Bangasen. 1984. *Bangasen*, 1st ed. India: Chaukhambha Sanskrit Bharati.

Barua, C.C., S. Sen, A.S Das, A. Talukdar, N.J. Hazarika, A.G. Barua & I.A. Barua. 2014. A comparative study of the in vitro antioxidant property of different extracts of *Acorus calamus* Linn. *Journal of Natural Product and Plant Resources* 4(1):8–18.

Batra, N., H. Honey, R. Bairwa & M. Bachwani. 2012. Pharmacognostical and preliminary phytochemical investigation of *Acorus calamus* Linn. *Asian Journal of Pharmaceutical Research* 2:39–42.

Bhuvaneswari, R.A. & C. Balasundaram. 2006. Traditional Indian herbal extracts used in vitro against growth of the pathogenic bacteria Acromonashydrophila. *Israeli Journal of Aquaculture – Bamidgeh* 58(2):89–96.

Chakraborty, R., S. Sen, F.A. Ahmed & H. Barakoti. 2020. Evaluation of in vitro anti-diabetic and antioxidant potential of Acorus calamus rhizome. *Indo Global Journal of Pharmaceutical Sciences* 10(3):35–40.

Chamlagai, D. & B. Singh. 2016. Study of in vitro anti-inflammatory activity of ethnomedicinal plants of Sikkim Viscum articulatum and Acorus calamus. *Asian Journal of Pharmaceutical and Clinical Research* 119–122.

Chaubey, P., A. Parki, O. Prakash, R. Kumar & A.K. Pant. 2018. Comparative study of chemical composition and antioxidant activity of essential oil extracted from *Acorus calamus* L. leaves. *Journal of Medicinal Herbs* 8(4):203–211.

Chellaiah, M., A. Muniappan, R. Nagappan & I. Savarimuthu. 2006. Medicinal plants used by traditional healers in Kancheepuram District of Tamil Nadu, India. *Journal of Ethnobiology and Ethnomedicine* 2:43.

Cho J., Y.H. Kim, J. Hong, C.H. Yang & C.G. Park. 2002. Protection of cultured rat cortical neurons from excitotoxicity by asarone, a major essential oil component in the rhizomes of Acorus gramineus. *Life Sciences* 71(5):591–599.

Chunekar, K.C. & G.S. Pandey. 1998. *Bhavprakash Nighantu*, 1st ed. India: Chaukhambha Bharati Academy.

Derymedvid, L.V., K.O. Kalko, A.V. Taran, V.E. Kashuta, O.O. Riabova, O.S. Shpychak & T.V. Shpychak. Prospects for the use of *Acorus calamus* extracts in medicine: generalized data from own research. *PharmacologyOnLine* 3:1831–1842.

Dethoup, T., P. Songkumarn, T. Sirirak & A. Kijjoa. 2019. Fungicidal activity of *Acorus calamus* L. extracts against plant pathogenic fungi. *Agriculture and Natural Resources* 53(5):527–532.

Devi, S.A. & D. Ganjewala. 2009. Antimicrobial activity of *Acorus calamus* (L.) rhizome and leaf extract. *Acta Biologicaszegediensis* 53(1):45–49.

Devi, S.A. & D. Ganjewala. 2011. Antioxidant activities of methanolic extracts of sweet-flag (*Acorus calamus*) leaves and rhizomes. *Journal of Herbs, Spices & Medicinal Plants* 17(1):1–11.

Dhivya, V., S.J. Nelson, K.S. Subramanian, Y.S.J.T. Edward, K. Rajamani, V.P. Santhanakrishnan & S. Sithanantham. 2019. Formulation of sweet flag oil (*Acorus calamus*) nanoemulsion by spontaneous emulsification method for the management of *Sitophilus oryzae*. *International Journal of Chemical Studies* 7:2072–2076.

Dinev, T., M. Tzanova, K. Velichkova, D. Dermendzhieva & G. Beev. 2021. Antifungal and antioxidant potential of methanolic extracts from *Acorus calamus* L., *Chlorella vulgaris* Beijerinck, *Lemna minuta* Kunth and Scenedesmus dimorphus (Turpin) Kützing. *Applied Sciences* 11(11):4745.

Elayaraja, A., M. Vijayalakshmi & G. Devalarao. 2010. In vitro free radical scavenging activity of various root and rhizome extracts of *Acorus calamus* Linn. *International Journal of Pharmacy and Biological Sciences* 1:301–304.

Gaidhani, S.N., G.S. Lavekar, A.S. Juvekar, S. Sen, A. Singh & S. Kumar. 2009. In vitro anti-cancer activity of standard extracts used in Ayurveda. *Pharmacognosy Magazine* 5: 425–429.

Ghosh, M. 2006. Antifungal properties of haem peroxidase from *Acorus calamus*. *Annals of Botany* 98(6):1145–1153.

Gilani, A.U., A.J. Shah, M. Ahmad & F. Shaheen. 2006. Antispasmodic effect of *Acorus calamus* is mediated through calcium channel blockade. *Phytotherapy Research* 20:1080–1084.

Govindarajan, R., A.K. Agnihotri, S. Khatoon, A.K.S. Rawat & S. Mehrotra. 2003. Pharmacognostical evaluation of an antioxidant plant-Acorus calamus Linn. *Natural Product Sciences* 9(4):264–269.

Gyawali, R. & K.S. Kim. Volatile organic compounds of medicinal values from Nepalese *Acorus calamus* L. *Kathmandu University Journal of Science, Engineering and Technology* 5:51–65.

Haghighi, S.R., M.H. Asadi, H. Akrami & A. Baghizadeh. 2017. Anti-carcinogenic and anti-angiogenic properties of the extracts of *Acorus calamus* on gastric cancer cells. *Avicenna Journal of Phytomedicine* 7(2):145.

Hanson, K.M., M. Gayton-Ely, L.A. Holland, P.S. Zehr & B.C. Söderberg. 2005. Rapid assessment of β-asarone content of *Acorus calamus* by micellar electrokinetic capillary chromatography. *Electrophoresis* 26(4–5):943–946.

Hao, Z.Y., Y.G. Cao, Y. Wang, C.L. Zhang, H. Luo, D. Liang & D.Q. Yu. 2021. Alkaloids from the rhizomes of *Acorus calamus* and their PPARα, PPARγ, and glucokinase-activating activities. *Phytochemistry Letters* 45:37–43.

Ignacimuthu, S., M. Ayyanar & S. Sivaraman. 2006. Ethnobotanical investigations among tribes in Madurai District of Tamil Nadu (India). *Journal of Ethnobiology and Ethnomedicine* 2:25.

Jadhav, T. 1994. *Sushrut Samhita*, 1st ed. India: Chaukhambha Surbharati Acadamy.

Jatisatiener, C., & A. Jatisatiener. 1999. An ethnomedico botanical survey of Parbativalley in Himachal Pradesh (India). *Acta Horticulturae* 501:87–93.

Jayaraman, R.T., T. Anitha, & V.D. Joshi. 2010. Analgesic and anticonvulsant effects of *Acorus calamus* roots in mice. *International Journal of Pharmaceutical Technology Research* 2:552–555.

Joshi, S. & S. Bashyal. 2018. Study on the chemical constituents and antibacterial activity of essential oil of *Acorus calamus* L. rhizomes of Rupendehi district (Nepal). *Journal of Institute of Science and Technology* 23(1):57–60.

Kalko, K.O., L.V. Derimedvid, L.A. Korang, O.Y. Toziuk, I.V. Okipniak, N.A. Domar & D.S. Zhurenko. 2021. Antiinflammatory, membraneprotective and capillary strengthening properties of sweet flag leaf extracts. *PharmacologyOnLine* 2:145–157.

Khanal, S., D.P. Bhandari, L. Bhandari & A. Adhikari. 2020. Chemical profiling and antioxidant activities of essential oil from rhizomes of *Acorus calamus* L. *Bibechana* 17:89–95.

Kim, H., T.H. Han, & S.G. Lee. 2009. Antiinflammatory activity of a water extract of *Acorus calamus* L. leaves on keratinocyte HaCaT cells. *Journal of Ethnopharma-cology* 122:149–156.

Kirtikar, K.R. & B.D. Basu. 1987. *Indian Medicinal Plants*, 4th ed. India: Mahendra Pal Sing Publishers.

Lad, Y.H., K.V. Waghmare, K.N. Methe, S.V. Kasabe & A.A. Kanase. 2010. Antiinflammatory effect of acetone extract of *Acorus calamus* in albino rats. *Electronic Journal of Pharmacology & Therapy* 3:5–7.

Lee, M.H., Y.Y. Chen, S.C. Wang, T. Watanobe & Y.C. Tsai. 2011. Inhibitory effect of β-asarone, a component of *Acorus calamus* essential oil, on inhibition of adipo-genesis in 3T3-L cells. *Food Chemistry* 126:1–7.

Lekshmi, J.L. & M. Kensa. 2020.Gc-Ms analysis of bioactive compounds in methanolic extract of *Acorus calamus* Linn.. *Cikitusi Journal for Multidisciplinary Research* 7:1–10.

Loying, R., R. Gogoi, N. Sarma, A. Borah, S. Munda, S.K. Pandey & M. Lal. 2019. Chemical compositions, in-vitro antioxidant, anti-microbial, antiinflammatory and cytotoxic activities of essential oil of *Acorus calamus* L. rhizome from North-East India. *Journal of Essential Oil Bearing Plants* 22(5):1299–1312.

MacGaw, L.J., A.K. Jagar & J. Van Staden. 2002. Isolation of β-asarone, an antibacte-rial and antihelmintic compound, from *Acorus calamus* in South Africa. *African Journal of Botany* 68:31–35.

Manikandan, S. & R.S. Devi. 2005. Antioxidant property of α-asarone against noise-stress-induced changes in different regions of rat brain. *Pharmacological Research* 52(6):467–474.

Manju, S., R.P. Chandran, P.K. Shaji & G.A. Nair. 2013. In vitro free radical scavenging potential of *Acorus calamus* L. rhizome from Kuttanad Wetlands, Kerala, India. *International Journal of Pharmaceutical Sciences* 5:376–380.

Mazza, G. 1984. Identification of oxidation products of beta-asarone in Acorus calamus L. by gas chromatography and mass spectrometry. *Sciences des Aliments (France)* 4(3):437–482.

Mazza, G. 1985. Gas chromatographic and mass spectrometric studies of the constituents of the rhizome of calamus: I. The volatile constituents of the essential oil. *Journal of Chromatography A* 328:179–194.

Meena, A.K., M.M. Rao, A. Singh & S. Kumar. 2010. Physiochemical and preliminary phytochemical studies on the rhizome of *Acorus calamus*. *International Journal of Pharmacy and Pharmaceutical Science* 2:130–131.

Mehrotra, S., K.P. Mishra, P. Maurya, C.C. Srimaln, V.S. Yadav, R. Pandey & V.K. Singh. 2003. Anticellular immunosuppressive properties of ethanolic extract of *Acorus calamus* rhizome. *International Immunopharmacology* 3:53–61.

Mikami, M., O. Takuya, Y. Yoshino, S. Nakamura, K. Ito, H. Kojima & H. Hara. 2021. *Acorus calamus* extract and its component α-asarone attenuate murine hippocampal neuronal cell death induced by L-glutamate and tunicamycin. *Bioscience, Biotechnology, and Biochemistry* 85:493–501.

Motley, T.J. 1994. The ethnobotany of sweet flag, *Acorus calamus* (Araceae). *Economic Botany* 48(4):397–412.

Muchtaromah, B., A. Hayat & E. Agustina. 2019. Phytochemical screening ana antibacterial activity of *Acorus calamus* L. extracts. *Jurnal Biodjati* 4(2):68–78.

Muchtaromah, B., M. Ahmad, R. Romaidi, L.A. Nazilah & N.A. Naja. 2018. Antibacterial activity of water and ethanol extract of *Allium sativum*, *Curcuma mangga*, and *Acorus calamus* combination. *Berkala Penelitian Hayati* 24(1):8–15.

Muthuraman, A. & N. Singh. 2012. Neuroprotective effect of saponin rich extract of *Acorus calamus* L. in rat model of chronic constriction injury (CCI) of sciatic nerve-induced neuropathic pain. *Journal of Ethnopharmacology* 142(3):723–731.

Muthusamy, A., P.S. Swathy, S. Tantry, S.A. Thorat, A. Kaniyassery & K.R. Kiran. 2022. Large-scale production of in vitro plantlets of sweet flag (*Acorus calamus* L.) and quantitative analysis of α- and β-asarone contents using high pressure liquid chromatography. *Vegetos* 35(1):122–132.

Naderi, G.A., M. Khalili, M. Karimi & M. Soltani. 2010. The effect of oral and intraperitoneal administration of *Acorus calamus* L. extract on learning and memory in male rats. *Journal of Medicinal Plants* 9(34):46–56.

Nadkarni, K.M. 1998. *Indian Materia Medica*, 1st ed. India: Popular Prakashan.

Nandakumar, S., S. Menon & S. Shailajan. 2013. A rapid HPLC-ESI-MS/MS method for determination of β-asarone, a potential anti-epileptic agent, in plasma after oral administration of *Acorus calamus* extract to rats. *Biomedical Chromatography* 27(3):318–326.

Nešić, M., N. Stojanović, P. Ranđelović & N. Randulović. 2018. Effects of acorenonetype *Acorus calamus* essential oil on rat gastric fundus contractions and intestinal transit. *Facta Universitatis, Series: Physics, Chemistry and Technology* 16(1):139.

Palani, S., S. Raja, R.P. Kumar, D. Venkadesan, K. Devi, A. Sivaraj & B.S. Kumar. 2009. Therapeutic efficacy of antihepatotoxic and antioxidant activities of *Acorus calamus* on acetaminophen-induced toxicity in rat. *International Journal of Integrative Biology* 7(1):39–44.

Palani, S., S. Raja, R.P. Kumar, P. Parameswaran & B.S. Kumar 2010. Therapeutic efficacy of *Acorus calamus* on acetaminophen-induced nephrotoxicity and oxidative stress in male albino rats. *Acta Pharmaceutica Sciencia* 52(1):89–100.

Pawar, R., S. Barve & V. Zambare. 2020. In vitro antibacterial activity of *Acorus calamus* extract on methicillin-resistant *Staphylococcus aureus* wound isolates and reduced invasion into mucosal fibroblasts. *Vegetos* 33(4):712–721.

Phongpaichit, S., N. Pujenjob, V. Rukachaisirikul & M. Ongsakul. 2005. Antimicrobial activities of the crude methanol extract of *Acorus calamus* Linn. Songklanakarin. *Journal of Science and Technology* 27:517–523.

Pokhrel, S. & K. Chaulagain. 2020. Phytoconstituents and biological analysis of *Acorus calamus* rhizome of Sindhupal chowk District, Nepal. *Bibechana* 17:104–109.

Porwal, P., M. Niladri, S. Yadav, A.K. Jha, P.Y. Patil & K. Shanawaz. 2021. Phytochemical screening, HPLC and GC-MS chromatography of two Indian traditional medicinal plants. *Drugs and Cell Therapies in Hematology* 10(3):371–379.

Prabha, S. & J. Kumar. 2021. Gas chromatographic and mass spectroscopic (GC-MS) analysis of rhizome of *Acorus calamus* Linn. for identification of potent antimicrobial bio-active compounds. *Journal of Scientific Research* 13(1):263–273.

Pradhan, S.N., K. Ray, R. Hazra & R.D. Ghua. 2007. Inhibitory role of *Acorus calamus* in ferric chloride-induced epileptogenesis in rat. *Human Experimental Toxicology* 27:947–953.

Prajapati, N.D., S.D. Purohit, D.D. Sharma & K. Tarun. 2003. *A Handbook of Medicinal Plants, Section II*, 1st ed. India: Agrobiaos.

Puthur, S., K.K. Raj, A.N. Anoopkumar, S. Rebello & E.M. Aneesh 2020. *Acorus calamus* mediated green synthesis of ZnONPs: a novel nano antioxidant to future perspective. *Advanced Powder Technology* 31(12):4679–4682.

Rabadia, J., U. Hirani, M. Dadhaniya & S. Bhensdadiya. 2014. Neuroprotective effect of *Acorus calamus* on a rat model of Parkinson's disease induced by 6-hydroxydopamine. *Asian Journal of Pharmacology and Toxicology* 2(3):16–14.

Radušienė, J., A. Judžentienė, D. Pečiulytė & V. Janulis. 2007. Essential oil composition and antimicrobial assay of Acorus calamus leaves from different wild populations. *Plant Genetic Resources* 5(1):37–44.

Rajput, S.B. & S.M. Karuppayil. 2013. Beta-asarone, an active principle of Acorus calamus rhizome, inhibits morphogenesis, biofilm formation and ergosterol biosynthesis in *Candida albicans*. *Phytomedicine* 20:139–142.

Rita, W.S., I.M.D. Swantara & G.A.P. Utami. 2019. Antimicrobial activity of *Acorus calamus* L. rhizome extract and its total flavonoid and phenolic contents. *AIP Conference Proceedings* 2155(1):020054.

Sabitha Rani, A., M. Satyakala, V. Sandya Devi & U. Suryanarayana Murty. 2003. Evaluation of antibacterial activity from the rhizome extract of Acorus calamus linn. *Journal of Scientific & Industrial Research* 62(6):623–625.

Sandeep, D. & C.K.K. Nair. 2010. Protection of DNA and membrane from γ-radiation-induced damage by the extract of *Acorus calamus* Linn.: an in vitro study. *Environmental Toxicology and Pharmacology* 29(3):302–307.

Shah, N.C. 2005. *Bharat Bhaishajya Ratnakar*. India: B. Jain Publishers Pvt. Ltd.

Shastri, K. & G. Pandey. 1997. *Charaka Samhita of Agnivesa*, 5th ed. India: Chaukhambha Sanskrit Sansthan.

Shukla, P.K., V.K. Khanna, M.M. Ali, R. Maurya, M.Y. Khan & R.C. Srimal. 2006. Neuroprotective effect of *Acorus calamus* against middle cerebral artery occlusion-induced ischaemia in rat. *Human & Experimental Toxicology* 25(4): 187–194.

Shukla, R., P. Singh, B. Prakash, & N.K. Dubey. 2013. Efficacy of *Acorus calamus* L. essential oil as a safe plant-based antioxidant, A flatoxin B 1 suppressor and broad-spectrum antimicrobial against food-infesting fungi. *International Journal of Food Science & Technology* 48(1):128–135.

Si, M.M., J.S. Lou, C.X. Zhou & J.N. Shen. 2010. Insulin-releasing and alpha glucosidase inhibitory activity of ethyl acetate fraction of *Acorus calamus* in vitro and in vivo. *Journal of Ethnopharmacology* 128:154–159.

Singh, R., P.K. Sharma & R. Malviya, R. 2011. Pharmacological properties and ayurvedic value of Indian buch plant (*Acorus calamus*): a short review. *Advances in Biological Research* 5(3):145–154.

Subamalani, S., A. Sasikumar, R. Vijayaragavan, R. Senthilkumar, S.M. Kumar, L.S. Raj, & I. Kannan. 2019. Effect of Acorus calamus Linn on histomorphometric changes in the CA1 and CA3 regions of Hippocampus in Wistar Albino rats. *Research Journal of Pharmacy and Technology* 12(7):3531–3536.

Sujitha, R., B.V. Bhimba, M.S. Sindhu & P. Arumugham. 2013. Phytochemical evaluation and antioxidant activity of *Nelumbo nucifera, Acorus calamus* and *Piper longum. International Journal of Pharmaceutical and Chemical Sciences* 2:1573–1578.

Susanah, R.W., K. Retno & S.I.M. Dira. 2018. Total phenolic and flavonoid contents and antimicrobial activity of *Acorus calamus* L. rhizome ethanol extract. *Research Journal of Chemistry and Environment* 22:65–70.

Tanaka, S., S. Yoichi, L. Ao, M. Matumoto, K. Morimoto, N. Akimoto & I.M. Said. 2001. Potential immunosuppressive and antiinflammatory activities of Malaysian medicinal plants characterized by reduced cell surface expression of cell adhesion molecules. *Phytotherapy Research* 15(8):681–686.

Tariq, R.M., S.N.H. Naqvi, M.I. Choudhary & A. Abbas. 2010. Importance and implementation of essential oil of Pakistanian *Acorus calamus* Linn., as a biopesticide. *Pakistan Journal of Botany* 42(3):2043–2050.

Thakur, A., N. Sharma, M. Bhatti, M. Sharma, A.V. Trukhanov & P. Thakur. 2020). Synthesis of barium ferrite nanoparticles using rhizome extract of *Acorus calamus*: characterization and its efficacy against different plant phytopathogenic fungi. *Nano-Structures & Nano-Objects* 24:100599.

Vaidyaratnam, P.S. 1994. *Varier's Indian Medicinal Plants*, 1st ed. India: Oriental Longman Ltd.

Vakayil, R., T. Abdul Nazeer & M. Mathanmohun, 2021. Evaluation of the antimicrobial activity of extracts from *Acorus calamus* rhizome against multidrug-resistant nosocomial pathogens. *Research Journal of Agricultural Sciences* 12:1613–1617.

Varde, A.B., S.S. Ainapure, S.R. Naik & S.R. Amladi. 1988. Antiinflammatory activity of coconut oil extract of *Acorus calamus, Ocimum sanctum* and *Ocimum basilicum* in rats. *Indian Drugs* 25(6):226–228.

Vohora, S.B., S.A. Shah & P.C. Dandiya. 1990. Central nervous system studies on an ethanol extract of *Acorus calamus* rhizomes. *Journal of Ethnopharmacology* 28(1):53–62.

Vohra, S.B., S.A. Shah, K. Sharma, S.A.H. Naqvi & P.C. Dandiya. 1989. Antibacterial, antipyretic, analgesic and antiinflammatory studies on *Acorus calamus* Linn. *Annals of National Academy of Medical Sciences* 25:13–20.

Wu, H.S., D.F. Zhu & C. Zhou. 2009. Insulin sensitizing activity of ethyl acetate fraction of *Acorus calamus* L. in vitro and in vivo. *Journal of Ethnopharmacology* 123:288–292.

7 Recent Update on Medicinal Properties of *Rotula aquatica* Lour.

Protha Biswas,[1] Ikbal Hasan,[1] Uttpal Anand,[2] Mimosa Ghorai,[1] Abdel Rahman Al-Tawaha,[3] Ercan Bursal,[4] Vartika Jain,[5] Devendra Kumar Pandey,[6] Mallappa Kumara Swamy,[7] Mahipal S. Shekhawat,[8] Tabarak Malik,[9] and Abhijit Dey[1,]*

[1] Department of Life Sciences, Presidency University, Kolkata, West Bengal, India
[2] Department of Life Sciences and the National Institute for Biotechnology in the Negev, Ben-Gurion University of the Negev, Beer-Sheva, Israel
[3] Department of Biological Sciences, Al-Hussein Bin Talal University, Maan, Jordon
[4] Department of Biochemistry, Mus Alparslan University, Turkey
[5] Department of Botany, Government Meera Girls' College, Udaipur, India
[6] Department of Biotechnology, Lovely Professional University, Punjab, India
[7] Department of Biotechnology, East West First Grade College of Science (Bangalore University), Bengaluru, Karnataka, India
[8] Plant Biotechnology Unit, KM Government Institute for Postgraduate Studies and Research, Puducherry, India
[9] Department of Biochemistry, College of Medicine & Health Science, School of Medicine, University of Gondar, Gondar, Ethiopia
* Corresponding author (Abhijit Dey): osthum.dbs@presiuniv.ac.in

7.1 INTRODUCTION

Rotula aquatica is a woody plant, rheophytic in nature, and is an aromatic medicinal bush common to South East Asia, India, Latin America, and Sri Lanka belonging to the Boraginaceae family. *R. aquatica* is a small shrub and is found in sandy and rocky beds throughout central, western, and southern India, often submerged in water. The plant is indigenous to India and a component of aquatic, lotic ecosystems having fragrant flowers. This plant is not very common; its presence is restricted to rocky riverbanks. Approximately 100 genera and 2000 species are represented by *Rotula* (Sivarajan and Balachandran 1994; Arya et al. 2022).

Around 80% of the world's population rely on plant-produced drugs for various ailments to a large extent. Various traditional treatment methods have employed plants as a source of medicines (Rouhi-Boroujeni et al. 2015). Herbal medicine was accepted as an alternative to treat diseases since it reduces the risks and costs associated with synthetic medicines. Ayurvedic practitioners or Vaidyas extensively use the plant, and the most important Ayurvedic drugs are made from the root tuber,

DOI: 10.1201/9781003256830-7

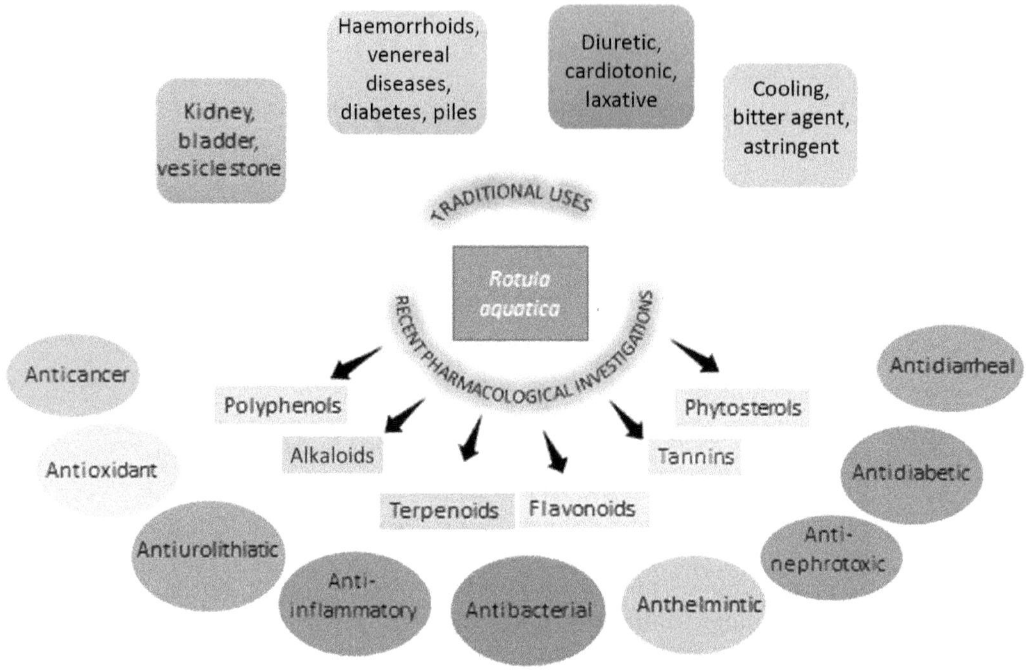

FIGURE 7.1 Traditional and pharmacological attributes of *R. aquatica*.

which is the medicinal part of the plant. The tubers containing rhabdiol and allantoin have traditionally been used to treat cardiovascular and blood disorders, as well as uterine diseases, dysuria, ulcers, bladder stones, and venereal diseases (Sivarajan and Balachandran 1994). The Ayurvedic system uses *R. aquatica* widely for dissolving kidney stones and urinary calculi, particularly in Karnataka and Kerala in India.

The stem decoctions of *R. Aquatica* are used as diuretics and sudorifics in the Philippines (Martin 2003). The aqueous extract from the root also exhibits anti-urolithiatic and antioxidant activity. The aqueous root extract shows crystal dissolving activity in monosodium urate monohydrate type of urinary calculi (Raut et al. 2008). In addition to treating piles, a decoction made from the root is utilized to treat bladder stones and venereal diseases. Roots are traditionally used to treat kidney stones, vesicle stones, haemorrhoids, venereal illnesses, and blood sugar levels. It is a laxative for treatment of piles, kidney stones, and bladder stones. In addition to its cooling, astringent, and bitter effects, it also exhibits cardiotonic and diuretic qualities. The stem extract also shows anti-proliferative properties against pancreatic cancer cell lines (Patil et al. 2004). Figure 7.1 presents the traditional and pharmacological attributes of *R. aquatica*.

A significant amount of consideration is given to the study of medicinal plants as sources of cures for several ailments, primarily to find less harmful and more cost-effective drugs. Among them, *R. aquatica* is an important medicinal plant that can treat various ailments and has shown no toxicity in acute toxicity studies (Mengi and Bakshi 2009). In this chapter, we will discuss the pharmacological properties of *R. aquatica* with recent updates on *in vitro* and *in vivo* studies.

7.2 PHYTOCHEMISTRY

R. aquatica's phytochemistry has yet to be comprehensively investigated and reported. The plant contains steroids, alkaloids, rhabdiols, and a diverse array of nutrients like amino acids and proteins, in addition to alkaloids, flavonoids, and phenolic compounds of medicinal value (Mamta and Abhishek 2010). In the qualitative detection of important phytochemicals from the methanolic extract of the root, various secondary metabolites such as alkaloids, flavonoids, phenols, tannins, phytosterols, reducing sugars, proteins, and terpenoids were detected, among which phenolic compounds were found as the principal component (Vysakh et al. 2020a). Analyses of the aqueous root extract revealed the presence of alkaloids, flavonoids, phenolics, saponins, tannins, terpenoids, anthraquinones, anthocyanin, and proteins and carbohydrates (Vijayakumari et al. 2013). ^{1}H nuclear magnetic resonance (NMR), infrared spectroscopy, mass spectrometry, and ^{13}C NMR techniques identified the bioactive components 3-*O*-acetyl-11-keto-β-boswellic acid, which shows anti-urolithiatic activity by inhibiting calcium oxalate (CaOx) crystallization (Vijayakumari et al. 2016a). The extract from methanol and chloroform *R. aquatica* root and leaves contained carbohydrates, proteins, amino acids, glycosides, alkaloids, tannin, and polyphenols. Both chloroform extracts possess steroids and triterpenoids, whereas methanolic extracts do not. Both methanol and chloroform and extracts of roots and leaves showed flavonoids, whereas quinoline was found in extracts of chloroform of both leaf and root. Only root and leaf methanol extracts contained saponins (Patil et al. 2003; Pallavi et al. 2016). Patil et al. (2003) detected tannins and steroids in the phytochemical screening test. Using elemental analysis, it was determined that the spot with a molecular weight of 388 had a molecular formula of $C_{26}H_{46}NO$. In the spectrum of the study, it was found that the compound contains an amide or nitrogen in the heterocyclic side chain and could be considered a steroidal alkaloid. It was found that a second compound, with a molecular weight of 387 and the formula $C_{25}H_{39}SO$, was identified as steroidal alcohol (Patil et al. 2004). Two components were isolated by developing high-performance thin layer chromatography (HPTLC) fingerprints and were categorized by ultraviolet (UV) spectroscopy, mass spectroscopy, and NMR.

HPTLC analysis of aqueous root extract from *R. aquatica*, revealed the existence of seven varied kinds of phenols with retention factor (Rf) values ranging from 0.01 to 0.97. A kaempferol-related Rf value in the sample indicates that the extract contains phenolic compounds. Twelve tannins were found with improved resolution, and one of the Rf values matched with gallic acid (standard) (Vijayakumari et al. 2016b). Three major phenolic compounds were identified by analysing the UV-visible spectra and retention times of the extracts: gallic acid, *p*-coumaric acid, and vanillic acid. The RP-HPLC data of methanolic *R. aquatica* extract was shown to have a total phenol content of 94.4 mg/ml/mg equivalent to gallic acid and 25.2 μg QE/g dry weight equivalent to quercetin. Polyphenols were also found in the extract (Aswathanarayan and Vittal 2013). Figure 7.2 illustrates the chemical structures of various phytochemicals.

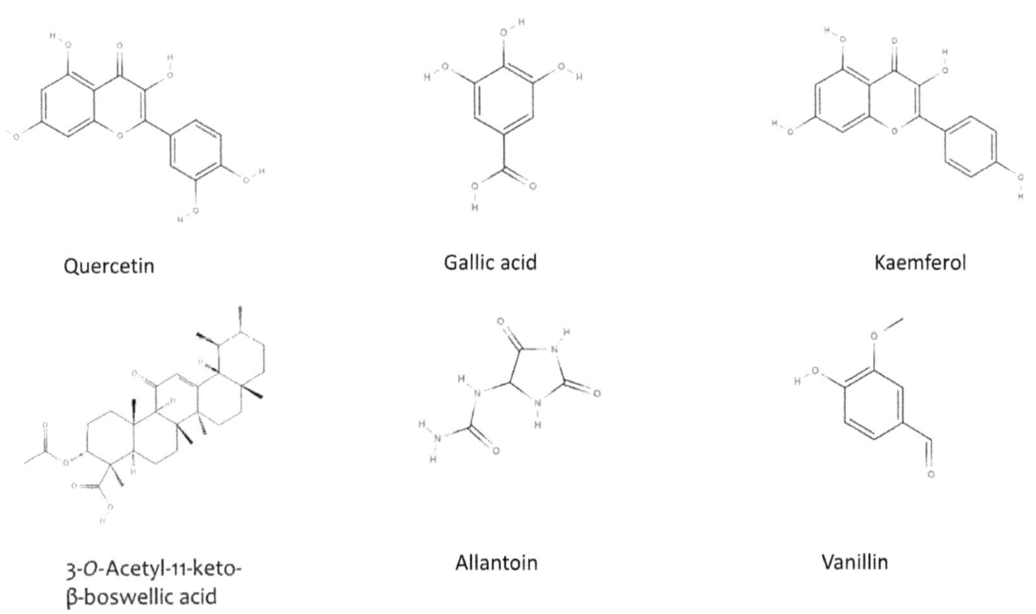

FIGURE 7.2 Structure of some pharmacological important phytochemicals of *R. aquatica*.

7.3 PHARMACOLOGICAL PROPERTIES

See Table 7.1.

7.3.2 Antioxidant Activity

The phosphomolybdenum technique was used to assess methanol extracts for total antioxidant activity and expressed as the equivalent ascorbic acid/gram (AA/g) dry extracts. *R. aquatica* has a total antioxidant capacity of 74.1 mg AA/g dry weight. *R. aquatica* methanolic extracts exhibited effectiveness as free radical scavengers with a half-maximal inhibitory concentration (IC_{50}) = 19.8 μg/ml in the DPPH test results. According to the ABTS free radical scavenging method, *R. aquatica* had a IC_{50} of 171.62 μg/ml for its antioxidant activity. The extract of *R. aquatica* stem showed antioxidant activity of 81.74% by showing the bleaching effect of β-carotene at 500 μg/ml concentration (Aswathanarayan and Vittal 2013). Pallavi et al. (2016) observed that both the root and methanolic leaf extracts had significant radical-scavenging efficacy in the 2,2-diphenyl-1-picryl hydrazyl (DPPH) test, with IC_{50} values at 54.68 and 77.01 μg/ml, respectively. Root and leaf chloroform extracts showed IC_{50} values of 125.31 and 159.74 μg/ml, respectively. The chloroform extracts of the leaves and roots had 50% inhibition at 198.01 and 140.64 μg/ml concentration, respectively. In contrast, the methanolic extracts showed IC_{50} values of 128.27 and 134.51 μg/ml, respectively, in superoxide-anion radical-scavenging efficacy assayed by the PMS/NADH system. Methanol and chloroform extract also showed nitric oxide, hydroxyl-radical (HO) scavenging, and ferrous-ion chelating efficacy, proving it a potential antioxidant agent (Pallavi et al. 2016). In another experiment, the administration of ethanol fraction at 50 and 100 mg/kg body weight (b.w.) doses, antioxidant levels, and rat kidney antioxidant enzyme status (CAT, SOD, GSH, GR, GPx,) were considerably restored. Ethanol fraction administration also reduced serum parameters like nitrate levels, reactive oxygen species (ROS), and malondialdehyde (MDA) (Vysakh et al. 2018). *R. aquatica* methanolic and aqueous stem extracts were assessed for *in vitro* antioxidant activity with superoxide radical scavenging capacity, DPPH, and reducing power assay (RPA). Methanolic and aqueous stem extracts had DPPH and superoxide IC_{50} values of 50.98 ± 0.19, 73.67 ± 0.12, 65.41 ± 0.42, and 71.97 ± 0.31 μg/ml respectively, and their reducing power enhanced as the concentration increased (Priya et al. 2013). Vijayakumari et al. (2013) also evaluated the *in vitro* antioxidant activity of *R. aquatica* by studying the non-enzymatic and enzymatic parameters. There were maximum levels of enzymatic and non-enzymatic antioxidants in the root sample including polyphenol oxidase, glutathione *S*-transferase, glutathione peroxidase, α-tocopherol ascorbic acid, and total polyphenol. However, stem and leaf samples also showed moderate activity (Vijayakumari et al. 2013). Considerable antioxidant activity with an IC_{50} value of 11.07 μg/ml was also observed in a DPPH free radical scavenging assay using an aqueous extract of *R. aquatica* (Mengi and Bakshi 2009). In another experiment, DPPH, ABTS, and superoxide scavenging assays were used to determine the antioxidant capacity of the methanolic extracts of roots. The IC_{50} values were 14.71 ± 0.03, 12.82 ± 0.08, and 36.02 ± 0.05 mg/ml, respectively, for the three tests (Vysakh et al. 2020a).

7.3.2 Anti-urolithiatic Activity

One of the oldest and most prevalent diseases in human history, urolithiasis refers to the presence of a kidney stone at any location in the urinary tract (Prasad et al. 2007). Approximately 80% of kidney stones are CaOx calculi. CaOx monohydrate and CaOx dihydrate are the two forms of CaOx crystals (Bensatal and Ouahrani 2008). Based on ionic changes in urine and serum, the decoctions of *R. aquatica* were tested for antilithic activity in male Wistar rats. Rats were fed feed containing 3% glycolic acid for 45 days to induce nephrolithiasis. Decoction treatment reduced oxalate and calcium ion concentrations in urine, supporting its stone-inhibitory properties; histopathological study of the kidney tissue also proved the same (Christina et al. 2002).

TABLE 7.1
Pharmacological Activities of *R. aquatica*

Biological activity	*In vitro/ in vivo* tests	Test organism	Plant extract (concentration/dose)	Result	References
Antibacterial effect	*In vitro* disc diffusion assay	*Escherichia coli*, *Klebsiella pneumoniae*, *Salmonella typhi*, *Enterobacter aerogenes* and *Shigella flexneri*; *P. aeruginosa* and *Staphylococcus aureus*; and *B. cereus*	Methanol extract of stem at 250, 500, and 750 µg	A dose-dependent increase in antibacterial activity	Aswathanarayan and Vittal 2013; Raj and Joseph 2016; Prashanthi et al. 2012
	In vitro well diffusion assay	*E. coli* BRL-17	Ethanolic extract, aqueous methanol, *n*-hexane, chloroform, ethyl acetate, and *n*-butanol fractions	MIC and MBC are 2.5 and 5 mg/ml, respectively	Vysakh et al. 2020
Anticancer activity	*In vitro* MTT assay	Pancreatic cancer cell line HPAF-II, CAPAN-2, and BxPC3	Aqueous extract 0, 0^{-4}, 10^{-3}, 10^{-2}, 10^{-1}, 10 mg/ml	Prevention of cell proliferation of the pancreatic cancer cell lines in a concentration-dependent manner	Patil et al. 2004
Antidiabetic activity	*In vivo*	Alloxan-induced diabetic male albino rats	400 mg/kg methanolic extract from the leaves	Reduction of serum glucose level, TC, TG, and TP levels, induced glucose-tolerance in OGTT	Priya et al. 2014
	In vivo	Streptozotocin-induced diabetes in rats	100, 200, 400 mg/kg b.w. of the methanol extract of roots	Blood glucose levels decreased significantly	Shyam et al. 2013; Balakrishnaiah and Satyanarayana 2017
Antidiarrhoeal activity	*In vivo*	Charcoal-meal, PGE2-induced diarrhoea, and castor oil in albino Wistar rats	Both 100 and 200 mg/kg alcohol extracts	Delayed defecation time by 186.980 minutes and 97.779 minutes, prevented enteropooling induced by PGE2	Rai et al. 2012
Anti-inflammatory potential	*In vivo*	Paw oedema-induced by carrageenin and granuloma by cotton pellet, inflammation induced by crystal in Wister rat	50, 100, 200 mg/kg b.w. petroleum ether, ethyl, methanolic, aqueous extract	Significant inhibition of paw oedema and inflammation compared to standard drug indomethacin, decrease in the number of neutrophils and monocytes, inhibition of IL-6 levels, reduction of weight of granuloma, antipyretic effect, significant increased body weight, proliferation of epithelium and collagen	Mengi and Bakshi 2009; Kamurthy et al. 2014; Vysakh et al. 2020; Agrawal et al. 2021

(continued)

TABLE 7.1 (Continued)
Pharmacological Activities of *R. aquatica*

Biological activity	*In vitro*/ *in vivo* tests	Test organism	Plant extract (concentration/dose)	Result	References
	In vivo	Tail immersion and hot plate tests, carrageenin-induced paw oedema in Wistar rat	Alcoholic and aqueous extracts of root at 200 mg/kg b.w. dose	Noteworthy analgesic, antipyretic, and anti-inflammatory efficacy with increased retention time	Gupta et al. 2011
	In vivo	Acute pyelonephritis in rat model	Methanolic extract at 100 and 200 mg/kg b.w.	TNF-α, NF-κB, TLR-4, IL-6, and THP levels were restored	Vysakh et al. 2022
Anthelmintic activity	*In vitro*	*P. posthuma*	25, 30, 50, 75 mg/ml root and bark aqueous extract	Caused paralysis and death in a dose-dependent way	Singh-Sunder et al. 2011a; Singh-Sunder et al. 2011b; Lakshmi et al. 2012
Anti-nephrotoxic activity	*In vivo*	Gentamicin-induced nephrotoxicity in Wistar rats	50 and 100 mg/kg b.w. ethanol fraction	Decreased ROS, MDA, and nitrate level, reduced NF-κB, IL-6 mRNA, TNF-α, and KIM-1 expression, thus ameliorated nephrotoxicity	Vysakh et al. 2018
Antioxidant activity	*In vitro* total antioxidant capacity, ABTS, and DPPH test, β-carotene bleaching, lipid-peroxidation, DNA-protection activity	–	500 µg/ml methanolic extract of stem	Effective antioxidant activity, total antioxidant capacity 74.1 mg AA/g dry weight, free radical scavenging IC_{50} in DPPH 19.8 µg/ml in ABTS IC_{50} 171.62µg/ml, β-carotene bleaching of 81.74%	Aswathanarayan and Vittal 2013
	In vitro DPPH test	–	Methanolic, aqueous extracts of roots and leaves	Potential antioxidant agent, nitric oxide and ferrous ion-chelating and hydroxyl radical scavenging activities; free radical scavenging potential with $IC_{50} = 11.07$µg/ml	Mengi and Bakshi 2009; Pallavi et al. 2016

	In vivo	Gentamicin-induced oxidative stress in Wistar rat	50 and 100 mg/kg b.w. ethanol fraction	Antioxidant levels and antioxidant enzymes level in rat kidney (CAT, GSH, GPx, SOD, GR) were significantly restored; reduced serum parameters like ROS, nitrate, and MDA levels	Vysakh et al. 2018
	In vitro DPPH, SOD scavenging capacity, RPA	–	Methanolic and aqueous stem extracts	DPPH and superoxide IC_{50} values of 50.98, 73.67, 65.41, and 71.97 µg/ml, respectively	Priya et al. 2013
	In vitro antioxidant assay	–	Leaf, stem, and root extract	Maximum levels of non-enzymatic and enzymatic antioxidants present in the root sample, leaf samples also showed moderate activity	Vijayakumari et al. 2013
Anti-urolithiatic activity	*In vitro* crystal dissolving assay	–	Aqueous, alcoholic, and hydroalcoholic petroleum-ether, methanol and chloroform extracts of root, stem and leaf 40, 100, 200, 300, 400, 500 µg/ml aqueous extracts, aqueous root extract (0.0%, 0.5%, 1%)	Dissolved monosodium urate monohydrate crystals, significant decreased in growth rate, volume, and depth of growth of struvite crystals in a dose-dependent way, inhibition of CaOx crystallization, up to 100% dissolution of CaOx and CaPO4, inhibited crystal formation and aggregation, reduced crystals number, reduced supersaturation and particle size	Raut et al. 2008; Chauhan et al. 2011; Gilhotra et al. 2013; Sasikala et al. 2013; Prasanthi et al. 2015
	In vivo	Ethylene glycol-induced urolithiasis in albino rats	100, 200 mg/kg b.w. extract orally along with 1% ethylene glycol	Urine volume increased, crystallization tendency reduced, urinary phosphate, calcium, oxalate, protein, creatinine, uric acid reduced, urinary magnesium increased, microcrystal in kidney	Kr and Christina 2011; Sulaiman et al. 2022
	In vivo	Glycolic acid-induced antilithic activity against male Wistar rats	Plant decoctions	Stone-inhibitory properties with reduced calcium and oxalate ion concentration in urine	Christina et al. 2002
	In vivo	Male Wistar albino rats	100 and 200 mg/kg b.w. aqueous root extract	Decreased urine-protein, phosphate, uric acid, calcium, oxalate, and creatinine, phosphate, serum-protein uric acid, calcium, blood urea, creatinine, oxalate, and nitrogen, renal-tissues calcium, oxalate and phosphate, reduced urolithiasis, regenerated kidney tissues	Vijayakumari et al. 2017

Analyses of aqueous, alcoholic, and hydroalcoholic extracts of *R. aquatica* were conducted to determine their anti-crystal efficacy towards basic calcium phosphate (BCP), calcium pyrophosphate (CPPD), and monosodium urate monohydrate (MSUM). An *in vitro* test was conducted with the extracts of *R. aquatica* in 24-well microplates to determine their effects on microcrystals. The aqueous extracts were shown to be effective in dissolving the crystals in MSUM (Raut et al. 2008). The efficacy of *R. aquatica* alcoholic extract was studied in albino rats in preventing ethylene glycol-induced urolithiasis. Administration of 200 mg/kg b.w. extract for 28 days of oral consumption of 1% ethylene glycol increased the urine volume, thus reducing the crystallization tendency. The urinary phosphate, calcium, protein, oxalate, creatinine, and uric acid levels were decreased, though urinary magnesium was enhanced. The histopathological studies confirmed that the extract reduced microcrystal deposition in the section of the kidney of rats with lithiasis (Kr and Christina 2011). The aqueous extract of roots of *R. aquatica* (0.0%, 0.5%, 1%) was also investigated for the inhibitory effect against struvite crystal *in vitro* using a single diffusion gel growth technique. The findings demonstrated that the extract considerably reduced the growth rate, volume, and depth of growth of the crystals in a dose-dependent way (Chauhan et al. 2011). Gilhotra et al. (2013) also experimented on the anti-urolithiatic efficacy of polyherbal formulation, which contains hydroalcoholic extract of *R. aquatica* and *osthumanpinnata* by the *in vitro* method. This study showed the inhibition of CaOx crystallization, which was beneficial in treating renal lithiasis (Gilhotra et al. 2013). Various *R. aquatica* extracts (chloroform, methanol, and petroleum ether extracts of root, stem, and leaf) were investigated for their effect on sodium oxalate solution-induced CaOx crystallization *in vitro*. Using a spectrophotometer, turbidity was measured at 620 nm in the absence or presence of the extract at different concentrations of the extract (100, 200, 300, 400, and 500 μg/ml). The study found that aqueous root extracts have the greatest effect on inhibiting crystal formation and aggregation compared to leaf and stem extracts in petroleum ether, methanol, and chloroform. Plant extracts reduced the number of crystals in the solution, thus reducing super-saturation and particle size (Sasikala et al. 2013). In another experiment, a systematic approach evaluated the anti-urolithiatic potential of roots using simple *in vitro* models. A semi-permeable membrane of the egg was used to screen successive solvent extracts for anti-urolithiatic activity [(formation of CaOx and calcium phosphate ($CaPO_4$)]. The aqueous extract had a high dissolving potential for CaOx and $CaPO_4$, with a value of 29.182% and 65.445%, respectively. There was 100% dissolution of CaOx and $CaPO_4$ at 38 and 20 mg of aqueous extract, respectively (Prasanthi et al. 2015). An aqueous plant root extract was assessed for its anti-urolithiatic efficacy in urolithiatic albino Wistar rats. The treatment with the extract at 100 and 200 mg/kg b.w. doses potentially reduced the calcium, urine protein, phosphate, uric acid, oxalate, and creatinine, as well as serum protein, phosphate, uric acid, creatinine, and calcium. The extract also decreased blood urea, oxalate, nitrogen, and renal tissues oxalate, calcium, and phosphate in urolithiatic rats after treatment for 28 days. This study reported that the 200 mg/kg *R. aquatica*'s aqueous root extract was more efficient at lowering urolithiasis and regenerating kidney tissues in rats. Using simple *in vitro* models, the present study evaluated the anti-urolithiatic potential of roots through a systematic approach (Vijayakumari et al. 2017). A recent investigation also revealed the anti-urolithiatic activity of *R. aquatica* in ethyl glycol-induced rats (Sulaiman et al. 2022).

7.3.3 Antibacterial Activity

Stem extracts in methanol of *R. aquatica* were tested at 1 mg/disc for their antibacterial activity. The extract showed significant antibacterial action against foodborne pathogens *Escherichia coli*, *Klebsiella pneumoniae*, *Salmonella typhi*, *Enterobacter aerogenes*, and *Shigella flexneri* (Gram-negative) with an inhibition zone between 17.46 and 10.10 mm (Aswathanarayan and Vittal 2013). Extracts of *R. aquatica* were also found to be effective against bacteria associated with urinary tract infection, such as *E. coli*, *K. pneumoniae*, *Pseudomonas aeruginosa*, and *Staphylococcus aureus*

(Raj and Joseph 2016). Another experiment exhibited the antibacterial activities of aqueous and different solvent extracts of *R. aquatica* roots against *Bacillus cereus* (Gram positive), *Salmonella abony* (Gram negative), *K. pneumoniae*, and *P. aeruginosa* in a manner dependent on dose at 250, 500, and 750 μg (Prashanthi et al. 2012). *R. aquatica* ethyl acetate fraction was investigated towards clinically distinct uropathogenic *E. coli* for anti-bacteria and anti-biofilm activity by the well diffusion assay. The inhibition zones were very clear against *E. coli*, and the minimum bactericidal concentration (MBC) and minimum inhibitory concentration (MIC) were 5 and 2.5 mg/ml, respectively. Scanning electron microscopy, fluorescence microscopy, and propidium iodide uptake assays also revealed the effectiveness of *R. aquatica* in killing bacteria (Vysakh et al. 2020b).

7.3.4 Anthelmintic Activity

The anthelmintic activity of *R. aquatica* extract was evaluated by evaluating its concentrations (10, 20, 30, 40, and 50 mg/ml) on the adult Indian earthworm *Phertima osthuman*. There was moderately significant anthelmintic activity in the *R. aquatica* methanol extract in a dose-dependent way and it showed spontaneous paralysis and killing of the earthworm, proving it a vermifuge and vermicidal agent (Lakshmi et al. 2012). In another study, the aqueous extract of *R. aquatica* roots was found to be effective against the adult *P. osthuman* in a dose-dependent manner (25, 50, 75 mg/ml) using 15 mg/ml piperazine citrate (standard drug). The aqueous extract induced paralysis in 16.95 minutes and death in 22.33 minutes at a dosage of 75 mg/ml, whereas 50 and 25 mg/ml extracts exhibited paralysis in 21.02 and 28.60 minutes, followed by death in 30.35 and 36.98 minutes, respectively against *P. postuma* (Singh et al. 2011a). Aqueous extracts of *R. aquatica* provide anthelmintic effects in a dose-dependent manner at 25, 50, and 75 mg/ml concentrations, with 75 mg/ml being the most effective dosage for the least time between paralysis and death. Against adult *P. postuma*, 75, 50, and 25 mg/ml *R. aquatica* aqueous extract caused paralysis in 16.99, 21.02, and 28.60 minutes and death in 22.33, 30.35, and 36.98 minutes, respectively (Singh et al. 2011b).

7.3.5 Anti-inflammatory Effect

To study the anti-inflammatory efficacy of *R. aquatica*, 100 and 200 mg/kg oral doses of methanol extract of root were used in carrageenan-induced paw oedema and dextran-induced paw oedema tests in female Wister rats, and the result showed inhibition of paw oedema formation. The histopathological analysis showed that both concentrations of methanol extract administrated to rat groups significantly suppressed inflammation, epithelium proliferation of collagen, and oedema of the epidermis compared to the treatments of dextran and carrageenan (Vysakh et al. 2020a). Researchers examined *R. aquatica* roots for their anti-inflammatory properties in rats suffering from acute and chronic inflammation. Rats' paw oedema caused by carrageenan, granuloma induced by cotton pellet, and inflammation caused by crystals were all tested against the plant's aqueous extract using doses of 50, 100, and 200 mg/kg orally. The extract showed (p >0.05) impediment of rat paw oedema in crystal-induced inflammation, a decrease in the number of neutrophils and monocytes, effective inhibition of interleukin (IL)-6 levels, and also in comparison to the control group there was a decrease in the weight of cotton pellet-induced granuloma (Mengi and Bakshi 2009). Alcoholic and aqueous extracts of the root of *R. aquatica* were employed to assess the analgesic, anti-inflammatory, and antipyretic activity in mice and rats by administration of a 200 mg/kg b.w. dose using the hot plate method, tail immersion test, and carrageenin-induced paw oedema. The results indicated that the alcoholic extract significantly increased reaction time at 3, 4, and 6 hours (p <0.01) compared to the standard drug pentazocine. The tail immersion and hot plate tests revealed the highest level of analgesia and antipyretic activity in alcoholic extract compared to aqueous extract, possibly due to blocking the lipoxygenase pathway (Gupta et al. 2011). Anti-inflammatory potential on acute and subacute inflammation was also investigated using paw oedema induced by carrageenan and cotton pellet granuloma test in rats. Petroleum ether or ethyl extract at 200 mg/kg b.w. dose considerably reduced inflammation ($p > 0.01$) as compared to indomethacin (standard drug), while ethanol extract exhibited more pronounced inhibition of the acute and sub-acute inflammation model ($p > 0.05$) (Kamurthy et al. 2014). The methanol extract of *R. aquatica* was investigated for its efficacy towards acute pyelonephritis-related inflammatory alterations in the rat model. The RT-PCR analysis revealed that mRNA level expressions of nuclear transcription factor kappa B (NF-κB), tumour necrosis factor-alpha (TNF-α), Toll-like receptor 4 (TLR-4), IL-6, and Tamm Horsfall protein (THP) were restored to normal by the treatment of methanolic extract at 100 and 200 mg/kg b.w. doses (Vysakh et al. 2022). In mice and rat models, a recent study also showed the anti-inflammatory, analgesic, and antipyretic effects of methanolic extract and aqueous extract of *R. aquatica* leaves. As determined by the hot plate method, the total methanolic extract at 200 mg/kg b.w. dose showed a significantly increase ($p > 0.01$) in reaction time at 3, 4, and 6 hours compared to pentazocine, the standard drug, but was somewhat less ($p > 0.05$) after 2 hours. Methanolic plant extract's analgesic properties compared to aqueous extract were also revealed in the tail immersion test. Significant ($p < 0.01$) antipyretic activity was also found in carrageenin-induced paw oedema, which might be due to potential lipoxygenase-pathway inhibition (Agrawal et al. 2021).

7.3.6 Anti-nephrotoxic Activity

The ethanol fraction of *R. aquatica* was found to inhibit gentamicin-induced nephrotoxicity, which causes high urea and creatinine levels and tubular necrosis. The fraction with a dose of 50 and of 100 mg/kg b.w. restored the changed antioxidant parameters caused by gentamicin; nitrate level, ROS, and MDA were reduced. Histological studies also exhibited that gentamicin-induced nephrotoxicity was ameliorated by the fraction. NF-κB, IL-6 mRNA, TNF-α, and KIM-1 expression were reduced in *R. aquatica* treated groups (Vysakh et al. 2018).

7.3.7 Antidiabetic Activity

A complete or partial absence of insulin secretion and an altered carbohydrate, fat, and protein metabolism constitutes diabetes mellitus, a chronic metabolic disorder. The antidiabetic properties of methanol extract of roots of *R. aquatica* was evaluated in streptozotocin-induced diabetic rats at 100, 200, and 400 mg/kg b.w. When a 400 mg/kg b.w. dose of methanol extract was given orally, blood glucose levels decreased in a more significant ($p > 0.01$) and highly significant ($p > 0.001$) manner after 4 and 8 hours, respectively, proving *R. aquatica* as a good antidiabetic agent (Shyam et al. 2013). In another experiment, extracts of the aerial portions in methanol of *R. aquatica* were studied for antidiabetic potential towards streptozotocin-induced diabetes. Effects of doses of 100, 200, and 400 mg/kg of plant extracts were compared with that of glibenclamide, the standard drug. Upon administration of 400 mg/kg b.w. of the methanolic extract, blood glucose levels decreased significantly ($p > 0.05$) at 2 and 18 hours, more significantly ($p > 0.01$) at 4 and 8 hours, and highly significantly ($p > 0.001$) at 8 hours (Balakrishnaiah and Satyanarayana 2017). The antidiabetic property of methanolic extract of *R. aquatica* leaves was evaluated in rats, where diabetes was induced by alloxan (120 mg/kg, orally). Diabetic rats were administered with an oral dose of the methanolic extract at 100–400 mg/kg (for 21 days). The blood samples were evaluated for serum glucose and lipid levels such as triglyceride (TG), total cholesterol (TC), and total protein (TP); rats were also studied for the oral glucose tolerance test (OGTT). The methanolic extract administered group (400 mg/kg, orally) demonstrated a significant ($p > 0.01$) decrease in blood glucose levels. Rats with 200 and 400 mg/kg methanol extract administration also demonstrated a substantial decrease in serum TC, TP, and TG levels, while substantial ($p > 0.01$) improvement was observed in OGTT glucose tolerance (Priya et al. 2014).

7.3.8 Antidiarrhoeal Activity

The diarrhoea problem is complicated for young children (under the age of five years), and children hospitalized in the paediatric unit had an approximately 17% mortality rate. A difference between the absorption and the intestinal tract's secretory systems results in diarrhoea and the associated faecal urgency and incontinence due to an imbalance along with hypermobility leading to an excess loss in fluid and electrolytes in faeces. The antidiarrhoeal effect of *R. aquatica* was assessed using the charcoal meal test, PGE2-induced diarrhoea, and castor oil-induced diarrhoea in albino Wistar rats using alcoholic extract in a dose of 100 mg/kg and 200 mg/kg b.w. When castor oil is used to induce diarrhoea, a 200 mg/kg dose delayed the defecation time by 186.980 minutes and a 100 mg/kg dose exhibited a lag in defecation by 97.779 minutes. In different models of antidiarrhoeal activity, 200 mg/kg of *R. aquatica* alcohol extract proved to be more effective than 100 mg/kg. Alcohol extracts of 100 and 200 mg/kg prevented enteropooling induced by PGE2, possibly inhibiting prostaglandin synthesis, which helps inhibit diarrhoea (Rai et al. 2012).

7.3.9 Anticancer Activity

The anticancer activity of *R. aquatica* was investigated in three cell lines for pancreatic cancer BxPC-3, CAPAN-2, and HPAF-II. MTT colorimetric assay found that the aqueous extract effectively inhibited cell proliferation in the cancer cells in a concentration-dependent manner. The root tip assay of *Allium cepa* was used to evaluate *R. aquatica* extract for preliminary antimitotic activity. It was found that, with an increase in aqueous extract concentration, the mitotic index of the root tips decreased significantly (Patil et al. 2004).

7.3.10 Acute Toxicity Test

It was observed that methanolic extracts of roots of *R. aquatica* at 2000 mg/kg b.w. dose produced no mortality, adverse effects, or unusual clinical symptoms during the 14-day study time in female Wistar rats. Comparing the untreated control group with the treated group, animal weight and food consumption remained unchanged. In the acute toxicity study, liver function markers (AST, ALP, ALT) did not differ significantly between the rats that received methanol extract and the control animals ($p > 0.05$). A histopathological examination of the important organs, such as the liver, kidneys, and heart, of animals treated with methanol extract showed no evidence of tissue impairment or toxicity impairment (Vysakh et al. 2020a).

7.4 CONCLUSION

A number of natural therapeutic drugs are becoming increasingly popular in primary medication, particularly in developing countries. It is thought that 80% of the population in remote areas depends on traditional medicine. Medicinal plants contain many safe bioactive compounds that can be used as self-medication without compromising health. *R. aquatica* is one of the important medicinal plants of the Indian Ayurvedic medicinal system belonging to the family Boraginaceae. As a traditional treatment for bladder and kidney stones, this plant has been an essential component in several Ayurvedic remedies from ancient times. The phytochemical investigations of the plant tissue showed the presence of secondary metabolites such as flavonoids, phenolics, alkaloids, reducing sugars, terpenoids, tannins, phytosterols, and proteins which are responsible for the pharmacological properties of the plant such as anti-urolithiatic, antioxidant, anthelmintic, anti-inflammatory, antidiabetic, and anticancer activities. It is still difficult to fully understand the pharmacological effects of phytochemical compounds and their mode of action, even though there is limited research conducted in certain areas. In most pharmacological studies, crude or poorly dissolved extracts are used. Therefore, bioassay-guided isolation of bioactive compounds should be conducted. To ensure the safe use of this plant as a therapeutic agent in modern medicine, more studies need to be conducted on phytochemical screening using different plant parts and isolated compounds for molecular mechanism investigations of different bioactivities.

REFERENCES

Agrawal, P., Mruthunjaya, K., Goyal, K., Ahuja, D. and Gupta, M.K. 2021. Analgesic, anti-inflammatory and antipyretic activity of *Rotula aquatica* Lour leave. *Research Journal of Pharmacy and Technology* 14(10): 5503–5507.

Arya, A.K., Durgapal, M., Bachheti, A., Joshi, K.K., Gonfa, Y.H., Bachheti, R.K. and Husen, A. 2022. Ethnomedicinal use, phytochemistry, and other potential application of aquatic and semiaquatic medicinal plants. *Evidence-Based Complementary and Alternative Medicine* 2022:4931556.

Aswathanarayan, J.B. and Vittal, R.R. 2013. *In vitro* evaluation of antioxidant and antibacterial activities of *Rotula aquatica* and *Ancistrocladus heyneanus*: antioxidant and antimicrobial activity of medicinal plants. *Journal of Pharmacy Research* 6(2): 313–317.

Balakrishnaiah, P. and Satyanarayana, T. 2017. Comparative study of antidiabetic activity among three medicinal plants against streptozotocin-induced diabetes rat model. *Journal of Pharmacognosy and Phytochemistry* 6(1): 278–282.

Bensatal, A. and Ouahrani, M.R. 2008. Inhibition of crystallization of calcium oxalate by the extraction of *Tamarix gallica* L. *Urological Research* 36(6): 283–287.

Christina, A.J.M., Mole, M.P. and Moorthy, P. 2002. Studies on the antilithic effect of *Rotula aquatica* lour in male Wistar rats. *Methods and Findings in Experimental and Clinical Pharmacology* 24(6): 357–360.

Chauhan, C.K., Joshi, M.J. and Vaidya, A.D. 2011. Growth inhibition of struvite crystals by the aqueous root extract of Rotula aquatica. *Indian J. Biochem. Biophys.* 48(3): 202–207.

Gilhotra, U.K., Mohan, G. and Christina, A.J.M. 2013. Antilithiatic activity of poly-herbal formulation tablets by in-vitro method. *Journal of Applied Pharmaceutical Science* 3(5): 43.

Gupta, M.G., Mruthunjaya, K., Saini, L., Garg, S.K., Agrawal, S.G. and Goyal, S. 2011. Analgesic, anti-inflammatory and antipyretic activity of *Rotula aquatica* lour root. *Inventi Rapid: Planta Activa* 2011(3): 037.

Kamurthy, H., Nampally, S. and Dontha, S. 2014. Anti-inflammatory activity of *Rotula aquatica* Lour. In albino rats. *International Journal of Pharmaceutical Sciences and Drug Research* 6(1): 48–51.

Kr, G.U. and Christina, A.J.M. 2011. Effect of *Rotula aquatica* Lour. On ethylene-glycol induced urolithiasis in rats. *International Journal of Drug Development & Research* 3(1): 273–280.

Lakshmi, V.K., Triveni, K.B., Anitha, S. and Shashidhara, S. 2012. *In vitro* anthelmintic activity of *Rotula aquatica* Lour bark. *Pharma Science Monitor* 3(4): 2332–2339.

Mamta, K. and Abhishek, B. 2010. Pharmacognostical evaluation on the root of *Rotula aquatica* Lour. *International Journal of Pharma and Bio Sciences* 1(2): 119.

Martin, K. 2003. Rapid *in vitro* multiplication and *ex vitro* rooting of *Rotula aquatica* Lour., a rare rhoeophytic woody medicinal plant. *Plant Cell Reports* 21(5): 415–420.

Mengi, S.A. and Bakshi, V.J. 2009. Evaluation of the aqueous extract of *Rotula aquatica* roots in acute and chronic inflammatory conditions in rats. *Pharmaceutical Biology* 47(6): 491–495.

Pallavi, M., Ramesh, C.K., Krishna, V., Channakeshava, G.H. and Jamuna, K.S. 2016. Total phenolics and antioxidant potentials of *Rotula aquatica* Lour. *Journal of Applied Pharmaceutical Science* 6: 169–174.

Patil, S., Jolly, C.I. and Narayanan, S. 2003. Free radical scavenging activity of Acacia catechu and *Rotula aquatica*: implications in cancer therapy. *Indian Drugs* 40(6): 328–332.

Patil, S., Narayanan, S., Eibl, G. and Jolly, C.I. 2004. Evaluation of antimitotic activity of *Rotula aquatica* (Lour): a traditional herb used in treatment of cancer. *Indian Journal of Experimental Biology* 42: 893–899.

Prasad, K.V.S.R.G., Sujatha, D. and Bharathi, K. 2007. Herbal drugs in urolithiasis – a review. *Pharmacognosy Reviews* 1(1): 175–178.

Prashanthi, P. and Anitha S, S. 2015. Effect of *Rotula aquatica* Lour. On experimental kidney stones. *International Journal of Pharmacognosy and Phytochemical Research* 7(6): 1142–6.

Prashanthi, P., Anitha, S. and Shashidhara, S. 2012. Studies on the antibacterial activity of the aqueous extract of the roots of *Rotula aquatica* (*R. aquatica*) Lour. *International Journal of Fundamental Applied Science* 4: 87–90.

Priya, B., Gahlot, M., Joshi, P., Zade, S. and Bagmar, U. 2014. Evaluation of antidiabetic activity of methanolic extract from the leaves of *Rotula aquatica* Lour in alloxan-induced diabetic rats. *Pharmacognosy Journal* 6(2): 110–114.

Priya, B., Zade, S.R., Shaikh, A., Gahlot, M. and Joshi, P. 2013. *In-vitro* antioxidant activity and determination of total phenolic, flavonoid contents of stems of *R. aquatica* Lour. *International Journal of Pharmaceutical Sciences and Research* 4(9): 3608.

Rai, A.K., Sharma, P., Barshiliya, Y., Sihare, M. and Negi, A. 2012. Antidiarrhoeal activity of *Rotula aquatica* in rats. *Asian Pacific Journal of Tropical Biomedicine* 2(1): S175–S177.

Raj, K.B. and Joseph, A.V. 2016. Antibacterial potential of phytochemicals against urinary tract infectious bacteria. *International Journal of Innovative Research and Development* 5(4): 179–183.

Raut, A.A., Sunder, S., Sarkar, S., Pandita, N.S. and Vaidya, A.D. 2008. Preliminary study on crystal dissolution activity of *Rotula aquatica, Commiphora wightii* and *Boerhaaviadiffusa* extracts. *Fitoterapia* 79(7–8): 544–547.

Raut, A.A., Sunder, S., Sarkar, S., Pandita, N.S. and Vaidya, A.D. 2008. Preliminary study on crystal dissolution activity of *Rotula aquatica, Commiphora wightii* and *Boerhaavia diffusa* extracts. *Fitoterapia* 79(7–8): 544–547.

Rouhi-Boroujeni, H., Gharipour, M., Mohammadizadeh, F., Ahmadi, S. and Rafieian-Kopaei, M. 2015. Systematic review on safety and drug interaction of herbal therapy in hyperlipidemia: a guide for internist. *Acta Bio-medica: Atenei Parmensis* 86(2): 130–136.

Sasikala, V., Radha, S.R. and Vijayakumari, B. 2013. *In vitro* evaluation of *Rotula aquatica* Lour. For antiurolithiatic activity. *Journal of Pharmacy Research* 6(3): 378–382.

Shyam, T., Ganapaty, S. and Balakrishnaiah, P. 2013. Antidiabetic activity of *Rotula aquatica* lour roots in streptozotocin-induced diabetic rats. *American Journal of Phytomedicine and Clinical Therapeutics* 1(7): 530–535.

Singh, S., Rai, A.K., Sharma, P. and Barshiliya, Y. 2011a. Comparative study of Anthelmintic activity between aqueous extract of *Areva Lanata* and *Rotula aquatica* Lour. *Asian Journal of Pharmacy and Life Science* 1(3): 211–215.

Singh, S., Rai, A.K., Sharma, P. and Barshiliya, Y. 2011b. Evaluation of aqueous extract of *Rotula aquatica* Lour for *in vitro* anthelmintic activity. *Journal of Advanced Pharmacy Education & Research* 2: 184–188.

Siverajan, V.V. and Balachadran, I. 1994. *Ayurvedic Drugs and Their Plant Sources*, India, New Delhi: Oxford and IBH Publishing Co. Pvt. Ltd., pp. 358–359.

Sulaiman C.T., Jyothi, C., Jinu Krishnan Unnithan G., Prabhukumar, K.M., and Balachandran, I. 2022. Phytochemical characterization and evaluation of antiurolithiatic activity of selected source plants of *Pashanabheda. Clinical Phytoscience* 8(1): 1–9.

Vijayakumari, B., Sasikala, V. and Radha, S.R. 2013. Preliminary phytochemical screening of the various extracts of *Rotula aquatica* Lour. *World Journal of Pharmacy and Pharmaceutical Science* 2: 6371–6380.

Vijayakumari, B., Sasikala, V., Radha, S.R. and Rameshwar, H.Y. 2016a. *In silico* analysis of aqueous root extract of *Rotula aquatica* Lour for docking analysis of the compound 3-*O*-acetyl-11-keto-β-boswellic acid contents. *Springer Plus* 5(1): 1–8.

Vijayakumari, B., Sasikala, V., Radha, S.R. and Rameshwar, H.Y. 2016b. HPTLC fingerprinting of aqueous root extract of *Rotula aquatica* for phenol, tannin and saponin contents. *Der Pharmacia Lettre* 8(9): 299–306.

Vijayakumari, B., Sasikala, V., Radha, S.R. and Rameshwar, H.Y. 2017. *Rotula aquatica* Lour aqueous extract as anti-urolithiatic agent in experimentally induced urolithiatic rat model. *International Journal of Pharmacognosy and Phytochemical Research* 9: 553–558.

Vijayakumari, B., Sumithra, S., Sasikala, V. and Radha, S.R. 2013. Evaluation of enzymatic and non-enzymatic antioxidant potential of *Rotula aquatica* Lour. *Journal of Pharmacy Research* 6(5): 442.

Vysakh, A., Abhilash, S., Kuriakose, J., Midhun, S.J., Jyothis, M. and Latha, M.S. 2018. Protective effect of *Rotula aquatica* Lour against gentamicin-induced oxidative stress and nephrotoxicity in Wistar rats. *Biomedicine & Pharmacotherapy* 106: 1188–1194.

Vysakh, A., Jayesh, K., Helen, L.R., Jyothis, M. and Latha, M.S. 2020a. Acute oral toxicity and anti-inflammatory evaluation of methanolic extract of *Rotula aquatica* roots in Wistar rats. *Journal of Ayurveda and Integrative Medicine* 11(1): 45–52.

Vysakh, A., Jayesh, K., Jisha, N., Vijeesh, V., Midhun, S.J., Jyothis, M. and Latha, M.S. 2022. *Rotula aquatica* Lour. mitigates oxidative stress and inflammation in acute pyelonephritic rats. *Archives of Physiology and Biochemistry* 128(1): 92–100.

Vysakh, A., Midhun, S.J., Jisha, N., Jayesh, K., Vijeesh, V., Jyothis, M. and Latha, M.S. 2020b. *Rotula aquatica* Lour. inhibits growth and biofilm formation of clinically isolated uropathogenic *Escherichia coli*. *Asian Pacific Journal of Tropical Biomedicine* 10(12): 547.

8 An Overview of Traditional Uses and Pharmacological Profile of *Sphaeranthus indicus*

Prakash Chandra Gupta and Nisha Sharma*
School of Pharmaceutical Sciences, Chhatrapati Shahu Ji Maharaj University, Kanpur, Uttar Pradesh, India
* Corresponding author (Prakash Chandra Gupta): herbalprakash@yahoo.com

8.1 INTRODUCTION

Humans have utilized herbal remedies since the dawn of civilization. Ancient Indians were well versed in the healing benefits of various plants as per Ayurveda, India's oldest traditional medical system. The flora of India is incredibly diversified due to the broad range of temperatures and geographical conditions in the country. Since the beginning of time, science has thoroughly investigated several ancient systems' rudimentary remedies.

In Indian traditional medicine, the medicinal plant *Sphaeranthus indicus* L. is frequently used to cure several ailments (Paranjape 2001, pp.148–149). It can be found in rice fields, desert wastelands, and cultivated plains throughout tropical India. It grows 1,200 feet above sea level in Australia, Africa, Sri Lanka, and India (Kapoor 1980). *Sphaeranthus hirtus* Willd. and *Sphaeranthus mollis* Roxb. are synonyms of *S. indicus* L.

Sphaeranthus L. is a genus of plants in the Asteraceae family. Most of the members of this genus are fragrant herbs that grow as weeds in temperate climates. *S. indicus* is the only plant in this genus that has been explored scientifically in terms of biological activity and phytoconstituents.

S. indicus L. (Asteraceae), locally named as Gorakhmundi, is an annual, aromatic, densely branched medicinal plant growing to about 30–60 cm in height. *S. indicus* is a popular ethnic medicinal plant in Ayurvedic and Unani treatments. Its use in Indian indigenous folk therapy dates back a long time. According to Unani specialists, it is a beneficial plant since it is used to heal a number of diseases. It is also known as Aab Hayat (Mohiuddin 2004, p.309; Nadkarni 2007, p.1163).

In ancient literature, it was thought to be one of the most potent plants in the world for enhancing intelligence. Its entire plant and blooms have long been regarded as having medicinal properties and utilized in Unani medicine to treat various illnesses. The plant is used both internally and externally. In Unani and other traditional medicinal systems, it is used as a circulatory cleanser, stimulant, eyesight, and vital organ tonic. As a result, traditional medicine uses it to treat integumentary, cardiorespiratory, gastrointestinal, urogenital, and locomotor disorders (Ghani 2004, pp.1263–1265; Mohiuddin 2004, p.309). This chapter aims to emphasize the traditional uses and therapeutic applications as well as the ethnomedicinal properties of *S. indicus* and to indicate its potential for managing various illnesses as mentioned in various ancient pieces of literature. It also surveys its bioactive components and pharmacological and clinical findings.

DOI: 10.1201/9781003256830-8

8.2 OVERVIEW OF TRADITIONAL USES OF *SPHAERANTHUS INDICUS*

People have used the *S. indicus* plant for thousands of years to cure many illnesses. The leaves of *S. indicus* are effective for treating jaundice and urethral and persistent skin issues (Saidapur 1978). The plant's extract is used as a cathartic and styptic. It is said to help with liver and stomach problems (Chadha 1976, pp.4–5).

The roots and seeds of the plant have anthelmintic and stomachic effects (Said 1956, pp.239–241).

The plant can also be used to treat skin problems by 'purifying' the blood (Kirtikar and Basu 1918). *S. indicus* leaves were used to treat smallpox (Latha et al. 2007). The bark part of this plant is helpful in treating piles. The root and leaf parts of *S. indicus* are used against cough (Kirtikar and Basu 1935, p. 1347). *S. indicus* is traditionally used to cure mouth ulcers. This plant extract is effective against leucorrhoea, elephantiasis, diarrhoea, and dysentery (Behera et al. 2006). Additionally, it has been proven that the plant extract possesses antibacterial properties and inhibits the activity of the hyaluronidase enzyme (Lichtenbelt et al. 1998).

S. indicus plant parts showed antimicrobial activities (Sasidharan et al. 2008). This plant contains erucic acid and polyunsaturated fatty acids, which work as antitumour and anti-eczema agents (Hamid et al. 2013). The plant's phytochemical examination indicated the existence of alkaloids, terpenoids, saponins, resins, gums, sugars, mucilage, and flavonoids (Kokate 1986, p.111; Garg and Dwivedi 2021).

The entire plant is used to improve memory. Its Arq or Sharbat (distillate or syrup) treats melancholia and epilepsy when combined with Kagzi lemon. In particular, Unani physicians characterized it for treating eye illnesses, and it was used to cure conjunctivitis and eye pain (Mohiuddin 2004, p.309). Its crushed floral juice can also help with discomfort, cataracts, and visual impairment. It is cooked as a vegetable and utilized in Balgham (phlegmatic) diseases and halitosis. It helps with cough and chronic tuberculosis; palpitations; flatulence; bleeding; haemorrhoids; correction of liver and intestine functions; burning micturition; urethral ulcer and gonorrhoea; and is useful in thyroid glandular enlargement. The plant is effective against urticaria, itching, eczema, and leprosy and it is also helpful for weight gain in underweight people (Kirtikar et al. 1981, pp.1346–1348; Nadkarni 2007, p.1163).

S. indicus is an excellent Unani herb for blood purification and immunomodulation, with a wide range of ethnobotanical applications and pharmacological effects (Sultana et al. 2021). The plant is used for tuberculous glands, anaemia, splenic disorders, elephantiasis, pain in the uterus and vagina, bronchitis, asthma, indigestion, biliousness, vomiting, dysentery, pain in the rectum, haemorrhoids, insanity, hemicrania, convulsion, urinary discharges, laxity of the breasts, leucoderma, etc. (Kirtikar and Basu 2003, p.1863; Ramachandran 2013). The plant is mentioned in Ayurveda as an alexipharmic, anthelmintic, fattening, digestible tonic and laxative.

8.2.1 MORPHOLOGY

S. indicus L. is an upright annual herb with branches and taproots that is heavily branched and aromatic (Ambavade et al. 2006). The roots are bulbous, 10–15 cm long, 0.1–0.4 cm in diameter, brown on the outside and light brown inside, with longitudinal stripes and transverse scars visible at regular intervals, and they have a characteristic odour. The stalk is greenish in colour (Makhija et al. 2011). Sessile, decurrent, obovate, serrate, sticky, granularly hairy, thin at the base, and greenish-brown in colour are the characteristics of the leaves. A violet inflorescence with short, thin, pointed bracts surrounds the single, globular, clustered flowers at the terminal ends. The outside of the flowers in each cap is female, few or numerous, and fertile, while the interior blooms are hermaphroditic, meristemetic, or sterile, with slender appendages and pale, hairy bracts. Hermaphroditic flower corollas are purple with a white tinge, funnel-shaped, with four to five teeth, serrated base anthers, pointed leaves or tails, stylized, filamentous, and sometimes spherical. The fruit is oval and contains compressed pods with no stem. The fragrance of the herb is modest; however, it fades after a lengthy period of storage (Galani et al. 2010).

8.2.2 Chemical Components

The whole plant, flower, and leaf extracts contain various proven chemical compounds. Preliminary phytochemical analysis of petroleum ether extract of roots and rhizomes of *S. indicus* showed the presence of steroids, fats, and oils. In contrast, the methanol extract contained carbohydrates, amino acids, tannins, proteins, phenols, steroids, fats, and oils, and the aqueous extract contained saponins and alkaloids (Ambavade et al. 2006; Shirwaikar et al. 2006). Essential oils obtained by steam distillation of the whole plant include ocimene, α-citral, α-terpinene, β-ionone, methylchavicol, geraniol, α- and β-ionone, stigmasterol, and sesquiterpene lactones (Basu and Lamsal 1946; Ambavade et al. 2006). Hentriacontan, sesquiterpene lactone, sesquiterpene glycoside sphaerantholide, flavone and isoflavone glycosides, stigmasterol, β-sitosterol, and flavone and isoflavone are all found in the alcoholic extract of Caputula powder (Yadav and Kumar 1998). Along with β-eudesmol and silicic acid, a sesquiterpene lactone named 7-α-hydroxyeudesm-4-en-6,12-olide and a brand-new sesquiterpene acid named 2-hydroxycostic acid were found in the acetone extract of the plant *S. indicus* L (Kirtikar and Basu 2003, p.1863). Methylchavicol, α-terpinene, geraniol, geranyl acetate, decadiene, *p*-methoxycinnmaldehyde, *n*-triacontanol, phenyl urethane, and *n*-pentacosane have been extracted from the oils and sphaerene, indicusene, and sphaeranthol as essential oils (Singh et al. 2019). Decadiene, α-ionone, α-terpinene, β-ionoe, *p*-methoxy cinnamaldehyde, geraniol, sphaerene, indicusene, sphaeranthol as essential oils and *n*-pentacosane, *n*-triacontanol, phenyl urethane were isolated from the oil (Singh et al. 2019). According to recent reports, *S. indicus* flower extract includes numerous beneficial components from natural sources. Important secondary metabolites from the glycoside, flavonoid, and phenolic groups have been identified by using high performance liquid chromatography–photodiode array (HPLC-PDA) and ultra-high performance liquid chromatography–mass spectrometry (UHPLC-MS) techniques (Ahmad et al. 2021).

8.3 PHARMACOLOGICAL ACTIVITY

8.3.1 Anxiolytic Properties

The elevated plus-maze, open field test, and foot-shock-induced aggressiveness test were used to assess the anxiolytic effects of petroleum ether, alcohol, and water extracts (10, 30, and 100 mg/kg, orally) from *S. indicus* flowers. In mice, the alcoholic and water extracts of *S. indicus* flowers had a significant anxiolytic effect. The open arm of the elevated plus-maze showed an increase in time spent, percentage entries, and total entries. The open field test demonstrated increased ambulation, activity in the centre, and overall locomotion, but the foot-shock-induced aggression test revealed fewer fighting bouts, indicating anxiolytic activity (Ambavade et al. 2006). In another study, anxiolytic effects were observed in a 100 mg/kg hydroalcoholic extract of the entire herb of *S. indicus* (Galani and Patel 2010).

8.3.2 Neuroleptic Activity

The neuroleptic properties of petroleum ether, alcohol, and aqueous extracts of *S. indicus* flowers were examined using a mouse model of apomorphine-induced cage climbing and catalepsy. The extracts were administered intraperitoneally at 30, 100, and 300 mg/kg doses. Only the petroleum ether extract (300 mg/kg) reduced the time it took to ascend apomorphine-induced cages. Aqueous (300 mg/kg) and alcoholic (300 mg/kg) extracts, but not petroleum ether extract, caused catalepsy (Mhetre et al. 2006). Catalepsy was generated by a hydroalcoholic extract of *S. indicus* whole herb (100, 200, and 500 mg/kg, orally) that potentiated haloperidol-induced catalepsy while antagonizing apomorphine-induced stereotypy (Galani and Patel 2009). A hydroalcoholic extract of the entire *S. indicus* plant has also been demonstrated to have neuroleptic properties.

8.3.3 Effects of Sedation

It was discovered that administering *S. indicus* entire plant hydroalcohol extract to rodents at oral doses of 100, 200, and 500 mg/kg increased pentobarbital-induced sleep and decreased locomotor activity and exploratory action (Galani and Patel 2009).

8.3.4 Immunomodulatory Properties

The immunomodulatory potential of *S. indicus* was examined in mice employing antibody titre, delayed-type hypersensitivity response, phagocytic activity, and cyclophosphamide-induced myelosuppression as evaluation criteria. Extract of methanol and its various fractions (100 and 200 mg/kg, orally) were found to have immunostimulant effects. Methanol extract, petroleum ether, chloroform, and the remainder methanol fraction of *S. indicus* L. flower heads were found to be effective in enhancing phagocytic activity, haemagglutination antibody titre, as well as delayed-type hypersensitivity in a mouse model with cyclophosphamide-induced myelosuppression. However, only the remaining methanol fraction was found to be active in normalizing total white blood cell levels in these mice (Bafna and Mishra 2004). Immunostimulant effects have been discovered in the sesquiterpene eudesmanolide from *S. indicus* (Shekhani et al. 1990).

8.3.5 Antioxidant Properties

An ethanolic extract of *S. indicus* at a concentration of 1000 g/m was used in an experiment conducted *in vitro*. This extract was able to neutralize the free radicals 2,2-azino-bis(3-ethylbenzothiazoline-6-sulphonic acid) (ABTS), 1,1-diphenyl-2-picrylhydrazyl (DPPH), superoxide, and nitric oxide to the greatest extent possible. The extract has a similar amount of iron chelation scavenging activity (Shirwaikar et al. 2006). An *in vivo* investigation found that a methanolic extract of *S. indicus* raised superoxide dismutase, catalase, and glutathione peroxide levels in rats while decreasing malondialdehyde levels (Tiwari and Khosa 2009). At 150 g/ml concentrations, the methanol extract of leaves and flowers demonstrated the best scavenging action (Krishna et al. 2013). According to Vijayalakshmi and Rao (2019), *S. indicus* has a strong antioxidant capacity, which may play a significant role in its therapeutic properties.

8.3.6 Anti-inflammatory Properties

Tumour necrosis factor (TNF)-α and interleukin (IL)-8 generation by *Propionibacterium acnes*, which are major mediators of proinflammatory cytokine-stimulated monocytes, were significantly inhibited by an aqueous extract of *S. indicus* root (Jain and Basal 2003).

The herb reduced the ability of *P. acnes* to create proinflammatory cytokines and reactive oxygen species. Polymorphonuclear leukocytes and monocytes were treated with *P. acnes* supernatant with or without *S. indicus* (5 and 50 g/ml). This resulted in a less severe reduction of reactive oxygen species, which was nonetheless significant. It was discovered in another investigation to have anti-inflammatory effects (Heinrich et al. 1998; Chakrabarti et al. 2012).

8.3.7 Analgesic and Antipyretic Activity

Analgesic and antipyretic activity was tested using Eddy's hot plate, tail immersion, and brewer's yeast-induced pyrexia protocols (using doses of 200 and 400 mg/kg, orally). The whole *S. indicus* plant was extracted with petroleum ether, benzene, chloroform, ethanol, and triple-distilled water. After 1 hour, the extracts of petroleum ether, chloroform, and ethanol worked as well as the standard painkiller diclofenac sodium at both doses. Compared to the standard drug paracetamol, chloroform and ethanol extracts showed antipyretic activity after 1 hour, while aqueous extracts showed this activity after 2 hours (Nanda et al. 2009). Another study found that ethanolic extracts of leaves and flowers at various dosages had a strong analgesic effect in several pain models (Ali et al. 2011).

8.3.8 Mast Cell Stabilizing Action

In sheep serum allergy tests and compound 48/80 hypersensitivity tests, ethanol and ethyl acetate extracts of *S. indicus* prevented mast cell degranulation better than ketotifen (Mathew et al. 2009).

8.3.9 Antihyperglycaemic Activity

The antihyperglycaemic action of *S. indicus* alcoholic extract was studied in rats with diabetes caused by nicotinamide (120 mg/kg, intraperitoneally) and streptozotocin (60 mg/kg, intraperitoneally). A hypoglycaemic effect was claimed for a 50% ethanolic extract of the plant (Dhar et al. 1968). Analyses were performed on diabetic and non-diabetic rats to determine their fasting plasma glucose levels, serum insulin levels, serum lipid patterns, magnesium thresholds, glycosylated haemoglobin, body weight fluctuations, and liver glycogen thresholds. Fasting normal rats fed *S. indicus* ethanol extract had better glucose tolerance. Oral *S. indicus* for 15 days reduced blood glucose and increased hepatic glycogen and plasma insulin (Prabhu et al. 2008).

8.3.10 Hepatoprotective Activity

When the methanolic preparations of *S. indicus* (300 mg/kg, orally) were administered, there was a major drop in liver function biomarkers like serum glutamate oxaloacetate transaminase (SGOT), serum glutamate pyruvate transaminase (SGPT), acid phosphatase (ACP), alkaline phosphatase (ALP), bilirubin, as well as total protein. It was also confirmed by analysing the histology of the livers of treated animals (Tiwari and Khosa 2009).

The hepatotoxic effects of acetaminophen on rats were compared with those of methanolic and water-based extracts of *S. indicus* flower heads. Additionally, following acetaminophen-induced hepatotoxicity, the methanolic extract of *S. indicus* improved the functions of antioxidant enzymes, e.g. superoxide dismutase, catalase, and glutathione peroxidase lowered the proportion of lipid peroxides in these animals (Nayak et al. 2007).

8.3.11 Skin Disease

In Guinea pig excised wound models, an ethanol extract of *S. indicus* showed 88% wound healing activity. A decrease of roughly 88% in the wound area was documented after a 15-day study. The cream, equivalent to neomycin, boosted wound rate closure and the duration of epithelialization (Sadaf et al. 2006). Numerous ointments comprising the ethanolic extract of *S. indicus* heads of flower in differing concentrations were evaluated for the healing of wounds in albino rats. Substantial epithelization of the wound, area of contraction, and wound closure activity in the test medication, was equivalent to the standard framycetin sulphate cream (Chopda et al. 2010). Six traumatized mice received extract-containing treatment once per day for 15 days *in vivo*. The formulation containing 2% (w/w) alcoholic extract of *S. indicus* flower head was better than the control and reference formulations in a study of wound healing activities. Compared to the control and standard formulations, the hydroxyproline content was higher in healed wounds (Jha et al. 2009).

8.3.12 Antimicrobial Activity

A crude ethanol extract of *S. indicus* flowers was found to have antibacterial action against both Gram-negative and Gram-positive bacteria (Shaikh et al. 1986).

Staphylococcus albus, *Staphylococcus aureus*, *Escherichia coli*, *Fusarium* spp., *Helminthosporium* spp., and several other microorganisms were susceptible to the bicyclic sesquiterpene obtained from the apical region of *S. indicus* (Singh et al. 1988). The plant's alcoholic and aqueous extracts were found to be highly efficient in inhibiting *Fusarium oxysporum*, *Alternoriasolani*, and *Penicillium pinophitum* from growing (Dubey et al. 2000). Antibacterial activity of the volatile oil extracted from the leaves of *S. indicus* was shown against *Salmonella paratyphi* A, B, C, and *Schigella flexneri*, *Salmonella enteritidis*, *Salmonella typhimurium*, *Shigella sonnei*, and *Vibrio cholera* (Garg and Kasera 1983). *S. indicus* fruits were revealed to have substantial antibacterial properties towards Gram-positive and Gram-negative microorganisms (Naqvi et al. 1998). Tandon et al. compared the antibacterial and antioxidant activities of the entire *S. indicus* plant and its different parts. They found that inflorescence extract could be a source of phytochemicals that have antibacterial and antioxidant properties (Tandon and Gupta 2020: Arya et al. 2022).

8.3.13 Antifungal Activity

Essential oil extracted from leaves possesses a high level of antifungal activity that has been demonstrated against *Rhizopus nodosus*, *Aspergillus niger*, *Curvularia prasadie*, *Trichoderma viride*, and *Trichophyton rubrum* (Garg and Kasara 1982). The antibacterial and antifungal activities of petroleum ether, acetone, methanol (90%), and water extracts of flowers were studied in bacterial and fungal colonies utilizing the diffusion method. Antibacterial and antifungal activities were present in all extracts (Lalla et al. 2005). In another study, the antibacterial and antifungal properties of *n*-hexane, benzene, chloroform, ethyl acetate, and acetone extracts from the flowers and the apical portions of the *S. indicus* plant were investigated using an *in vitro* disc diffusion technique at doses of 5, 2.5, and 1.25 mg/disc. *S. aureus* and *Candida albicans* were both killed by *n*-hexane floral extract (Duraipandiyan et al. 2009).

8.3.14 Antiviral Activity

S. indicus methanol extract suppressed mouse corona and herpes simplex viruses at 0.4 g/ml (Vimalanathan et al. 2009). Antiviral activity was also shown in the plant extract against the vaccinia and Ranikhet viruses (Dhar et al. 1968).

8.3.15 Action Against Nematodes

The dried fruits' methanol extract was shown to have nematocidal activities (Nisha et al. 2007). At concentrations of 100–500 ppm, it proved hazardous to the *Calex quinquefasciatar* mosquito's second and fourth instar larvae. The larvae in the fourth instar were much more vulnerable than the larvae in the second instar (Tiwari and Saxena 2003).

It was discovered that the root extract's larvicidal potency was markedly more significant than the leaf extract (Hameed and Shah 2003). Acetone extracts of roots and leaves were shown to kill more than half of a common Indian mosquito species that transmits the filarial worm (at doses of 750 and 1000 ppm). As a larvicidal agent, the purified fraction of *S. indicus* acetone extract is effective against mosquitos. *Tribolium castaneum* was a repellent and feeding deterrent to a methanol extract of *S. indicus* at 1% concentration. At a 5 ml dose, complete feeding deterrent efficacy was established, whereas repellent activity was found at a 4 ml dose (Dhar et al. 1968).

A worm motility test was used to evaluate the *in vitro* macrofilaricidal ability of a methanol extract of *S. indicus* (1–10 mg/ml) against the adult stage of the wild filarial worm *Setaria digitata*.

It had macrofilaricidal efficacy with a concentration of less than 4 mg/ml and an incubation period of 100 minutes (Hameed and Shah 2003).

8.3.16 Effect of Bronchodilation

The methanol extract and fractions in solvents in increasing order of polarity, petroleum ether, benzene, chloroform, and ethyl acetate, dramatically reduced histamine-induced bronchospasm in guinea pigs (Sharma and Saxena 1996). In guinea pigs, an investigation was conducted to determine the bronchodilatory prospects of a methanolic extract of the entire plant of *S. indicus* L. and its various fractions (87 and 174 mg/kg, orally) for histamine-induced acute bronchospasm.

8.3.17 Activity Against Hyperlipidaemia

Following treatment with an extract from *S. indicus*, there was a sizeable increase in the number of high-density lipoprotein molecules (Ali et al. 1991). In rats with hyperlipidaemia caused by an atherogenic diet, an alcoholic extract of the floral heads of *S. indicus* L. was investigated for its potential to lower blood cholesterol levels. Body weight, total cholesterol, triglycerides, low-density lipoprotein, and very low-density lipoprotein all decreased significantly when *S. indicus* extract (500 mg/kg/day, orally, for eight days) was administered.

8.3.18 Renoprotective Properties

The ethanolic extracts of *S. indicus* were given to rats with acute kidney damage caused by gentamicin, and then it was assessed for nephroprotective screening. Because of the renal impairment caused by gentamicin, there was an increase in biochemical markers such as blood urea and serum creatinine and a decrease in total protein and serum albumin. Acute tubular necrosis was discovered on histopathology. The ethanol extract of *S. indicus*, given at a concentration of 300 mg/kg, restored kidney function to near-normal levels and normalized the biochemical indicators stated above, as shown microscopically (Sarpate et al. 2020).

8.3.19 Unrelated Activities

Entamoeba histolytica is also resistant to the plant's anticancer and antiprotozoal capabilities (Dhar et al. 1968). Hyaluronidase is inhibited by the extract of *S. indicus* (Pande and Dubey 2009). Alcoholic *S. indicus* floral extract has been demonstrated to have hypotensive, peripheral vasodilatory, and cathartic effects (Srinivasan et al. 2008).

8.4 DISCUSSION

As per the Unani philosophy, removing toxins and harmful metabolites from the blood is known as blood purification, resulting in its optimum level, and strengthening the intrinsic heat. The physical and digestive activities improve significantly. As a result, aberrant humours will not be formed, and poison will not be produced. Both impurities and contamination will be avoided (Alam et al. 2020).

S. indicus (known as Mundi in Unani medicine) has the property of Musaff-i-Dam (blood cleanser), as well as additional properties according to a literature review of Unani and ethnomedicinal plant writings (tonic for vital organs) (Kirtikar and Basu 2003, p.1863; Ghani 2004, pp.1263–1265; Mohiuddin 2004, p.309; Kabir al-Din 2007, pp.581–583).

Blood purifier medications work by causing required changes in the blood and removing waste materials, thereby cleaning it of contaminants. Their moderate warm, cold, dry, and wet qualities create necessary alterations and maintain blood composition. Therefore, blood purifiers serve as a counterbalance, preserving blood quantity, quality, and homeostasis. Additionally, they normalize the blood composition by balancing the heat of the disproportionate blood due to their alternative, moderately warm, and dry temperament. Further, blood cleansing medications strengthen the body's defences, boost immunity, shield against contaminants, and repair sluggish liver and kidney malfunctioning to normal operating conditions (Mushir et al. 2019).

Flavonoids, found in Mundi, are a type of phytoconstituent that offer cellular protection due to their potent antioxidant qualities. These properties either stop or repair the damage caused by free radicals to red cells. Glycosides and saponins have been found in studies to boost the ability of bone marrow cells to proliferate. Due to the presence of these phytoconstituents, blood purification medications may demonstrate the haematinic potential of Mundi.

Antioxidant properties are seen in medications that contain flavonoids, saponins, and alkaloids (Mushir et al. 2019). Triterpene saponins stimulate lymphocyte proliferation to produce interferon, haematopoietic growth factors, and cytokines. Alkaloids aid in the repair of haematopoiesis.

Mundi has been shown to have neuroprotective, anticonvulsant, sedative, anxiolytic, antitussive, bronchodilator, anti-asthmatic, anti-inflammatory, and antitussive properties in recent experimental research. Experts in Unani medicine believe that the Mundi can treat various conditions, including memory loss, epilepsy, melancholy, cough, chronic tuberculosis, worm infestation, flatulence, skin problems, liver disorders, renal disorders, and intestinal disorders. Additionally, it helps with genital infections in both men and women (leucorrhoea, gonorrhoea, uterine inflammation, infertility).

According to existing research, Mundi shows antioxidant, antihelminthic, nephroprotective, hepatoprotective, antiulcer, wound healing, and antimicrobial activities; it also has antihypertensive, immunomodulatory, antioxidant, antihyperlipidaemic, antioxidant, antipyretic, and antibacterial properties, making it effective in diabetes, hyperlipidaemia, infections, and other conditions. Its effectiveness in several ocular illnesses is uncertain due to a paucity of scientific research; consequently, experimental and clinical trials are needed.

8.5 CONCLUSION

S. indicus is a plant that the Unani system of medicine has discovered as a blood cleanser and immunomodulator with a wide range of qualities that can help with various diseases. The above-mentioned pharmacological and clinical research studies substantiate *S. indicus*'s potential advantages in numerous illnesses indicated in traditional Unani literature. It is advised that prospective randomized clinical trials be conducted to demonstrate both its effectiveness and safety. Furthermore, rather than employing plant materials in their natural state, they can be made into a beautiful completed product with established data, making it simple for patients and pharmacists to distribute.

REFERENCES

Ahmad, H.I., M.F. Nadeem, H.M. Shoaib Khan, M. Sarfraz, H. Saleem, U. Khurshid, M. Locatelli, M. Ashraf, N. Akhtar & A. Alghamdi. 2021. Phytopharmacological evaluation of different solvent extract/fractions from *Sphaeranthus indicus* L. flowers: from traditional therapies to bioactive compounds. *Frontiers in Pharmacology* 12:708618.

Alam, A., J.I. Siddiqui, M.H. Kazmi, I. Ahmad & M.S. Moin. 2020. Tasfiyaal-dam (blood purification) in Unani perspective: a comprehensive review. *International Journal of Herbal Medicine* 8(2):100–105.

Ali, M.A., M. Mikage, F. Kiuchi, Y. Tsuda & K. Kondo. 1991. Screening of crude drugs used in Bangladesh for nematocidal activity on the larva of Toxocaracanis. *Journal of the Pharmaceutical Society of Japan* 45(3):206–214.

Ali, A., S.B.S. Naqvi, S. Gauhar & R. Saeed. 2011. Anti-inflammatory and analgesic activities of ethanolic extract of Sphaeranthus indicus Linn. *Pakistani Journal of Pharmaceutical Sciences* 24(3):405–410.

Ambavade, S.D., N.A. Mhetre, V.D. Tate & S.L. Bodhankar. 2006. Pharmacological evaluation of the extracts of *Sphaeranthus indicus* flowers on anxiolytic activity in mice. *Indian Journal of Pharmacology* 38(4):254–259.

Arya, A.K., M. Durgapal, A. Bachheti, K.K. Joshi, Y.H. Gonfa, R.K. Bachheti & A. Husen 2022. Ethnomedicinal use, phytochemistry, and other potential application of aquatic and semiaquatic medicinal plants. *Evidence-Based Complementary and Alternative Medicine* 2022:4931556.

Bafna, A.R. & S.H. Mishra. 2004. Immunomodulatory activity of methanol extract of flower heads of *Sphaeranthus indicus* Linn. *Ars Pharmaceutica* 45(3):281–291.

Basu, N.K. & P.P. Lamsal. 1946. Chemical investigation of *Sphaeranthus indicus*. *Journal of the American Pharmacist Association* 35(9):274–275.

Behera, S.K., A. Panda, S.K. Behera, & M.K. Misra. 2006. Medicinal plants used by the Kandhas of Kandhamal district of Orissa. *Indian Journal of Traditional Knowledge* 5(4):519–528.

Chadha, Y.R. 1976. *The Wealth of India. The publications and Information Directorate*. New Delhi: CSIR.

Chakrabarti, D., A. Suthar, G. Jayaraman, B. Muthuvelan, S. Sharma & M. Padigaru. 2012. A standardized extract from *Sphaeranthus indicus*, inhibits inflammatory, migratory and proliferative activity in keratinocytes and immune cells. *Pharmacology and Pharmacy* 3(2):178–194.

Chopda, M., S.S. Patole & R.T. Mahajan. 2010. Wound healing activity of Sphaeranthus indicus (Linn) in Albino rats. In: *Bioresources for Rural Livelihood*. G.K. Kulkarni, B.N. Pandey and B.D. Joshi, editors. New Delhi: Narendra Publishing House, pp. 239–244.

Dhar, M.L., M.M. Dhar, B.N. Dhawan, B.N. Mehrotra & C. Ray. 1968. Screening of Indian plants for biological activity. *Indian Journal of Experimental Biology* 6(4):232–247.

Dubey, K.S., A.H. Ansari & M. Harduha. 2000. Antimicrobial activity of the extract of *Sphaeranthus indicus*. *Asian Journal of Chemistry* 12(2):577–578.

Duraipandiyan, V., P. Kannan & S. Ignacimuthu. 2009. Antimicrobial activity of *Sphaeranthus indicus* L. *Ethnobotanical Leaflets* 2009(3):320–325.

Galani, V.J. & B.G. Patel. 2010. Effect of hydroalcoholic extract of *Sphaeranthus indicus* against experimentally induced anxiety, depression and convulsions in rodents. *International Journal of Ayurveda Research* 1(2):87–92.

Galani, V.J. & B.G. Patel. 2009. Psychotropic activity of *Sphaeranthus indicus* Linn. in experimental animals. *Pharmacognosy Research* 1(5):307–313.

Galani, V.J., B.G. Patel & D.G. Rana. 2010. *Sphaeranthus indicus* Linn.: a phytopharmacological review. *International Journal of Ayurveda Research* 1(4):247–253.

Garg, S.C. & H.L. Kasera. 1983. *In vitro* antibacterial activity of the essential oil of *Spharanthus indicus* L. *Fitoterapia* 54:37–39.

Garg, S.C. & H.L. Kasera. 1982. Antifungal activity of the essential oil of *Sphaeranthus indicus* Linn. *Pafai Journal* 4:23–24.

Garg, M. & N. Dwivedi. 2021. Physicochemical and phytochemical studies on *Sphaeranthus indicus* Linn. with HPTLC fingerprinting. *Journal of Drug Delivery and Therapeutics* 11:100–107.

Ghani, N. 2004. *Khazainul Advia*. New Delhi: Idarae Kitabus Shifa.

Hameed, S.V.S. & D.S. Shah. 2003. Effect of aqueous extracts of *S. indicus* against *Calex fatigans*, Weid. (Diptera; Culicidae). *Journal of Experimental Zoology* 6:279–284.

Hamid, R., A. Masood, I.H. Wani & S. Rafiq. 2013. Lectins: proteins with diverse applications. *Journal of Applied Pharmaceutical Science* 1:93–103.

Heinrich, M.M., J.E. Robles, B.R. West, B.R. Ortiz de Montellano & E. Rodriguez. 1998. Ethnopharmacology of Mexican Asteraceae (Compositae). *Annual Review of Pharmacology and Toxicology* 38:539 565.

Jain, A. & E. Basal. 2003. Inhibition of *Propionibacterium acnes* induced mediators of inflammation by Indian herbs. *Phytomedicine* 10(1):34–38.

Jha, R.K., N. Garud & R.K. Nema. 2009. Excision and incision wound healing activity of flower head alcoholic extract of *Sphaeranthus indicus* Linn. in albino rats. *Global Journal of Pharmacology* 3(1):32–37.

Kabir al-Din, M. 2007. *Makhzan al-Mufridat*. New Delhi: Idarae Kitabus Shifa.

Kapoor, S.L. & L.D. Kapoor. 1980. Medicianl plant wealth of the Karimnagar district of Andhra pradesh. *Bull Med Ethnobot Res.* 1:120–144.

Kirtikar, K.R. & B.D. Basu. 1918. *Indian Medicinal Plants*. Vol. 3. Deharadun: Bishen Singh Mahendra Pal Singh and Periodical Experts.

Kirtikar, K.R. & B.D. Basu. 1935. *Indian Medicinal Plants, II*. Allahabad: Lalit Mohan Basu Publication.

Kirtikar, K.R. & B.D. Basu. 1981. *Indian Medicinal Plants*, 2nd ed. Vol. 2. E. Blatter, J.F. Caius & K.S. Mhaskar, editors. Allahabad: Lalit Mohan Basu.

Kiritikar, K.R. & B.D. Basu. 2003. *Indian Medicinal Plants*, 2nd ed. Vol. 6. Dehradun: Oriental Enterprises.

Kokate, C. 1986. *Preliminary Phytochemical Analysis. Practical Pharmacognosy*, 1st ed. New Delhi: Vallabh Prakashan.

Krishna, T.M., S.P. Thota, M. Jadhav, K.M. Kamal, A. Venuganti, D. Mrunalini, S. Vadiari & G. Mittapelli. 2013. Studies on *in vitro* antioxidant and antibacterial activities of *Sphaeranthus indicus* (Linn). *International Journal of Pharmaceutical Research and Biomedical Analysis* 2(1):1–9.

Lalla, P.M., A.R. Deshpande, M. Musaddip, N.H Shahare & C. Sukesh. 2005. Antimicrobial activity of medicinal plant Sphaeranthus indicus. In: *International Conference on Modern Trends in Plant Sciences with Special Reference to the Role of Biodiversity in Conservation*. Maharashtra: Amravati, p. 96.

Latha, L.Y., S. Sasidharan, Z. Zuraini, S. Suryani, L. Shirley & S. Sangeetha. 2007. Activity and toxicity of crude extract of the *Psophocarpus tetragonolobus*. *African Journal of Traditional Complementary and Alternative Medicine* 4:59–63.

Lichtenbelt, B., R. Crane & S. Naqvi. 1998. *Introduction to Volume Rendering*. Englewood Cliffs, NJ: Prentice Hall, Inc.

Makhija, K., L. Richard, S.P. Kirti, K. Saleemullah, M. Jessy & A. Shirwaikar. 2011. *Sphaeranthus indicus*: a review of its chemical, pharmacological and ethnomedicinal properties. *International Journal of Pharmacology* 7:171–179.

Mathew, J.E., K.K. Srinivasan, V. Dinakaran & A. Joseph. 2009. Mast cell stabilizing effects of *Sphaeranthus indicus*. *Journal of Ethnopharmacology* 122(2):394–396.

Mhetre, N.A., S.D. Ambavade & S.L. Bodhankar. 2006. Neuroleptic activity of extract of *Sphaeranthus indicus* in mice. *Indian Journal of Natural Product* 22:24–27.

Mohiuddin, G. 2004. *Fasihuddin. Rahnuma-i-Aqaqeer*, Vol. II. New Delhi: I'jaz Publishing House.

Mushir, A., N. Jahan & G. Sofi. 2019. Exploration of Unani concept of blood purifying drugs and likely models for its screening. *Bangladesh Journal of Medical Science* 18(2):312–321.

Nadkarni, A.K. 2007. *Materia Medica*, 3rd ed. Vol. 1. Bombay: Popular Prakashan Private Limited.

Nanda, B.K., J. Jena, B. Rath & B.R. Behera. 2009. Analgesic and antipyretic activity of whole parts of *Sphaeranthus indicus* Linn. *Journal of Chemical Pharmaceutical Research* 1(1):207–212.

Naqvi, B.S., K. Hashmi, D. Sheikh & A. Mahdi. 1998. Antibacterial activity in fruits and vegetables. *Pakistan Journal of Pharmacology* 15(2):7–11.

Nayak, S.S., T.K. Maity & B.C Maiti. 2007. Hepatoprotective activity of *Sphaeranthus indicus* Linn. *International Journal of Green Pharmacy* 1(S2):32–36.

Nisha, M., M. Kalyanasundaram, K.P. Paily, Abidha, P. Vanamail & K. Balaraman. 2007. *In vitro* screening of medicinal plant extracts for macrofilaricidal activity. *Parasitology Research* 100(3):575–579.

Pande, V.V. & S. Dubey. 2009. Antihyperlipidemic activity of *Sphaeranthus indicus* on atherogenic diet induced hyperlipidemia in rats. *International Journal of Green Pharmacy* 3(2):159–161.

Paranjape, P. 2001. Indian medicinal plants. In: *Forgotten Healer: A Guide to Ayurvedic Herbal Medicine*. P. Paranjape, editor. Delhi: Chaukhamba Sanskrit Pratisthan.

Prabhu, K.S., R. Lobo & A. Shirwaikar. 2008. Antidiabetic properties of the alcoholic extract of *Sphaeranthus indicus* in streptozotocin-nicotinamide diabetic rats. *Journal of Pharmacy & Pharmacology* 60(7):909–916.

Ramachandran, S. 2013. Review on *Sphaeranthus indicus* Linn. (Koṭṭaikkarantai). *Pharmacognosy Review* 7:157–169.

Sadaf, F., R.M. Saleem, S.L. Ahmed & Navaid-ul-Zafar. 2006. Healing potential of cream containing extract of *Sphaeranthus indicus* on dermal wounds in guinea pigs. *Journal of Ethnopharmacology* 107(2):161–163.

Said, H.M. 1956. *Medicinal Herbal: A Textbook for Medical Students and Doctors*. Karachi: Madinat Al-Hikmah, Bait Al-Hikmah.

Saidpur, S.K. 1978. Follicular atresia in the ovaries of non mammalian vertebrates. *International Review of Cytology* 54:225–244.

Sarpate, R.V., T.K. Deore & S.V. Tupkari. 2020. Bronchodilatory effect of *Sphaeranthus indicus* Linn against allergen induced bronchospasm in guinea pigs. *Zhongguo Yao Li Xue Bao* 7:651–654.

Sasidharan, S., Z. Zuraini, L.L. Yoga, S. Sangetha & S. Suryani. 2008. Antimicrobial activities of *Psophocarpus tetragonolobus* (L.) DC extracts. *Food Borne Pathogen and Disease* 5:303–309.

Shaikh, D., B.S. Naqui & R. Shaikh. 1986. The antimicrobial principles of *Sphaeranthus indicus*: Isolation, purification and antimicrobial action. *Pakistan Journal of Scientific and Industrial Research Series* 29:366–371.

Sharma, M. & R.C. Saxena. 1996. *Sphaeranthus indicus* as mosquito larvicide. *Journal of Applied Zoologist Research Association* 7:87–88.

Shekhani, M.S., P.M. Shah, A.Yasmin, R. Siddiqui, S. Perveen & K. Mohammed. 1990. An immune stimulant sesquiterpene glycoside from *Sphaeranthus indicus*. *Phytochemistry* 29:2573–2576.

Shirwaikar, A., K.S. Prabhu & I.S. Punitha. 2006. *In-vitro* antioxidant studies of *Sphaeranthus indicus* (Linn). *Indian Journal of Experimental Biology* 44:993–996.

Singh, S., B.C. Semwal & P. Upadhaya. 2019. Pharmacognostic study of *Sphaeranthus indicus* Linn. A review. *Pharmacog Journal* 11(6):1376–1385.

Singh, S.K., K.M. Saroj, V.J. Tripathi, A.K. Singh & R.H. Singh. 1988. An antimicrobial principle from *Sphaeranthus indicus* L. (family Compositae) *International Journal of Crude Drug Research* 26:235–239.

Srinivasan, V.M., K.K. Jessy & E.M. Anand Alex. 2008. Effect of *Sphaeranthus indicus* Linn. on gentamicin induced acute renal failure in rats. *Indian Journal of Pharmacology* 40:71–72.

Sultana, A., M. Fazmiya & K. Rahman. 2021. Mundi (*Sphaeranthus indicus* Linn): the best blood purifier and immunomodulatory Unani herb with versatile ethnomedicinal uses and pharmacological activities. *Cellmed* 11:10.1–10.10.

Tandon, D. & A.K. Gupta. 2020. Comparative assessment of antimicrobial and antioxidant activity between whole plant and parts of *Sphaeranthus indicus* Linn. (Asteraceae). *Clinical Phytoscience* 6:1–15.

Tiwari, A. & R.C. Saxena. 2003. Repellent and feeding deterrent activity of *Sphaeranthus indicus* against *Triboliumcastanum*. *Bio-research Bulletin* 1:179–284.

Tiwari, B.K. & R.L. Khosa. 2009. Hepatoprotective and antioxidant effects of *Sphaeranthus indicus* against acetaminophen induced hepatotoxicity in rats. *Journal of Pharmaceutical Science and Research* 1:26–30.

Vijaylakshmi, N. & M.R.K. Rao. 2019. The antioxidant studies of two medicinal plants, *Sphaeranthus indicus* and *Psophocarpus tetragonolobus*. *Asian Journal Pharmaceutical and Clinical Research* 12:321–327.

Vimalanathan, S., S. Ignacimuthu & J.B. Hudson. 2009. Medicinal plants of Tamil Nadu (Southern India) are a rich source of antiviral activities. *Pharmaceutical Biology* 47:422–429.

Yadava, R.N. & S. Kumar. 1998. Chemical examination of the leaves of *Sphaeranthus indicus*. *Asian Journal of Chemistry* 10:764–766.

9 Phytochemistry and Ethnopharmacological Importance of *Pistia stratiotes* L. (Hemsl) A. Grey (Araceae)

Maan B. Rokaya,[1,2,*] *Saroj Bashyal,*[3] *and Binu Timsina*[2]
[1] Institute of Botany, Czech Academy of Sciences, Průhonice, Czech Republic
[2] Global Change Research Institute, Czech Academy of Sciences, Brno, Czech Republic
[3] Central Department of Botany, Tribhuvan University, Kathmandu, Nepal
[*] Corresponding author (Maan B. Rokaya): rokayamaan@gmail.com

9.1 INTRODUCTION

Pistia stratiotes L. (Hemsl) A. Grey is commonly known as water lettuce and belongs to the family Araceae. It is a free-floating hydrophyte with a rosette of obovate to spatulate short-haired leaves (Langeland and Burks 1998). It grows in water bodies such as slow-moving rivers and reservoirs, irrigation channels, ponds, lakes, canals, and ditches (Cilliers 1991; Adebayo et al. 2011; Hussner et al. 2014; Arya et al. 2022). It also invades rice paddies in Asia as well as other wetland habitats. Because of seed survival and germination ability, *P. stratiotes* can endure drying and reinfest ephemeral waters subject to seasonal drying.

P. stratiotes is widely distributed in tropical and subtropical regions in the world. It is considered as native to south America (Langeland and Burks 1998). It has a widespread distribution in 40 countries (Holm 1991) in Africa (Witt and Luke 2017), Asia (Zhengyi et al. 2010), Europe (DAISIE 2013), North America (González-Torres et al. 2012), and Oceania (PIER 2013), where it is considered as an introduced or an invasive species. However, this plant is reported as a noxious weed or invasive aquatic weed in several countries such as the USA (USDA-NRCS 2012), Slovenia (Šajna et al. 2007), Papua New Guinea (PIER 2013), Argentina (Holm 1991), China (Zhengyi et al. 2010), Indonesia, and Malaysia (Waterhouse 1993) (Figure 9.1 and Table 9.1).

This aquatic weed negatively impacts the economy/livelihood and environment (Langeland and Smith 1993; Economic Community of West African States 1995). The oxygen and light limitations caused by *P. stratiotes* may have killed native plants, as in Slovenia (Šajna et al. 2007) or fish and wildlife (Florida Department of Environmental Protection 2007). The mode of spread of species is through international shipping routes (Bartram and Harper 1942, 1943), aquaculture, escape from confinement or garden, hitchhiker, intentional release, landscape improvement, nursery trade, ornamental purposes and pet trade (Rivers 2002; Gherardi 2007).

DOI: 10.1201/9781003256830-9

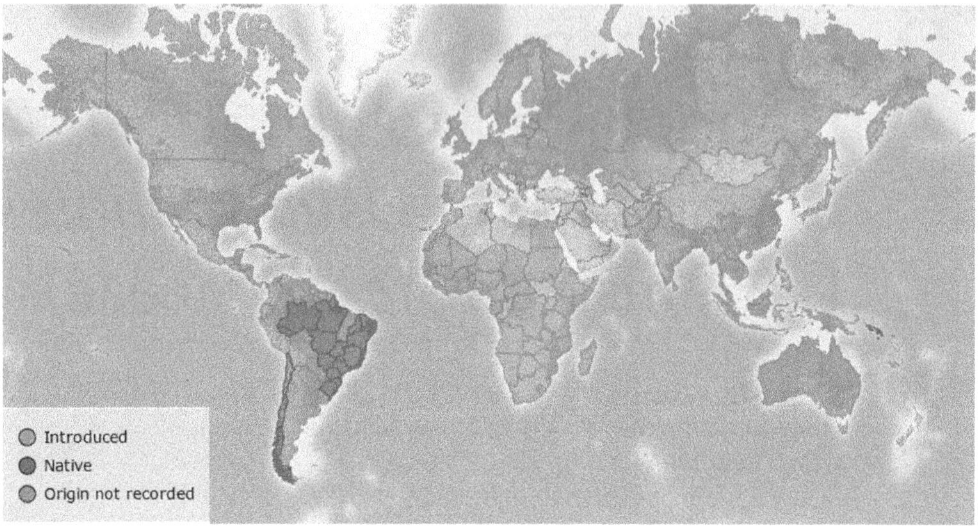

FIGURE 9.1 Distribution of *P. stratiotes* in the world. (From CABI (2022).)

TABLE 9.1
Distribution of *P. stratiotes* According to Regions and Countries in the World

Range	Region	Countries
Native range	South America	Bolivia, Brazil, Paraguay (considered native to the Pantanal region of South America)
Introduced range	Europe	Austria, Belgium, the Czech Republic, France, Germany, Hungary, Israel, Italy, Morocco, the Netherlands, Norway, Portugal, Romania, Russia, Slovenia, Spain, the UK
	Africa	Angola, Burkina Faso, Cameroon, Central African Republic, Chad, Comoros, Democratic Republic of Congo, Egypt, Equatorial Guinea, Gabon, Gambia, Guinea, Guinea-Bissau, Lesotho, Liberia, Madagascar, Mauritania, Mauritius, Morocco, Mozambique, Namibia, Niger, Nigeria, Rwanda, Senegal, Seychelles, Sierra Leone, Somalia, South Africa, Sudan, Swaziland, Togo
	Asia	Afghanistan, Bangladesh, India, Indonesia, Japan, Kazakhstan, Laos, Myanmar, Nepal, Pakistan, Singapore, Sri Lanka, Vietnam
	North America	Canada (Ontario), USA (Arizona, California, Colorado, Connecticut, Delaware, Georgia, Idaho, Illinois, Kansas, Louisiana, Maryland, Michigan, Minnesota, Mississippi, Missouri, New Jersey, New York, North Carolina, Ohio, Rhode Island, South Carolina, Texas)
	Central America and the Caribbean	Antigua and Barbuda, Belize, Costa Rica, Dominican Republic, El Salvador, Guadeloupe, Guatemala, Haiti, Honduras, Jamaica, Martinique, Montserrat, Nicaragua, Panama, Saint Lucia, Saint Vincent, Trinidad and Tobago.
	Oceania	Australia, New Zealand, Papua New Guinea, Solomon Islands
Invasive range	Africa	Benin, Botswana, Burundi, Côte d'Ivoire, Ethiopia, Ghana, Kenya, Malawi, Réunion, Tanzania, Uganda, Zambia, Zimbabwe
	Asia	Brunei, Cambodia, China, Malaysia, Philippines, Taiwan, Thailand
	Europe	Canary Islands
	North America	USA (Alabama, Florida, Hawaii)
	Central America and the Caribbean	Cuba, Puerto Rico,
	Oceania	Cook Islands, French Polynesia, Guam, New Caledonia, Palau, Vanuatu

(Ward 2017; CABI 2022.)

FIGURE 9.2 *P. stratiotes* plant growing in low land in the central Nepal. (A) Habit. (B) Roots. (C) Leaf. (D) Flower.

P. stratiotes is a floating herb (Figure 9.2A) with rosettes of grey-green leaves; rosettes occur singly or connected to others by short stolon. Roots are numerous, feathery, long, and fibrous (Tiwari et al. 2005), hanging unbranched 50–80 cm below the water (Figure 9.2B). Leaves are grey-green, often spongy near the base, densely soft-pubescent with prominent parallel veins, slightly broader than long, and widest at the apex (Figure 9.2C); the leaf blade 15 × 10 cm. Flowers are white or yellowish-white, inconspicuous (Figure 9.2D), and clustered on small fleshy stalks nearly hidden in leaf axils. There is a whorl of male flowers above a single female flower. The fruit arises from the female flower as a green berry with many viable seeds (Langeland and Burks 1998).

Despite its invasive nature, *P. stratiotes* is sold in the market as an ornamental plant for aquariums and water gardens (Rixon et al. 2005). It can potentially be used as biofuel (Mishima et al. 2008) and in water remediation (Lu et al. 2010). Because of its crude fibre, carbohydrate, and crude protein contents, this plant can be used as quality forage (Parsons and Cuthbertson 2001) for pig and duck fibrous (Tiwari et al. 2005).

Khan et al. (2014) presented a comprehensive review of the phytochemical, pharmacological, medicinal, and bioremediation potential, allelopathy, utilization, and management of water lettuce. Tripathi et al. (2010) updated information on the phytochemical and pharmacological activities of the plant. However, the focus of the previous reviews was mainly on diuretic, antidiabetic, antidermatophytic, antifungal, and antimicrobial properties, whereas the current review provides detailed and up-to-date information on traditional uses, proximate composition, phytochemistry, and biological activities of the plant including toxicity/cytotoxicity, which are lacking in previous reviews.

9.2 METHODS

We searched published literature on *P. stratiotes* from databases such as Google Scholar, Medicine, PubMed, Science Direct, Scopus, and SciFinder to collect information. We used keywords such as pharmacological, chemical constituents, phytochemistry, ethnobotany, ethnomedicine, traditional use or indigenous use, morphology, inorganic compound, proximate composition, and toxicity or the cytotoxicity of *P. stratiotes*.

9.3 RESULTS AND DISCUSSION

9.3.1 Ethnomedicinal Uses of *Pistia stratiotes*

The folkloric usage of *P. stratiotes* is presented in Table 9.2. *P. stratiotes* is used in folklore medicine in countries like India, Nigeria, Nepal, and Pakistan. Leaves or whole plant parts are used as medicine against many diseases ranging from external skin infections/disorders/diseases to complex internal diseases such as ulcers, tuberculosis, and thyroid problems. Paste, juice, ash, decoction, and poultice are forms of use. It is taken orally or applied externally to treat diseases. Sometimes, instead of taking it for a specific type of disease, whole plant decoction is used for bathing for three months in Lagos State, Nigeria (Sharaibi et al. 2014).

9.3.2 Chemical Composition of *Pistia stratiotes*

Generally, plant parts such as aerial parts or shoots, leaves, roots, and whole plants contain different types of compounds; in total, 74 different compounds have been isolated from *P. stratiotes* (Table 9.3). Phytochemicals such as alkaloids, flavonoids, phenols, saponins, tannins, and terpenoids, are present in acetone, aqueous, ethanolic, ethereal extract, ethyl ether, and methanolic extracts of the plant. The analysis was carried out using chromatography, column chromatography, gas chromatography–mass spectrometry analysis, high-performance liquid chromatography, and spectroscopic analysis. Flavonoids were the most common type of compounds, followed by steroids and fatty acid compounds.

Analysis of trace metals in the roots and leaves of *P. stratiotes* by Wasagu et al. (2013) also revealed the presence of sodium, potassium, phosphorus, calcium, and magnesium.

9.3.3 Proximate Composition of *Pistia stratiotes*

Proximate analysis of the shoot, leaves, roots, and whole plant of *P. stratiotes* by different studies have revealed the presence of carbohydrates (K.A. Ali et al. 2011; Wasagu et al. 2013; Dethe et al. 2014; Akter and Sultana 2019), crude fibre, moisture, total ash, crude protein, and crude fat in leaf

TABLE 9.2
Traditional Uses of *P. stratiotes*

Sl. no.	Ailments/use	Plant part	Mode of use	Region/country	References
1.	Arthritis	Lvs	Unspecified	Panjab, Pakistan; north-west, Nigeria	Salihu et al. 2018; M. Ali et al. 2020
2.	Asthma	Lvs	The leaves are used in the treatment	West Bengal, India	Datta and Banerjee 1978
		Wh pl	Juice is used	Bihar, India	Singh et al. 2013
		Lvs, Wh pl	Decoction and juice used	Punjab province, Pakistan	Umair et al. 2019
		Lvs, Wh pl	Poultice mixed with rose water and sugar and given	Uttar Pradesh, India	Uniyal et al. 2011
3.	Boils	Wh pl	Pound until fine and paste applied externally	Tamilnadu, India	Senthilkumar et al. 2014
		Wh pl	Plant paste is applied externally	Manipur, India	Jain et al. 2007
4.	Burn	Wh pl	Plant paste is applied externally	Manipur, India	Jain et al. 2007
5.	Cough	Lvs	The leaves are used in the treatment	West Bengal, India	Datta and Banerjee 1978
		Wh pl	Juice is used	Bihar, India	Singh et al. 2013
		Lvs, Wh pl	Poultice mixed with rose water and sugar and given	Uttar Pradesh, India	Uniyal et al. 2011
		Lvs, Wh pl	Decoction and juice used	Punjab province, Pakistan	Umair et al. 2019
6.	Digestion	Wh pl	Water from the braise plant is taken orally	Tamilnadu, India	Senthilkumar et al. 2014
7.	Dysentery	Lvs	The leaves are used in treatment	West Bengal, India	Datta and Banerjee 1978
		Wh pl	Water from the braise plant is taken orally	Tamilnadu, India	Senthilkumar et al. 2014
		Lvs, Wh pl	Poultice mixed with rose water and with coconut milk	Uttar Pradesh, India	Uniyal et al. 2011
8.	Ear, nose, throat	Lvs	Leaf extract is used as drops	Bengal, India	Chowdhury and Das 2015
		Lvs, Wh pl	Plant juice is used	Uttar Pradesh, India	Uniyal et al. 2011
9.	Eczema	Lvs	The leaf paste is used in eczema	Bengal, India	Chowdhury and Das 2015
		Lvs	Leaf juice is boiled with coconut oil and applied externally	Bengal, India	Chowdhury and Das 2015
		Lvs, Wh pl	Decoction and juice used	Punjab province, Pakistan	Umair et al. 2019
		Lvs	Unspecified	Uttar Pradesh, India	Uniyal et al. 2011
10.	Fracture/ dislocation of bone	Lvs	Boil in water and drink used for 7–14 days	Benue State, Nigeria	Uleh et al. 2017
11.	Gonorrhoea	Wh pl	Pound until fine and paste applied externally	Tamilnadu, India	Senthilkumar et al. 2014
12.	Headache	Lvs	Unspecified	Punjab, Pakistan	M. Ali et al. 2020
13.	Haemorrhoid	Lvs, Wh pl	Poultice applied externally	Uttar Pradesh, India	Uniyal et al. 2011

(continued)

TABLE 9.2 (Continued)
Traditional Uses of *P. stratiotes*

Sl. no.	Ailments/use	Plant part	Mode of use	Region/country	References
14.	Inflammation/ swelling	Wh pl	Pound until fine and paste applied externally	Tamilnadu, India	Senthilkumar et al. 2014
		Lvs, Wh pl	Decoction and juice used	Punjab province, Pakistan	Umair et al. 2019
		Lvs, Wh pl	Unspecified	Ondo State, Nigeria; Punjab, Pakistan; Morang, Nepal	Acharya (Siwakoti) and Pokhrel 2006; Abiodun and Tunji 2019; M. Ali et al. 2020
15.	Leprosy	Lvs	Leaf juice is boiled with coconut oil and applied externally	Bengal, India; Uttar Pradesh, India	Uniyal et al. 2011; Chowdhury and Das 2015
		Wh pl	Unspecified	Ondo State, Nigeria	Abiodun and Tunji 2019
		Lvs, Wh pl	Decoction and juice used	Punjab province, Pakistan	Umair et al. 2019
16.	Menstrual flow	Wh pl	Water from the braise plant is taken orally	Tamilnadu, India	Senthilkumar et al. 2014
17.	Painful joint	Lvs, Wh pl	Decoction and juice used	Punjab province, Pakistan	Umair et al. 2019
18.	Piles	Lvs	The leaf paste is used in piles	Bengal, India	Chowdhury and Das 2015
		Lvs, Wh pl	Unspecified	Ondo State, Nigeria; Uttar Pradesh, India	Uniyal et al. 2011; Abiodun and Tunji 2019
		Lvs, Wh pl	Decoction and juice used	Punjab province, Pakistan	Umair et al. 2019
19.	Respiratory diseases	Wh pl	Unspecified	Niger State, Nigeria	Mann et al. 2008
20.	Rheumatism	Wh pl	Whole plant is boiled and tied on the swollen parts	Bengal, India	Chowdhury and Das 2015
21.	Ringworm	Lvs	Unspecified	Punjab, Pakistan	M. Ali et al. 2020
		Lvs, Wh pl	Plant ashes are applied externally	Uttar Pradesh, India	Uniyal et al. 2011
22.	Skin infection/ disorders/ diseases	Lvs, Wh pl	Unspecified	Ondo State, Nigeria, Iloilo, Philippines; Uttar Pradesh, India	Uniyal et al. 2011, Tantiado 2012; Abiodun and Tunji 2019
		Lvs	The leaf paste is used in leprosy	Bengal, India	Chowdhury and Das 2015
		Lvs	Leaf juice is boiled with coconut oil and applied externally	Bengal, India	Chowdhury and Das 2015
		Lvs	Juice is used against skin diseases	Tamilnanu, India	Vanila et al. 2008
		Lvs	The leaves are used in skin diseases	West Bengal, India	Datta and Banerjee 1978
23.	Stomach disorders	Wh pl	Unspecified	Ondo State, Nigeria	Abiodun and Tunji 2019

		Wh pl	Pound until fine and paste applied externally	Tamilnadu, India	Senthilkumar et al. 2014
24.	Syphilis	Wh pl	Pound until fine and paste applied externally	Tamilnadu, India	Senthilkumar et al. 2014
25.	Thyroid	Lvs	Dry leaves added to mustard oil made in to paste apply externally	Amrabad Tiger Reserve Forest, India	Ramesh et al. 2019
26.	Tuberculosis	Wh pl	Unspecified	Niger State, Nigeria	Mann et al. 2008
27.	Ulcers	Lvs	The leaf paste is used in ulcers	Bengal, India	Chowdhury and Das 2015
		Lvs, Wh pl	Unspecified	Ondo State, Nigeria; Uttar Pradesh, India	Uniyal et al. 2011; Abiodun and Tunji 2019
28.	Urinary problems	Wh pl	Two teaspoonfuls of plant extract are applied twice daily	Bengal, India	Chowdhury and Das 2015
		Wh pl	Water from the braise plant is taken orally	Tamilnadu, India	Senthilkumar et al. 2014
		Wh pl	Pound until fine and paste applied externally	Tamilnadu, India	Senthilkumar et al. 2014
		Lvs, Wh pl	Decoction and juice used	Punjab province, Pakistan	Umair et al. 2019
29.	Wound	Lvs	Unspecified	Punjab, Pakistan	M. Ali et al. 2020
30.	Others	Wh pl	The decoction is used for bathing for 3 months	Lagos State, Nigeria	Sharaibi et al. 2014

Lvs, leaves; Wh pl, whole plant.

TABLE 9.3
Compounds Isolated from *P. stratiotes*

Chemicals	Chemical group	Plant part	Extract type	Method	PubChem CID	Molecular formula	References
(12*R*,9*Z*,13*E*,15*Z*)-12-Hydroxy-9,13,15-octadecatrienoic acid	Fatty acid	Plant	Ethyl ether extract	Chromatography	–	–	Aliotta et al. 1991
(24*R*)-ergosta-7,22-diene-3β,5α,6β-triol	Steroid	Whole plant	Ethanolic extract	Spectroscopic analysis	–		Liu et al. 2008
(3*S*,5*R*,6*S*,7*E*,9*R*)-3-Hydroxy-5,6-epoxy-β-ionyl-3-*O*-β-D-glucopyranoside	Terpene glucoside	–	Methanol extract	Spectroscopic method (Mosher method)	–	–	Greca et al. 1991
(9*S*,10*E*,12*Z*,15*Z*)-9-Hydroxy-10,12,15-octa-decatrienoic acid	Fatty acid	Plant	Ethyl ether extract	Chromatography	–	–	Aliotta et al. 1991
11α-Hyroxy-14*S*-ethyl-5α-cholest-22en-3,6-dione	Steroid	Plant	–	Electron ionization mass spectroscopy	–	–	Monaco and Previtera 1991
12,15-Octadecadiynoic acid, methyl ester	Fatty acid ester	Leaves	Ethanolic extract	GC-MS analysis	538453	$C_{19}H_{30}O_2$	Tyagi and Agarwal 2017b, Lawal et al. 2020
1-Monolinoleoylglycerol trimethylsilyl ether	Steroid	Leaves, root, shoot	Ethanolic extract	GC-MS analysis	250006068	–	Tyagi and Agarwal 2017a, 2017b
1-*O*-Galloyl-β-D-glucose	Tannins	–	Aqueous extract	HPLC analysis	–	–	Ganesh and Sharma 2018
2,6,10,14,18,22-Tetracosahexane	Squalene	Aerial part, leaves	Methanol extract	GC-MS analysis	57417215	$C_{24}H_{38}$	Nisha et al. 2018, Lawal et al. 2020
(24*R*)-Ergosta-7,22-diene-3β,5α,6β-triol	Ergosterol	–	–	Column chromatography	–	–	Liu 2007
24*S*-Ethyl-4,22-cholestadiene-3,6-dione	Cholesterol	Plant	Ethyl ether extract	Chromatography	–	–	Aliotta et al. 1991
2-Hexadecan-1-01,3,7,15-tetramethyl (R-[R*,R*-*E*])	Alkane	Leaves	Methanol extract	GC-MS analysis	–	–	Nisha et al. 2018
2-Hexadecan-1-ol, 3,7,11,15-tetramethyl (*R*-[*R**,*R**-*E*])-(CAS)	Acyclic diterpenoid	Aerial part	–	GC-MS analysis	–	–	Lawal et al. 2020
7-β-Hydroxy-4,22-stigmastadien-3-one	Steroid	Plant	–	Spectroscopic methods—ultraviolet	–	–	Ayyad 2010

7β-Hydroxysitosterol	Steroid	Whole plant	Ethanolic extract	Spectroscopic analysis	135117690	$C_{21}H_{28}N_4O_2$	Liu 2007; Liu et al. 2008
9,12,15-Octadecacatrienoic acid, methyl ester (*Z,Z,Z*)	Steroid	Leaves	Ethanolic extract	GC-MS analysis	–	–	Tyagi and Agarwal 2017b; Lawal et al. 2020
9,12-Octadecadienoic acid ethyl ester	Polyenoic fatty acid	Leaves, root, shoot	Ethanolic extract	GC-MS analysis	–	–	Tyagi and Agarwal 2017b
Annuionone E	Oxepane	–	–	Column chromatography	101260349	$C_{13}H_{22}O_3$	Liu 2007
Bis (2-ethylhexyl) phthalate	Plasticizer compound	Roots	Ethanolic extract	GC-MS analysis	8343	$C_{24}H_{38}O_4$	Tyagi and Agarwal 2017a
Catechin	Flavonoid	–	Aqueous, methanolic extract	HPLC analysis	9064	$C_{15}H_{14}O_6$	Ganesh and Sharma 2018
Chrysoeriol 4′-*O*-β-D-glucopyranoside	Flavonoid	Whole plant	Ethanolic extract	Spectroscopic analysis, column chromatography	14406834	$C_{22}H_{22}O_{11}$	Liu 2007; Liu et al. 2008
Cyanidin-3-glucoside	Flavonoid	Plant	Methanol extract	Chromatography	441667	$C_{21}H_{21}O_{11}^+$	Zennie and McClure 1977
Daucosterol	Lignan	Whole plant	Ethanolic extract	Spectroscopic analysis, column chromatography	5742590	$C_{35}H_{60}O_6$	Liu 2007; Liu et al. 2008
δ-Linolenic acid	Fatty acid	Plant	Ethyl ether extract	Chromatography	–	–	Aliotta et al. 1991
Diisooctyl phthalate	Plasticizer compound	Leaves	Ethanolic extract	GC-MS analysis	33934	$C_{24}H_{38}O_4$	Tyagi and Agarwal, 2017a, 2017b
Docosanoic acid ethyl ester (ethyl docosanoate)	Fatty acid ester	Leaves	Ethanolic extract	GC-MS analysis	22199	$C_{24}H_{48}O_2$	Tyagi and Agarwal, 2017a, 2017b
E-11-Hexadecanoic acid, ethyl ester	Stearic acid	Leaves, root, shoot	Ethanolic extract	GC-MS analysis	–	–	Tyagi and Agarwal 2017a, 2017b
Epicatechin	Polyphenol	–	Aqueous extract	HPLC analysis	72276	$C_{15}H_{14}O_6$	Ganesh and Sharma 2018
Epicatechin gallate	Polyphenol	–	Aqueous extract	HPLC analysis	107905	$C_{22}H_{18}O_{10}$	Ganesh and Sharma 2018
Ethane, 1,1-diethoxy	Ether	Roots	Ethanolic extract	GC-MS analysis	7765	$C_6H_{14}O_2$	Tyagi and Agarwal 2017b
Ethyl iso-allocholate	Steroid	Roots	Ethanolic extract	GC-MS analysis	6452096	$C_{26}H_{44}O_5$	Tyagi and Agarwal 2017b
Formic acid, 1-methylethyl ester (isopropyl formate)	Carboxylic acid esters	Leaves	Ethanolic extract	GC-MS analysis	12257	$C_4H_8O_2$	Tyagi and Agarwal 2017b; Lawal et al. 2020
Gallic acid	Phenolic acid	–	Acetone, Aqueous, methanolic extract	HPLC analysis	370	$C_6H_2(OH)_3COOH$	Ganesh and Sharma 2018
Glycerol 1-palmitate	Monoacylglycerol	Aerial part, leaves	Methanol extract	GC-MS analysis	14900	$C_{19}H_{38}O_4$	Nisha et al. 2018; Lawal et al. 2020

(*continued*)

TABLE 9.3 (Continued))
Compounds Isolated from *P. stratiotes*

Chemicals	Chemical group	Plant part	Extract type	Method	PubChem CID	Molecular formula	References
Hexadecanoic acid ethyl ester	Palmitic ester	Leaves, roots	Ethanolic extract	GC-MS analysis	–	–	Tyagi and Agarwal 2017b; Lawal et al. 2020
Hexadecanoic acid, 2-hydroxy-1-(hydroxymethyl) ethyl ester	Palmitic acid glycerol ester	Leaves	Ethanolic extract	GC-MS analysis	985	$C_{16}H_{32}O_2$	Tyagi and Agarwal 2017a, 2017b; Lawal et al. 2020
Inosine	Purine nucleoside			Column chromatography	135398641	$C_{10}H_{12}N_4O_5$	Liu 2007
Isobutyl alcohol	Alcohol	Leaves, roots	Ethanolic extract	GC-MS analysis	6560	$(CH_3)_2CHCH_2OH$	Tyagi and Agarwal 2017b; Lawal et al. 2020
Kaempferol	Flavonoid	–	Aqueous extract	HPLC analysis	5280863	$C_{15}H_{10}O_6$	Ganesh and Sharma 2018
L-Glutamine	Amino acid	Leaves	Ethanolic extract	GC-MS analysis	5961	$C_5H1_0N_2O_3$	Tyagi and Agarwal 2017b; Lawal et al. 2020
Linoleic acid	Fatty acid	Plant	Ethyl ether extract	Chromatography	5280450	$C_{18}H3_2O_2$	Aliotta et al. 1991
Linolelaidic acid methyl ester (methyl linolelaidate)	Palmitic ester	Leaves	Ethanolic extract	GC-MS analysis	5362793	$C_{19}H_{34}O_2$	Tyagi and Agarwal 2017b; Lawal et al. 2020
Linolenic acid, ethyl ester (ethyl linolenate)	Linoleic acid ester	Leaves, root, shoot	Ethanolic extract	GC-MS analysis	5367460	$C_{20}H_{34}O_2$	Tyagi and Agarwal 2017a
Lucenin	Flavonoid	Plant	Methanol extract	Chromatography	–	–	Zennie and McClure 1977
Luteolin	Flavonoid	Whole plant	Ethanolic extract	Spectroscopic analysis, column chromatography	5280445	$C_{15}H_{10}O_6$	Liu 2007; Liu et al. 2008
Luteolin-7-glycoside	Flavonoid	Plant	Methanol extract	chromatography	–	–	Zennie and McClure 1977
n-Hexadecanoic acid	Fatty acid	Leaves, root, shoot	Ethanolic extract	GC-MS analysis	985	$C_{16}H_{32}O_2$	Tyagi and Agarwal 2017a, 2017b; Lawal et al. 2020
Orientin	Flavonoid	Plant	Methanol extract	Chromatography	5281675	$C_{21}H_{20}O_{11}$	Zennie and McClure 1977
Palmitic acid	Fatty acid	–	–	Chromatography	–	–	Ling et al. 1999
Palmitic acid, ethyl ester (ethyl palmitate)	Stearic acid	Leaves, root, shoot	Ethanolic extract	GC-MS analysis	12366	$C_{18}H_{36}O_2$	Tyagi and Agarwal 2017b
Phytol [(*E*,7*R*,11*R*)-3,7,11,15-tetramethylhexadec-2-en-1-ol]	Diterpene	Leaves, root, shoot	Ethanolic extract	GC-MS analysis	5280435	$C_{20}H_{40}O$	Tyagi and Agarwal 2017b

Phytol acetate	Acyclic diterpenoid	Aerial part, leaves	Methanol extract	GC-MS analysis	6428538	$C_{22}H_{42}O_2$	Nisha et al. 2018; Lawal et al. 2020
Propane,1,1-diethoxy-2-methyl	Ether	Leaves, roots	Ethanolic extract	GC-MS analysis	519415	$C_8H_{18}O_2$	Tyagi and Agarwal 2017a, 2017b; Lawal et al. 2020
Rutin	Flavonoid	–	Acetone, Aqueous, methanolic extract	HPLC analysis	5280805	$C_{27}H_{30}O_{16}$	Ganesh and Sharma 2018
Sitoindoside I	Steroid glycosides	Whole plant	Ethanolic extract	Spectroscopic analysis, column chromatography	9832350	$C_{51}H_{90}O_7$	Liu 2007; Liu et al. 2008
Sitosterol-3-*O*-[2,4-di-*O*-acetyl-6-*O*-stearyl]-β-D-glucopyranoside	Saponin	Whole plant	Ethanolic extract	TCL	–	–	Singh et al. 2014
Sitosterol-3-*O*-[2',4'-*O*-diacetyl-6'-*O*-stearyl]-6'-β-D-glucopyranoside	Sitosterol acylglycosides	Plant	Ethereal extract	Spectroscopic methods	–	–	Greca et al. 1991
Sitosterol-3-*O*-[4'-*O*-stearyl]-β-D-xylopyranoside	Sitosterol acylglycosides	Plant	Ethereal extract	Spectroscopic methods	–	–	Greca et al. 1991
Soya-cerebroside I	Glycosphingolipid	Whole plant	Ethanolic extract	Spectroscopic analysis, column chromatography	6444148	$C_{40}H_{75}NO_9$	Liu 2007; Liu et al. 2008
Stearic acid ethyl ester (ethyl stearate)	Fatty acid ester	Leaves, root, shoot	Ethanolic extract	GC-MS analysis	8122	$C_{20}H_{40}O_2$	Tyagi and Agarwal 2017b, 2017a
Stigmasta 4,22-dien-3-one	Stigmastane	Plant part	–	Chromatography	6442194	$C_{29}H_{46}O$	Ling et al. 1999
Stigmasta-5,22-diene-3,7-diol	Stigmastane	–	–	Column chromatography	146159416	$C_{29}H_{48}O_2$	Liu 2007
Stigmasterol	Stigmastane	Leaves	–	Chromatography	5280794	$C_{29}H_{48}O$	Ling et al. 1999
Stigmasteryl stearate	Triterpenoid	–	–	Chromatography	13828744	$C_{47}H_{82}O_2$	Ling et al. 1999
Tetradecane	Alkane	Aerial part, leaves	Methanol extract	GC-MS analysis	12389	$C_{14}H_{30}$	Nisha et al. 2018; Lawal et al. 2020
Trans-2-fluoro-3-(trimethylsily) Octanocide	Fatty acid	Leaves	Methanol extract	GC-MS analysis	–	–	Nisha et al. 2018
Trideutetro methyl ethyl ether	Dialkyl ether	Aerial part, Leaves	Methanol extract	GC-MS analysis	–	–	Nisha et al. 2018; Lawal et al. 2020
Vanillic acid	*m*-Methoxybenzoic acids and derivatives	–	Acetone extract	HPLC analysis	8468	$C_8H_8O_4$	Ganesh and Sharma 2018

(continued)

TABLE 9.3 (Continued)
Compounds Isolated from *P. stratiotes*

Chemicals	Chemical group	Plant part	Extract type	Method	PubChem CID	Molecular formula	References
Vicenin	Flavonoid	Plant	Methanol extract	Chromatography	5280441	$C_{21}H_{20}O_{10}$	Zennie and McClure 1977
Vitexin	Flavonoid	Plant	Methanol extract	Chromatography	5280441	$C_{21}H_{20}O_{10}$	Zennie and McClure 1977
α-Asarone	Phenylpropanoid	Plant	Ethyl ether extract	Chromatography	348286817	$C_{12}H_{16}O_3$	Aliotta et al. 1991
α-Glyceryl linolenate	Fatty acid ester	Leaves, root, shoot	Ethanolic extract	GC-MS analysis	381122762	$C_{21}H_{36}O_4$	Tyagi and Agarwal 2017b
β-Sitosterol	Sterol	Whole plant	Ethanolic extract	Spectroscopic analysis, column chromatography	222284	$C_{29}H_{50}O$	Liu 2007; Liu et al. 2008

CID, compound identifier; GC-MS, gas chromatography–mass spectrometry; HPLC, high-performance liquid chromatography; TLC, thin-layer chromatography.

and roots (Wasagu et al. 2013). In addition, vitamins A, C, and E are also found, and in another study, volatile oils and resins are also reported (Tyagi and Agarwal 2017a).

The anti-nutrients such as alkaloids, anthocyanin phenol, anthraquinone glycoside, anthraquinone, cardiac glycoside, cyanogenic glycoside, flavonoids, glycosides, phenols, quinones, saponins, steroids, tannin, and terpenoid are also reported from this plant (Kumar et al. 2010; K.A. Ali et al. 2011; Wasagu et al. 2013; Dethe et al. 2014; Tyagi and Agarwal 2017a; Akter and Sultana 2019).

9.3.4 Biological Activities of *Pistia stratiotes*

9.3.4.1 Analgesic Activity

The *in vivo* tail immersion method and acetic acid-induced writhing method were used to evaluate analgesic activity in Swiss albino mice of either sex (20–25 g) by oral administration of petroleum ether and ethanolic extract of *P. stratiotes* leaves (Kumar et al. 2011).

In the tail immersion method, the control group received sodium lauryl sulphate (SLS) 0.5% (3 ml/kg), the standard control group received the standard drug diclofenac sodium (45 mg/kg) suspended in 0.5% (w/v) SLS, and the test group received petroleum ether extract of 200 and 400 mg/kg suspended in 0.5% (w/v) SLS, and ethanolic extract of 200 and 400 mg/kg suspended in 0.5% (w/v) SLS, separately. The time taken to withdraw the tail was noted as the reaction time that was measured at 0, 15, 30, 45, and 60 minutes, respectively. Thirty minutes after drug administration, reaction time increased significantly for the test and standard controls when compared to the control. The test extracts and the drug produced a dose-dependent increase in reaction time at various observation time intervals.

In the acetic acid-induced writhing method, the control group received 2% Tween 80 (3 ml/kg), the positive control group received aspirin (50 mg/kg), and the test groups received 200 and 400 mg/kg of petroleum ether extracts and 200 and 400 mg/kg of ethanolic extracts, separately. All the test doses were given 1 hour prior to intraperitoneal (i.p.) acetic acid injection (0.1 ml of 6% v/v) and then after 10 minutes, writhing movement was recorded for 20 minutes. Results showed that there was a dose-dependent reduction in writhing of the body when extracts were administered. The maximum percentage of inhibition of writhing response was exhibited by ethanolic extract at 400 mg/kg (47.34%), like the standard drug aspirin at 50 mg/kg (47.75%).

Rahman et al. (2011) tested the *in vivo* antinociceptive activity of methanolic extract of leaves of *P. stratiotes* in young Swiss albino mice (weighing 20–30 g) of either sex. Writhing was induced by using acetic acid which was administered at 250 and 500 mg/kg body weight and was compared with diclofenac sodium (25 mg/kg of body weight) as standard and control mice that were given 10 ml/kg body weight of vehicle (1% Tween 80 in water). Doses of 250 and 500 mg/kg body weight of the extract produced about 35.80% and 47.94% writhing inhibition in test mice, respectively, and were comparable to the standard drug diclofenac sodium which showed 60% writhing inhibition.

Methanolic acetate fractions of *P. stratiotes* were administered to Swiss albino mice (weighing 25–35 g) at 200 and 400 mg/kg doses to test analgesic activity by observing acetic acid-induced writhing and formalin-induced licking and biting performance (Hussain et al. 2018). Vehicle (1% Tween 80; control) or indomethacin (10 mg/kg; standard) were administered orally 30 minutes before the i.p. administration of 1% carboxylic acid. Then the mice were observed for specific contractions of the body termed 'writhing' for the succeeding 20 minutes and the number of 'writhes' in every treated group was compared to the activity produced with standard indomethacin (10 mg/kg; positive control). Plant extract caused a dose-dependent reduction in abdominal constrictions and stretching compared to the control and standard groups (indomethacin caused 78.62% inhibition of writhing).

A 20 μL dose of 2.5% formalin was injected into the dorsal surface of the right hind paw 30 minutes after the administration of the test extract, and indomethacin (10 mg/kg) was used as a positive control. The mice were observed for 30 minutes after the injection of formalin, and the total time of licking and biting the injected hind paw was recorded. The control group received 2.5% formalin for the formalin-induced licking and biting test. At a 400 mg/kg dose, there was a 71.25% reduction in hind paw licking. However, at a 200 mg/kg dose, there was a slightly lower, non-significant reduction in hind paw licking compared to the reference standard, indomethacin.

9.3.4.2 Anthelmintic Activity

A study investigated the anthelmintic activity of ethanolic extract of *P. strategies* against *Pheretima posthuma* and nematode *Ascaridia galli* (Kumar et al. 2010). Out of six groups with six worms each, group one was treated as control with worms placed in normal saline; test groups were treated with 10, 20, and 50 mg/ml of extracts in 1% gum acacia in normal saline; and reference standard groups with piperazine citrate (15 mg/ml) in normal saline and albendazole (20 mg/ml) in normal saline. The observation of time taken to set paralysis and death showed the anthelmintic activity in a dose-dependent manner giving the shortest time of paralysis (*P. posthuma* 3.15 ± 0.13 minutes, *A. galli* 1.92 ± 0.32 minutes) and death (*P. posthuma* 14.75 ± 0.33 minutes, *A. galli* 9.16 ± 0.58 minutes) with 50 mg/ml concentration, like the standard drug piperazine citrate but did not cause death. However, albendazole caused both paralysis and death of worms, but the time taken was much longer.

Karim et al. (2015) tested *in vitro* anthelmintic activity by taking *Pheretima posthuma* (4–8 × 0.2–0.4 cm) that was placed in a Petri dish containing 25 ml each of 10, 25, and 50 mg/ml methanolic and petroleum ether fractions of leaf and roots with distilled water containing 1% Tween 80, where piperazine citrate (10 and 20 mg/ml) and 1% Tween 80 in distilled water were used as positive and negative controls, respectively. The experiment showed that both extracts caused paralysis and death in a dose-dependent manner. However, petroleum root extract paralysed the animal for a short period of time (6.5 minutes) with a dose of 50 mg/ml, similar to the positive control piperazine citrate.

9.3.4.3 Anti-allergic Activity

To analyse anti-allergic properties, two tests (tear secretion test and tear film stability test) were carried out by Abokyi et al. (2014) *in vivo* in ICR mice of either sex (10–12 weeks old, 26–30 g). A test group of mice was treated with the aqueous leaf extract of *P. stratiotes* (100 mg/kg) for seven consecutive days. Test mice groups were compared with the control group treated with normal saline (2 ml/kg) and groups treated with cetirizine (5 mg/kg) and prednisolone (10 mg/kg).

First, tear film secretion and tear film stability tests were carried out before and after treatments using the phenol red thread test and the fluorescein tear break-up time (FTBUT) by using a 1 mm portion of the phenol red thread inserted at the conjunctival cul-de-sac to observe wetting and colour change to red. Next, the stability of the tear film on the ocular surface was performed by FTBUT where 1 μL of 1% sodium fluorescein solution was instilled into the conjunctival cul-de-sac and measurements of FTBUT were carried out using a cobalt blue filter. The study showed that there was no significant effect ($p > 0.05$) on tear secretion and tear film stability due to the plant extract. It was comparable to normal saline treatment, but tear secretion and tear film stability declined significantly with cetirizine treatment, while tear secretion reduced significantly when treated with prednisolone. The results mentioned above showed that *P. stratiotes* does not affect tear secretion and tear film stability. Hence, its use is not likely to be associated with adverse effects, which conventional anti-allergic drugs have, during the management of allergic conjunctivitis.

9.3.4.4 Antiarthritic Activity

Kyei et al. (2012) determined the antiarthritic effect of aqueous and ethanolic leaf extracts of *P. stratiotes* in arthritis induced in male and female six–eight-week-old Sprague Dawley mice (180–200 g). Arthritis was induced using 0.1 ml of complete Freund's adjuvant (mycobacterial components), which caused paw oedema. Nine days after induced arthritis, mice were treated with aqueous and ethanolic leaf extracts (30, 100, or 300 mg/kg) of *P. stratiotes*. They were compared with groups treated with control drugs (0.3 mg/kg of methotrexate) given i.p. every four days and 0.43 mg/kg of diclofenac given i.p. daily and 1 mg/kg of dexamethasone give i.p. every other day, and a normal group treated with 1 ml/kg normal saline given orally daily. The experiment also included a non-arthritic control group consisting of incomplete arthritis that was induced by intraplantar injection of 0.1 ml of sterile paraffin oil (incomplete Freund's adjuvant) and a normal control group had no induction of arthritis. Arthritic indices for injected paw and non-injected paw were recorded on day 28, and finally, mice were sacrificed and their blood samples analysed. It was found that the 30, 100, and 300 mg/kg doses of aqueous extract and the 30 and 100 mg/kg doses of ethanolic extract reduced the swelling of injected paw oedema, like the control drugs methotrexate, dexamethasone, and diclofenac, whereas only the 30 mg/kg dose of aqueous extract reduced the swelling in non-injected paw. Overall results showed that arthritic indices were reduced by all drug doses, except for the 100 and 300 mg/kg doses of ethanolic extract. Blood analysis showed that white blood cell levels decreased in arthritic mice treated with the 30 mg/kg dose of aqueous leaf extract and those treated with methotrexate. Erythrocyte sedimentation rate and C-reactive protein levels were lower in all the treatment groups except for the mice treated with aqueous leaf extract 300 mg/kg and ethanolic leaf extract 100 and 300 mg/kg doses. The arthritic animals treated with 30 mg/kg of the aqueous extract showed no inflammatory changes in the injected paw, while the non-injected paw showed only foci of mild chronic inflammatory changes like the reference drug treatment used in histopathological studies. Thus, it is concluded that aqueous and ethanolic leaf extracts of *P. stratiotes* have antiarthritic activity.

9.3.4.5 Antidiabetic Activity

Lawal et al. (2019) induced diabetes in Wister mice (150–200 g) by i.p. administration of alloxan monohydrate (160 mg/kg body weight) mixed with normal saline. Normal control and untreated diabetic control mice received the vehicle, diabetic mice groups received glibenclamide (5 mg) and 200 mg/kg of aqueous extract of the whole plant of *P. stratiotes*. The normal extract control mice group received 200 mg/kg of aqueous extract. Weekly fasting blood glucose (FBG) was monitored for three weeks, and after three weeks, all mice were sacrificed. Blood was collected by cardiac puncture for analysis of high-density lipoprotein (HDL), triacylglycerols (TG), total cholesterol (TC), serum aspartate aminotransferase (AST), alanine aminotransferase (ALT), direct and total bilirubin, urea, and creatinine. It was found that oral administration of aqueous extract significantly reduced FBG by 68.6% compared to untreated diabetic mice and was similar to glibenclamide (60.6%). Blood examinations showed that treatment of aqueous extract prevented the elevation of TC, TG, LDL, and HDL reduction and increased serum levels of urea, creatinine, ALT, AST, and total and direct bilirubin showing antidiabetic properties.

Lawal et al. (2020) studied molecular interactions of 19 bioactive compounds reported by Tyagi and Agarwal (2017a) against diabetic targets enzymes; glucose-6-phosphatase (G6PC, PDB ID: 1VNF) and sodium-glucose transporter-1 (SGLT1, PDB ID: 3DH4) by using the screening tool AutoDock 4.2.6 and the program PyRx version 0.8 docking software. It was found that out of 19 compounds, five compounds (tetracosahexaene, glycerol-1-palmitate, stigmasterol, diisooctyl phthalate, and phytol acetate) showed excellent inhibitory activity for SGLT1(3DH4) protein and G6PC (1VNF) protein. The evaluation of absorption, distribution, metabolism, and excretion-toxicity (ADME-T) showed that stigmasterol, glycerol-1-palmitate, diisooctyl phthalate, and phytol acetate were likely to be of potential use against diabetes.

9.3.4.6 Antidiarrhoeal Activity

Rahman et al. (2011) tested *in vivo* antidiarrhoeal activity of methanolic extract of leaves of *P. stratiotes* in young Swiss Albino mice (20–30 g) of either sex. Diarrhoea in mice was induced by using castor oil. Mice with diarrhoea were administered with extract of 250 and 500 mg/kg body weight and they were compared with loperamide (50 mg/kg of body weight) as a positive control and control mice that were administered with 10 ml/kg body weight of vehicle (1% Tween 80 in water). Diarrhoeal initiation time and the number of stools excreted by mice in 4 hours showed that the extract caused a significant delay of the onset of diarrhoeal episode by 250 mg/kg body weight (27.78% of diarrhoeal inhibition) and 500 mg/kg body weight (41.67% of diarrhoeal inhibition) and was comparable to loperamide (71.22% of diarrhoeal inhibition). The latent period for the initiation of stool excretion was delayed by 1.1 hours and 0.34 hours compared to control mice showing anti-diarrhoeal activity of plant extract.

Antidiarrhoeal activity was studied in three–four-week-old Swiss albino mice of both sexes (around 20–25 g) by Karim et al. (2015). Diarrhoea in mice was induced when they were kept under fasting for 24 hours and then fed with 0.5 ml of castor oil and magnesium sulphate (2 g/kg). In both cases, four different fractions (methanolic and petroleum ether fractions of leaves and roots each) of *P. stratiotes* at the doses of 200 and 400 mg/kg, loperamide (3 mg/kg) as a positive control and only vehicle (1% Tween 80 in water) as a negative control, were administered orally. Different extracts showed a dose-dependent antidiarrhoeal activity but much higher than the positive control and vehicle. Leaf extracts of both methanol and petroleum ether were more effective than other parts, and root extracts, too, possess antidiarrhoeal activity that was higher in a dose of 400 mg/kg.

9.3.4.7 Antifungal Activity

In vitro antifungal activity of methanolic leaf extracts of *P. stratiotes* against skin disease-causing fungi (*Trichophyton mentagrophytes*, *Microsporum gypseum*, *Microsporum nanum*, and *Epidermophyton floccosum*) was tested by using by the microplate dilution method where the minimum inhibitory concentration (MIC) and minimum fungicidal concentration (MFC) values were determined (Premkumar and Shyamsundar 2005). For determining MIC and MFC, 100 μl of serially diluted *P. stratiotes* extract ranging from 500 μg/ml to 7.8 μg/ml was added in each well, with the first well having 500 μg/ml concentration and the seventh well having a concentration of 7.8 μg/ml. The negative control contained 200 μl of Sabouraud's dextrose broth (SDB) without extract and inoculum. Miconazole was used as a positive control in a concentration range of 96 to 1.5 μg/ml prepared in SDB, containing fungal inoculum. The results showed that extract was the most active against the dermatophytes *T. rubrum*, *T. mentagrophytes* and *E. floccosum* with MIC and MFC values of 250 μg/ml, while against *M. gypseum* and *M. nanum*, the values were 125 μg/ml. Miconazole at a concentration of 3 μg/ml inhibited all fungi and the negative control showed no fungal growth.

10 and 20 μl of the crude ethanol extracts of the aerial part of *P. stratiotes* were used against *Candida albicans* and *Rhodotorula rubra* using the disc diffusion method (K.A. Ali et al. 2011). The inhibition zones were 18 mm (10 μl/disc) for *C. albicans* and 28 mm (20 μl/disc) for *R. rubra*. These zones of inhibitions were similar to nystatin (30 μg/disc), a positive standard control.

The next study for antifungal activity used a fresh crude methanol and ethanol extract of roots and of leaves of *P. stratiotes* against *Aspergillus niger* and *Candida albicans* using the agar-well diffusion method (Tyagi and Parashar 2017). To find out the MIC of both extracts, extracts were used at different concentrations (20, 40, 60, 80, 100, 120, 160, 200, 250, 300 mg/ml) and MIC test plates were incubated for 72 hours at 28°C. Only the concentration of 300 mg/ml showed antifungal activity and leaf extract showed higher inhibitory effect against *A. niger* (inhibition zone of 9.67 ± 0.47 mm) and *C. albicans* (inhibition zone of 10.00 ± 0.00 mm) than root extract acting on *A. niger* (inhibition zone of 8.33 ± 0.47 mm) and *C. albicans* (inhibition zone of 9.66 ± 0.47 mm). The zone of inhibition due to the standard drug fluconazole (5 mg/ml) against both fungi was similar to the above-mentioned results.

9.3.4.8 Anti-inflammatory Activity

The carrageenan-induced hind paw oedema model and the cotton-pellet granuloma model were used to evaluate anti-inflammatory activity in male albino mice (120–150 g) with oral administration of petroleum ether and ethanolic extract of the whole *P. stratiotes* plant (Kumar et al. 2011). In the carrageenan-induced hind paw oedema model, a control group of male mice was administered with 1 ml of vehicle propylene glycol, and another group with 12.5 mg/kg of diclofenac sodium as the standard drug suspended in propylene glycol; one set of test groups was administered with 200 and 400 mg/kg of petroleum ether extract suspended in propylene glycol, and another set of test groups with 200 and 400 mg/kg of ethanolic extract suspended in propylene glycol. After 1 hour of drug treatment, the animals were administered with subcutaneous injection (0.05 ml) of 2% w/v carrageenan solution and contralateral hind paws were injected with 0.1 ml of saline as control. The paw volume was measured plethysmographically at 0, 1, 2, and 3 hours after injection of oedematogenic agent. The result indicated that extracts of all doses showed a significant decrease in oedema-induced hind paw oedema, similar to standard drugs. However, the overall results showed that ethanolic extract has better anti-inflammatory activity than petroleum ether.

In the cotton-pellet granuloma model, two sterilized cotton pellets weighing 10 ± 1 mg were implanted subcutaneously by incision on either side of the back of the mice under ether anaesthesia. The control group was given 1 ml of vehicle (propylene glycol), and the standard test group was given 100 mg/kg of diclofenac sodium (100 mg/kg, orally) suspended in propylene glycol; the first set of test groups was given 200 and 400 mg/kg of petroleum ether extract suspended in propylene glycol and the next set of test groups was given 200 and 400 mg/kg of ethanolic extract suspended in propylene glycol. Extracts and standard drugs were administered daily orally for seven consecutive days and finally mice were sacrificed by cervical dislocation and the pellets together with the granuloma tissues were carefully removed, dried in an oven at 60°C for 24 hours, weighed, and compared with control. The results showed that in granuloma-induced sub-chronic inflammation, the ethanolic extract in doses of 200 and 400 mg/kg had more significant anti-inflammatory activity than petroleum ether extract. After drug administration, 400 mg/kg of ethanolic extract showed 40.33% and diclofenac sodium showed 52.99% inhibition of granuloma when compared to the control group.

Aqueous and ethanolic extracts of leaves of *P. stratiotes* were tested in eight-week-old Sprague Dawley mice of either sex (205 ± 20 g) with acute inflammation that was induced via sub-plantar injection of carrageenan (0.2 mg), histamine (0.1 mg), serotonin (0.1 mg), prostaglandin E (0.2 ml), and bradykinin (0.2 ml of 10 nmol) in the right hind paws (Koffuor et al. 2012). In all cases, aqueous and ethanolic extracts were administered in three doses (30, 100, and 300 mg/kg, orally), and distilled water (1 ml/kg, orally) was given to the control group. The next group was given reference anti-inflammatory drugs. Diclofenac sodium (0.93 mg/kg, i.p.) was given to carrageenan-induced mice 1 hour after induction of oedema, chlorpheniramine (0.35 mg/kg, orally) to histamine-induced mice, granisetron (28.5 mcg/kg, orally) to serotonin-induced mice, and again diclofenac sodium (0.93 mg/kg, i.p.) was given to prostaglandin-induced mice. As an indication of inflammation, paw thickness was measured plethysmographically. In carrageenan-treated mice, paw oedema was measured at 1, 2, and 3 hours post-treatment; in histamine-treated mice, paw oedema was measured at 30-minute intervals for 2.5 hours post-treatment; in serotonin-treated mice, paw oedema was measured at 30-minute intervals for 2.5 hours post-treatment; in prostaglandin E-treated mice, paw oedema was measured at 0.5, 1, 1.5, 2, and 2.5 hours post-treatment; and in bradykinin-treated mice was measured as for carrageenan-treated mice. Results showed that the extracts at all doses seriously reduced paw thickness in all the models of inflammation, except the 300 mg/kg doses in carrageenan and serotonin-induced inflammation. The extracts' anti-inflammatory effects were comparable to the different reference drugs.

Anti-inflammatory activity of methanolic acetate fractions of leaves of *P. stratiotes* was assessed in Swiss albino mice (weighing 25–35 g) by administering 200 and 400 mg/kg doses once acute inflammation was induced by injecting 0.1 ml of (1%) carrageenan in the plantar surface of one of the hind paws (Hussain et al. 2018). The paw volume was measured at 0, 1, 2, and 3 hours. Controls, normal saline (1 ml/kg), or ibuprofen (10 mg/kg) were administered 30 minutes before carrageenan administration. The plant extract at both doses exerted a significant (p <0.05) anti-inflammatory effect from 0 minutes to 3 hours after administration, and results were comparable to control and reference standard groups.

9.3.4.9 Antimicrobial Activity

Crude ethanol extracts of the aerial parts of *P. stratiotes* were used in volumes of 10 and 20 µl against Gram-positive (*Bacillus cereus*, *B. subtilis*, *Staphylococcus aureus*) and Gram-negative (*Escherichia coli*, *Salmonella typhi*, *Proteus mirabilis*) bacteria by using the disc diffusion method (K.A. Ali et al. 2011). The extract showed antimicrobial activity against *B. cereus* with 14 mm (10 µl/disc) and 23 mm (20 µl/disc) zones of inhibition; *B. subtilis* with 16 mm (10 µl/disc) and 21 mm (20 µl/disc) zones of inhibition; *Staph. aureus* with 25 mm (10 µl/disc) and 34 mm (20 µl/disc) zones of inhibition; *E. coli* with 16 mm (10 µl/disc), and 24 mm (20 µl/disc) zones of inhibition; *S. typhi* with 9 mm (10 µl/disc) and 19 mm (20 µl/disc) zones of inhibition; and *P. mirabilis* with 11 mm (10 µl/disc) and 13 mm (20 µl/disc) zones of inhibition. All these results were similar to the standard results of the standard controls: vancomycin (30 µg/disc) and tetracycline (30 µg/disc).

In vitro antimicrobial activity of the crude ethanol extract was investigated in Gram-positive (*Staph. aureus*, *Pseudomonas aeruginosa*) and Gram-negative (*E. coli*, *B. subtilis*) bacteria by using the agar-well diffusion method (Tyagi and Parashar 2017). To determine the minimum inhibitory concentration (MIC), fresh methanol and ethanol extracts were used at different concentrations (20, 40, 60, 80, 100, 120, and 160 mg/ml) and MIC test plates were incubated for 24 hours at 37°C. When antibacterial activities of 160 mg/ml ethanol extracts of leaves and roots were compared, leaf extract showed inhibitory effect in all bacteria with the maximum zone of inhibition in *B. subtilis* (9.33 ± 0.47 mm) and minimum in *Staph. aureus* (8.00 ± 0.00 mm) and root extract showed an inhibitory effect in all bacteria except in *E. coli* with the maximum zone of inhibition in *Staph. aureus* (9.67 ± 0.47 mm). The zone of inhibition was similar to that of the standard drug, ciprofloxacin (5 mg/ml), in all the bacteria.

Mukhtar and Tukur (2001) also tested aqueous and ethanolic extracts of *P. stratiotes* (0.5, 1.0. 2.5. 5.0, 10, 20, 40,60, 80, and 100 mg/ml) against different bacteria (*Staph. aureus*, *S. typhi*, *E. coli*, *Klebsiella* sp. *Proteus* sp., and *P. aeruginosa*) by using the agar diffusion and tube method. After incubation for 18–24 hours at 37°C ± 1°C, the aqueous extract inhibited the growth of *Staph. aureus* at a disc concentration range of 2.0–20 mg/ml (MIC range 8.5–16 mm), *S. typhi* at a disc concentration range of 40–100 mg/ml (MIC range 6.5–7 mm), and *E. coli* at a disc concentration range of 80–100 mg/ml (MIC range 6.5–7 mm).

Inhibitory activities of the ethanolic and water extracts of roots of *P. stratiotes* were evaluated against eight microbial isolates, including Gram-positive bacteria (*B. subtilis* and *Staph. aureus*) and Gram-negative bacteria (*E. coli*, *K. pneumoniae*, *P. aeruginosa*, *Salmonella* sp., and *Shigella* sp.) (Daboor and Haroon 2012). The agar-well diffusion method was used for the antimicrobial screening with streptomycin and nystatin as standard antibiotics. The results showed that the extracts of *P. stratiotes* showed no zone of inhibition against selected bacterial strains. Ethanolic extract showed similar inhibitory effects for only *B. subtilis* and *E. coli* without any significant difference with inhibition zones of values between 16.133 ± 0.115 and 16.30 ± 0.10 mm, respectively. No inhibitory activity was found for the water extract against all the tested microorganisms. However, ethanol and ethyl acetate extracts in low doses (1 and 2 mg/kg) of the aerial part of *P. stratiotes* tested against *Staph. aureus*, *S. typhi*, and *Lactobacillus* spp. showed no effect against them (Hossain et al. 2018).

9.3.4.10 Antioxidant Activity

Antioxidant activities of leaf and root methanol extracts were tested to determine the ability to scavenge 2,2-diphenyl-2-picryllhydrazyl (DPPH) radical (Tyagi and Parashar 2017). DPPH radical scavenging effect of leaf extract (0.8–3.4 mg/ml) and root extract (1.15–4.6 mg/ml) was shown in terms of IC_{50} (micromolar concentration required to inhibit DPPH radical formation by 50%). It was found that the methanolic extracts of roots showed higher antioxidant activity with IC_{50} values of 4.098 ± 0.03 mg/ml than leaves with IC_{50} values of 2.463 ± 0.018 mg/ml of minimum activity in the DPPH method. Similarly, Tyagi et al. 2017 tested the radical scavenging of DPPH by using methanol extract from leaves and roots. IC_{50} of the extract showed that roots had more antioxidant activity (4.098 ± 0.03 mg/ml) than leaves (2.463 ± 0.018 mg/ml).

In a separate study (Lestari et al. 2021), the antioxidant activity of different fractions of methanol extract subjected to the DPPH radical scavenging effect showed that out of seven fractions, the highest activity was in the third fraction with a half-maximal inhibitory concentration (IC_{50}) value of 131.66 parts per million (ppm). The fifth fraction with the IC_{50} was about 184.62 ppm, and all other fractions were relatively weak with the IC_{50} at more than 200 ppm but were much more potent than the control of vitamin C with 4.14 ppm.

A study by Karim et al. (2015) included the test for the antioxidant property of four different fractions (methanolic and petroleum ether fractions of leaves and roots) by using DPPH, nitric oxide scavenging, and ferric ion (Fe^{3+}) reducing methods. They found that the leaf methanolic extract scavenged nitric oxide (NO) and DPPH radicals in a dose-dependent manner and the petroleum ether fraction of root was found to have the highest efficacy in Fe^{3+} reducing power assay. However, the total antioxidant capacity of the different fractions was equivalent to the standard (ascorbic acid).

Jha et al. (2012a), when testing methanolic extract of *P. stratiotes* leaves, found a significant scavenging effect on the DPPH radical in a dose-dependent manner of concentrations (10–100 μg/ml). However, the effect of the extract was lower than that of ascorbic acid used as a positive control in the same concentrations as extracts.

Methanolic extract of *P. stratiotes* leaves at different doses (10, 25, 50, 75, 100, and 125 μg) were tested for an antioxidant properties by using DPPH free radical scavenging, superoxide (H_2O_2) anion scavenging, and nitric oxide radical scavenging effects (Jha et al. 2010). All of them showed antioxidant activity in a dose-dependent manner, meaning reducing power increased with an increase in the concentration of the extract and were similar to ascorbic acid and butylated hydroxytoluene (BHT) that were used as standards.

9.3.4.11 Antipyretic Activity

The brewer's yeast-induced pyrexia method was used to evaluate antipyretic activity in Swiss albino mice of either sex (20–25 g) by oral administration of petroleum ether and ethanolic extract (Kumar et al. 2011). Fever was induced by injecting 2.0 ml/kg of 20% aqueous suspension of brewer's yeast in normal saline. Then 18 hours after yeast injection, the control group was administered 1 ml of vehicle suspended in 2% gum acacia and the standard group with paracetamol (30 mg/kg suspended in 2% gum acacia); test groups were administered with 200 and 400 mg/kg of ethanolic extract suspended in 2% gum acacia, and with 200 and 400 mg/kg of petroleum ether extract suspended in 2% gum acacia. The rectal temperature was recorded by clinical thermometer at 0, 1, 2, and 3 hours after drug administration, and the results showed all extracts as antipyretic. However, 400 mg/kg ethanolic extract showed more antipyretic activity than petroleum ether extract.

9.3.4.12 Anti-spermatogenic Activity

Singh et al. (2014) investigated anti-spermatogenic activity in male albino mice (25–40 g) by using petroleum ether, chloroform, ethanolic, and aqueous extracts of whole *P. stratiotes* plant and saponin (sitosterol-3-*O*-[2,4-di-*O*-acetyl-6-*O*-stearyl-β-D-glucopyranoside). The control group received 0.5% carboxy methyl cellulose (CMC), the next group received arachis oil daily for 45 days, and all other test groups received different extracts in two oral doses (100 and 200 mg/kg body weight) for 45 days. In addition, saponin was given at a different oral dose of 50 mg/kg body weight/day. At the end of the experiment, all the mice were sacrificed and body weights of all the animals, before and after experimentation, were compared, sperm counts performed, sperm head morphology and histology of testis were studied, level of testosterone was measured, and sperm viability was calculated. It was found that there was a significant decrease in the weight of reproductive organs (testis, epididymis, and seminal vesicle) due to the administration of extracts. The sperm count, sperm viability, and serum testosterone levels were significantly lowered compared to the control group. However, ethanolic extract, at an oral dose of 200 mg/kg body weight was the most active among all investigated extracts.

Similarly, saponin showed a significant decrease in the weight of reproductive organs, including testis, epididymis, seminal vesicle sperm count, sperm viability, and serum testosterone level, and a significant increase in the number of abnormal spermatozoa in comparison to the control group. The histological study of testis indicated that the ethanolic extract- and the saponin-treated mice displayed degradative changes in the seminiferous tubules, which were not observed in control groups. From this study, it can be concluded that *P. stratiotes* can potentially act as a male contraceptive.

9.3.4.13 Antitumour Activity

Antitumour activity in mice that were irradiated with 4 gray radiation in three fractions by Co-60 ϒ radiation source was tested using different compounds (1-*O*-galloyl-β-D-glucose, catechin, epicatechin, epicatechin gallate, gallic acid, kaempferol, rutin, and vallinic acid) and unknown fractions obtained from aqueous, acetone, and methanolic fractions of *P. stratiotes* (Ganesh and Sharma 2018). The antitumour activity was recorded from day 1 when the palpable tumour developed until day 20. The tumour volume was measured between day 1 and day 5, day 11 and day 15, and day 16 and day 20 in two doses (200 and 500 mg/kg body weight). All tested compounds showed antitumour properties by inhibiting tumour volume but rutin isolated from methanolic and aqueous fractions revealed the maximum reduction of the tumour volume of 0.024 cm^3.

Jha et al. (2012a) tested the antitumour activity of the methanolic extract of *P. stratiotes* leaves on the two types of melanoma cell lines (B16F1 and B16F10) by using two different methods known as the trypan blue exclusion (TBE) assay and the microculture tetrazolium (MTT) assay. In the TBE method, plant extract in different dilutions (10–100 µg/ml) were used against cells for 24 hours by keeping them in a CO_2 incubator at 37°C. In the end of experiment, trypan blue dye was used to stain cells and it was found that the extract significantly inhibited cell proliferation in a dose-dependent manner. The percentage inhibition by the extract showed more pronounced efficacy against B16F1 than B16F10 cell lines. The extract at the dose of 100 µg/ml showed its best activity in B16F1 cell lines, and it was like the activity of the standard drug doxorubicin.

In the MTT assay method, extracts of different concentrations (10–100 µg/ml) were used against cell lines and were incubated again for 24 hours in a CO_2 incubator at 37°C. The percentage of inhibition was calculated, and the IC_{50} values of extract for B16F1 were at the dose of 5.09 µg/ml and for B16F10 was 8.05 µg/ml. Thus, in conclusion, among the two cell lines studied, the extract exhibited anticancer activity (Jha et al. 2012a).

9.3.4.14 Blood Pressure Alteration Activity

Wistar strain mice (200–300 g, of either sex) were anaesthetized with 50% urethane (0.3 ml/l00 g rat weight, i.p.), and the trachea was incised to insert a windpipe (Achola et al. 1997). The carotid artery was cannulated for blood pressure recordings, and the femoral vein was cannulated for administering the plant extract (10–60 µg). It was found that systolic, diastolic, and mean blood pressure extract was significantly higher in mice without extract administration than in those administered with the extract. This shows that the plant extract helps in reducing blood pressure.

9.3.4.15 Diuretic Activity

The diuretic activity of ethanol extract of *P. stratiotes* was studied in male Wistar albino mice (180–230 g/kg body weight) by Sahu et al. (2009). All mice received normal saline (25 ml/kg body weight) orally prior to the start of the experiment. The control group received normal saline, the standard control group received urea (1 kg/kg) and furosemide (5 mg/kg), while the test groups received the ethanol extracts (200 and 400 mg/kg body weight). At 5 and 24 hours after the last administration dose, urine volume was collected and measured (ml/day). It was found that the urine output volume increased at both doses (200 and 400 mg/kg body weight), like the controls of urea and furosemide, exhibiting the significant diuretic activity of the plant.

Methanolic and chloroform extracts of *P. stratiotes* leaves were orally administered to male Wistar mice (150–200 g) for screening of diuretic activity (Tripathi et al. 2011). The administration of normal saline, furosemide (4 mg/kg body weight), urea (1000 mg/kg), methanolic extract (200 mg/kg), and chloroform extract (200 mg/kg) to different test mice groups and recording of urine excretion after 5 and 24 hours were performed. Urine was subjected to the analysis of sodium and potassium contents, and it was found that the Na^+/K^+ ratio was 2.3 ± 0.07 for methanolic extract and 2.4 ± 0.09 for chloroform extract, showing a significant increase in sodium ion excretion for both extracts, as the normal Na^+/K^+ ratio is 2.05–2.83. These values are slightly less than that of furosemide (2.25 ± 0.008), but both extracts possess potent diuretic activity.

9.3.4.16 Hepatoprotective Activity

The hepatoprotective activity of ethanolic extract (200 mg/kg) of *P. stratiotes* leaves was investigated against paracetamol (750 mg/kg every 72 hours for 21 days)-induced liver damaged male Wistar albino mice (150–200 g) by Somasundaram et al. (2017). The results after 21 days showed that the hepatic enzymes serum glutamate oxaloacetate transaminase (SGOT), serum glutamate pyruvate transaminase (SGPT), serum alkaline phosphate (SALP), and total bilirubin were increased, while total protein activities decreased in paracetamol-treated mice when compared to the vehicle (0.1% CMC)-treated control group. In the reference control group of mice treated with silymarin (50 mg/kg), the levels of serum enzymes, total bilirubin, and total protein activities were reversed in the paracetamol-induced hepatic injury by reducing the levels of serum hepatic enzymes and total bilirubin, and restoring the decreased total protein level. The extract produced a similar effect to silymarin against paracetamol-induced hepatic damage in mice, therefore showing that ethanolic leaf extract of *P. stratiotes* has a significant hepatoprotective activity against paracetamol-induced hepatotoxicity.

9.3.4.17 Muscular Relaxation Activity

Achola et al. (1997) used methanol extract of *P. stratiotes* for smooth muscle relaxation of the intestine obtained from New Zealand white rabbits (2–2.5 kg), bronchiolar trachea obtained from guinea pigs (300–450 g), and diaphragm obtained from Wistar strain mice of either sex (200–300 g). For this, the calcium channel-blocking activity of the extract was demonstrated using isolated segments of rabbit jejunum, guinea pig trachea, and mice phrenic nerve diaphragm.

For jejunum, a piece (3 cm) of tissue was taken from the abdomen and mounted in a 20 ml double-walled tissue bath containing Krebs–Henseleit (KH) solution and was allowed to equilibrate, then Krebs solution was replaced with Krebs without calcium. Calcium solution (10, 30, and 50 µm) was added to the tissue bath at about 2-minute intervals to restore spontaneous contractions. Tissue was treated with plant extract (1 mg/ml) for about 3 minutes and calcium solution (30 and 50 µM). Finally, tissue was treated with verapamil (1 µM) for about 3 minutes and then calcium solution (30 and 50 µM) to restore spontaneous contractions. Results showed that, first, intrinsic contractions of jejunum were abolished when exposed to calcium-free Krebs solution and were restored by calcium in about 2 minutes. Second, when the tissues were pretreated with the plant extract, calcium did not restore the spontaneous movement of jejunum; however, after 5 minutes intrinsic movements started to appear. Third, calcium did not restore spontaneous contractions of jejunum when the tissues were pretreated with verapamil. However, the plant extract and verapamil inhibited calcium activity in the tissues.

About 3 cm of trachea tissue was equilibrated for 30 minutes in KH solution. Then doses of the plant extract (0.25 and 0.5 µg/ml bath) were investigated against histamine (0.5 µg/ml bath) and a

35-minute tissue–drug contact time cycle. The dilation of tissue was found to be dose dependent, showing plant extract to be a bronchodilator on isolated guinea pig trachea.

For the phrenic nerve diaphragm, tissue was placed in a Petri dish containing KH solution, attached to an electrode, and placed in a 50 ml double-walled tissue bath containing KH solution (95% O_2 and 5% CO_2 at 37°C) and was allowed to stabilize for about 30 minutes. The time of the onset of neuromuscular blockade of the plant extract was found to have a dose-dependent blockade of the mice phrenic nerve system as neuromuscular blockade of 8 mg/ml bath concentration of the plant extract was approximately half that of a 4 mg/ml bath concentration.

In conclusion, the methanol extract of the plant has a muscle-relaxing property of smooth (vascular, intestinal, and bronchiolar) muscles, possibly via calcium channel blockade.

9.3.4.18 Neuropharmacological Activity

For assessing neuropharmacological activities, central nervous system (CNS) depressant tests and antinociceptive tests were performed in Swiss albino mice (three–four weeks old, 20–25 g) by using ethanolic extract of the aerial part of *P. stratiotes* (K.A. Ali et al. 2011).

Four types of tests were performed for CNS depressant tests (hole cross test, open field test, beam walking test, and thiopental sodium-induced sedative test). For three tests (hole cross, open field, and beam walking tests), diazepam (1 mg/kg body weight) was used in the positive control group, and extracts of 250 and 500 mg/kg body weight were given to two test groups, respectively, and vehicle (1% Tween 80 in water) to the control group. For the hole cross test, the number of passages was counted for 5 minutes starting after oral administration of the extract at 0, 30, 60, 90 and 120 minutes. Results of extracts of 250 and 500 mg/kg body showed significant decrease of movement from its initial value at 0 to 120 minutes and were similar to those of diazepam.

In the open field test, the number of squares passed by the mice was counted for 3 minutes after oral administration of test extracts started at 0, 30, 60, 90 and 120 minutes. Extracts of 250 and 500 mg/kg body weight showed a significant decrease in movement from its initial value 0 to 120 minutes.

For the beam walking test, the number of foot slips depicts the insufficiency in motor coordination and only extract at 500 mg/kg body weight showed changes in motor coordination deficit.

In thiopental sodium-induced sedative test, thiopental sodium plus diazepam was given to the positive control group, thiopental sodium plus ethanolic extracts of 250 and 500 mg/kg body weight were given to test groups, and thiopental sodium was given to the control group. All mice were injected with thiopental sodium (10 mg/kg) 15 minutes after administration of either diazepam (1 mg/kg body weight) or extracts (250 and 500 mg/kg body weight). Sleeping time was measured by observing the time interval between losing and regaining of righting reflex and extracts at the doses 250 and 500 mg/kg body showed a significant decrease in onset of action and increased the sleeping time, which was comparable to the positive control group.

The hot plate test, tail immersion test, and acetic acid-induced writhing test were carried out for antinociceptive activity. For the hot plate test and tail immersion test, the positive control group received nalbuphine (10 mg/kg body weight), test groups received 250 and 500 mg/kg body weight extracts, and the control group received only vehicle (1% Tween 80 in water (10 mg/kg body weight)). In the hot plate test, after oral administration of the extracts, reaction time was recorded at 0, 30, 60, and 90 minutes, and test groups produced a dose-dependent increase in latency time relative to the control group. In the tail immersion test, the latent period of the tail-flick response was determined before and 0, 30, 60 and 90 minutes after the administration of extracts and administration of the extract was found to increase in a dose-dependent manner, like the reference drug nalbuphine.

In the acetic acid-induced writhing test, extracts given to test groups and control groups were the same as the hot plate test and tail immersion tests except the positive control group was treated with diclofenac sodium (25 mg/kg body weight). To induce pain in the mice, 0.7% acetic acid solution was used, and 30 minutes allowed for absorption. After an interval of 5 minutes, the number of writhings were counted for 30 minutes. Results showed that in the two test groups treated with the extract, the writhing response was significantly inhibited and the protection increased with the increase in dose. In conclusion, all the tests showed that plant extract exhibited antinociceptive activity.

CNS depressant activities were tested in hole cross and open field tests by using methanolic acetate fractions of *P. stratiotes* that were administered to Swiss albino mice (weighing 25–35 g) (Hussain et al. 2018). For the hole cross test, the first control group was given vehicle (1% Tween 80) at 10 ml/kg body weight, the second control group diazepam at 1 mg/kg body weight, and two test groups were treated with 200 and 400 mg/kg body weight, respectively, of methanolic acetate fractions of *P. stratiotes*. After the administration, the number of mice that passed through the hole from one chamber to the other was counted for 3 minutes at 0, 30, 60, 90, and 120 minutes. The number of movements through the hole was reduced in a dose-dependent manner for two doses of extracts compared to controls and standard.

For the open field, the case floor was marked with a grid, and the control group was treated with distilled water (10 ml/kg), the standard group with diazepam (1 mg/kg body weight), and test groups with 200 and 400 mg/kg of plant extract. The number of fields crossed by each mouse was observed over a timed interval of 3 minutes (at 0, 15, 30, 45, and 60 minutes after administration). Test extracts decreased the movements of the test animals at both dose levels.

9.3.4.19 Wound Healing Activity

The wound healing property of different compounds (1-*O*-galloyl-β-D-glucose, catechin, epicatechin, epicatechin gallate, gallic acid, kaempferol, rutin, and vanillic acid) and unknown fractions obtained from aqueous, acetone, and methanolic fractions of *P. stratiotes* were studied in 24 adult male mice (F1 hybrid from Swiss albino and C57BL male) (Ganesh and Sharma 2018). They were applied topically with the dose of 1.5% and the final wound was measured on day 15 by the physical evaluation of the wound by vernier calliper and re-epithelization property. Different compounds and unknown fractions showed re-epithelization and wound repair between 15 and 20 days. The most significant reduction of the wound was caused by vanillic acid (0.88 mm^3/15th day) and all the constituents were better than control drug povidone iodine.

Swiss albino mice of either sex (24 ± 2 g) were treated with methanolic extract of *P. stratiotes* leaves to analyse wound healing activity (Jha et al. 2012b). In the study, the standard group was treated with Betadine as standard, and the test groups with extract at 5% and 10% (w/w) in a base of Vaseline (pure petroleum jelly) on the wound created in the posterior side of mice. The wound measurements on 4, 6, 8, 10, 12, 14, and 16 days showed a significant improvement in wound contraction by extract when compared with the control group.

9.3.5 Toxicity and Cytotoxicity

Acute toxicity of crude ethanol extract of aerial part of *P. stratiotes* was evaluated by expressing its value as the amount of a material, given all at once, which causes the death of 50% (one half) of test group animals (LD_{50}) by K.A. Ali et al. (2011). First, the extract was given to mice by i.p. injection at different doses (10, 100, 500, 1000 mg/kg), and signs of toxicity and death within 24 hours were determined. Second, specific doses were administered depending on the results of the first phase, and final LD_{50} values were calculated. The maximum dose for LD_{50} was found to be 850 mg/kg body weight i.p.

Acute toxicity of petroleum ether and ethanolic extracts was evaluated in albino female mice (20–25 g) by Kumar et al. (2011). First, mice were fasted before the acute test, then different groups with six mice each were orally given extracts of 1000, 2000, 2500, 3000, 3500 and 4000 mg/kg, respectively, and the control group was given only propylene glycol (vehicle). Mortality and general behaviours of mice were observed for an initial 4 hours, and intermittently for 6 hours and then again at 24 hours and 48 hours after the first dose. In the end, it was found that up to 4000 mg/kg of ethanolic and petroleum ether extracts do not show any toxic effects. However, LD_{50} doses of 2000 mg/kg and above were categorized as unsuitable for administration.

Koffuor et al. (2012) tested toxicity in eight-week-old Sprague Dawley mice (205 ± 20 g) of either sex by administering aqueous and ethanolic extracts of *P. stratiotes* (10, 30, 100, and 300 mg/kg) to test groups; the control group (non-treatment group) received 1 ml/kg distilled water. Clinical and behavioural symptoms of toxicity and mortality were observed hourly for 24 hours and then daily for 14 days. Blood and urine samples were collected 24 hours after treatments and then ten days post-treatment for haematological assessment and biochemical parameters. There were no clinical or behavioural changes in mice. Haematological assessments revealed reductions in red blood cell count in all treatment groups over the ten days, and urine analysis of treated and untreated rats showed an increase in only urobilinogen and microalbumin in all treatment groups ten days post-treatment. Thus, as there is acute and delayed toxicity due to the administration of aqueous and ethanolic extracts, care should be taken, and haematological profile, as well as liver and kidney function, should be monitored.

The toxicity of crude aqueous extract of *P. stratiotes* was investigated by observing the change in body weight and renal cell function, such as concentration of creatinine, urea, potassium, and sodium in the serum of white albino mice (150–164 g). Mice of different groups were administered orally 100, 200, and 300 mg/kg body weight of the extract, respectively, while the control group received vehicle for five days. There was a dose-dependent decrease in body weight loss in mice following the administration of the extract. There was an increase in levels of serum creatinine, urea, and potassium whereas there was a decrease in levels of serum sodium at concentrations of 200 and 300 mg/kg showing nephrotoxicity effects (Wasagu et al. 2014).

Cytotoxic activities of ethanol extract of aerial part of *P. stratiotes* were tested by the brine shrimp lethality bioassay (K.A. Ali et al. 2011). Ten living larvae (nauplii) were kept in each test tube with 20, 40, 60, 80, 100, 120 and 140 µg/ml concentrations of extract with a duplicate of each concentration; and seven tubes were used for the control test. Alive nauplii were counted after 16 hours. The plot of the percentage of mortality versus log concentration of the extract gave an approximately linear correlation, and the graph showed the concentration at which there was 50% and 90% mortality (LC_{50} and LC_{90}) of nauplii. The percentage mortality of nauplii was found to be similar with both doses of 80 and 100 µg/ml of the extract. LC_{50} and LC_{90} were 1.8 and 2.07 µg/ml, respectively, proving that the extract possesses cytotoxic activity.

The cytotoxicity of petroleum ether and methanolic extracts were determined by the brine shrimp lethality bioassay (Karim et al. 2015), which showed median lethal concentration (LC_{50}) values of the test samples that were calculated after 24 hours. It was found that the cytotoxic action increased with the increment of doses. Specifically, petroleum ether fractions of leaf and root extracts were more toxic (LC_{50} values of 121.99 and 45.80 µg/ml for leaf and root, respectively) than methanolic fractions (LC_{50} values of 424.58 and 6217.75 µg/ml for leaf and root, respectively).

9.4 CONCLUSION

Regardless of the introduced or invasive nature of *P. stratiotes*, it is medically useful against different diseases. Furthermore, many of the traditional uses such as for skin infections, wound healing, anti-inflammatory, antimicrobial, diarrhoeal, and antiarthritic uses, and for ulcer and urinary problems have been supported by scientific investigations. However, more pharmacological activities (analgesic, antidiabetic, anthelmintic, antipyretic, anti-spermatogenic, alternation of blood pressure, eye anti-allergy, hepatoprotective, muscular relaxation, and neuropharmacological activities) have been investigated. Some bacteria are resistant to *P. stratiotes* extracts, e.g. *Staph. aureus*, *S. typhi*, and *Lactobacillus* spp. are resistant to the ethanol and ethyl acetate extracts in low doses (1 and 2 mg/kg) of *P. stratiotes*.

Toxicological reports in this review have shown some side effects like nephrotoxicity even at concentrations of 200 and 300 mg/kg; therefore, care must be taken when using the plant to prevent potential toxicity. In addition, as there is acute and delayed toxicity due to the administration of aqueous and ethanolic extracts of the plant, it is recommended to be cautious as there might be alteration in liver and kidney function (Wasagu et al. 2014).

Overall, all the elicited activities of *P. stratiotes* discussed in this review will serve as baseline information for further clinical and pharmacological investigation. Although many bioactive compounds are present in *P. stratiotes*, only eight different compounds (1-*O*-galloyl-β-D-glucose, catechin, epicatechin, epicatechin gallate, gallic acid, kaempferol, rutin, and vanillic acid) were found to have anticancer activity and also help in wound healing (Ganesh and Sharma 2018). Likewise, five compounds (tetracosahexaene, glycerol-1-palmitate, stigmasterol, diisooctyl phthalate, and phytol acetate) were found to have antidiabetic potential (Tyagi and Agarwal 2017a). However, more research on biological activities is recommended for the future for global acceptance of this plant in traditional and modern medicine.

Although several scientific studies have been carried out on *P. stratiotes*, there should be further research to achieve both effectiveness and safety of use for human beings in the future. Further, plant species considered introduced or invasive can be harvested and utilized for compound isolation or for use in herbal medicine so that their invasiveness can be managed for human benefit.

REFERENCES

Abiodun, A.A. & B.H. Tunji. 2019. Medicinal weed diversity and ethno-medicinal weeds in Odigbo local government area, Ondo State, Nigeria. *Journal of Medicinal Plants Studies*, 7(5), 81–85.

Abokyi, S., G.A. Koffuor, E.A. Ofori, A.K. Thomford, & K.A. Osei. 2014. Evaluating the effect of an aqueous extract of *Pistia stratiotes* Linn (Araceae) on tear secretion and tear film stability in ICR mice. *Journal of Applied Pharmaceutical Science*, 4(8), 1–5.

Acharya (Siwakoti), E. & B.B. Pokhrel. 2006. Ethno-medicinal plants used by Bantar of Bhaudaha, Morang, Nepal. *Our Nature*, 4, 96–103.

Achola, K.J., A.A. Indalo & R.W. Munenge. 1997. Pharmacological activities of *Pistia stratiotes*. *International Journal of Pharmacognosy*, 35(5), 329–333.

Adebayo, A., E. Briski, O. Kalaci, M. Hernandez, S. Ghabooli, B. Beric, F. Chan, A. Zhan, E. Fifield, T. Leadley & H. MacIsaac. 2011. Water hyacinth (*Eichhornia crassipes*) and water lettuce (*Pistia stratiotes*) in the Great Lakes: playing with fire? *Aquatic Invasions*, 6(1), 91–96.

Akter, P. & B. Sultana. 2019. Allelopathic effects, yields and qualitative phytochemical screening of root exudates of five weed species. *Malaysian Journal of Sustainable Agriculture*, 3(1), 44–48.

Ali, K.A., P. Paul, I. Md. Torequl, B.N. Nath, S. Shamir & A.A. Khan. 2011. Cytotoxicity, antimicrobial and neuropharmacological evaluation of ethanolic extract of *Pistia stratiotes* L. *International Research Journal of Pharmacy*, 2(2), 82–92.

Ali, M., I.M. Iqbal, A. Shabbir, Z.-U.-D. Khan & M.T.A. Khan. 2020. Ethnomedicinal studies on aquatic plants of tehsil Shakargarh, Punjab, Pakistan. *Journal of Medicine Plants Studies*, 8(1), 15–19.

Aliotta, G., P. Monaco, G. Pinto, A. Pollio & L. Previtera. 1991. Potential allelochemicals from *Pistia stratiotes* L. *Journal of Chemical Ecology*, 17(11), 2223–2234.

Arya, A.K., M. Durgapal, A. Bachheti, K.K. Joshi, Y.H. Gonfa, R.K. Bachheti & A. Husen 2022. Ethnomedicinal use, phytochemistry, and other potential application of aquatic and semiaquatic medicinal plants. *Evidence-Based Complementary and Alternative Medicine* 2022, 4931556.

Ayyad, S.N. 2010. A new cytotoxic stigmastane steroid from *Pistia stratiotes*. *Die Pharmazie*, 57(3), 212–214.

Bartram, J. & F. Harper. 1942. *Diary of a journey through the Carolinas, Georgia, and Florida from July 1, 1765, to April 10, 1766*. Philadelphia, PA: Transactions of the American Philosophical Society.

Bartram, J. & F. Harper. 1943. *Travels in Georgia, and Florida, 1773–74. A report to Dr. John Fothergill*. Philadelphia, PA: Transactions of the American Philosophical Society.

CABI. 2022. Pistia stratiotes (water lettuce) In: *Invasive Species Compendium*. [online]. Wallingford, UK: CABI. Available from: www.cabi.org/isc/datasheet/41496 (accessed 4 Apr 2022).

Chowdhury, A. & A.P. Das. 2015. Ethnopharmacological survey of wetland plants used by local ethnic people in sub-Himalayan Terai and Duars of West Bengal, India. *American Journal of Ethnomedicine*, 2(2), 1–13.

Cilliers, C.J. 1991. Biological control of water lettuce, *Pistia stratiotes* (Araceae), in South Africa. *Agriculture, Ecosystems & Environment*, 37(1), 225–229.

Daboor, S.M. & A.M. Haroon. 2012. In vitro: antimicrobial potential and phytochemical screening of some Egyptian aquatic plants. *The Egyptian Journal of Aquatic Research*, 38(4), 233–239.

DAISIE. 2013. *Delivering Alien Invasive Species Inventories for Europe (DAISIE).* Brussels: Research Institute for Nature and Forest.

Datta, S.C. & A.K. Banerjee. 1978. Useful weeds of West Bengal rice fields. *Economic Botany*, 32(3), 297–310.

Dethe, U.L., S.S. Joshi & V.T. Aparadh. 2014. Screening of bioactive compounds of *Sesbania grandiflora* and *Pistia stratiotes*. *Indian Journal of Advances in Plant Research*, 1(1), 27–30.

Economic Community of West African States (ECOWAS). 1995. *Control of Floating Weeds in the ECOWAS Member Countries. Main Report.* Lagos, Nigeria: ECOWAS.

Florida Department of Environmental Protection. 2007. *Status of the Aquatic Plant Maintenance Program in Florida Public Waters: Annual Report Fiscal Year 2005–2006.* Tallahassee, FL: Florida Department of Environmental Protection, Bureau of Invasive Plant Management.

Ganesh, N. & P. Sharma. 2018. In vivo determination of wound healing and antitumor properties of HPLC eluted fractions of water weeds (*Eichhornia crassipes* and *Pistia stratiotes*). *Research Journal of Life Sciences, Bioinformatics, Pharmaceutical and Chemical Sciences,* 4(2), 355–367.

Gherardi, F. 2007. *Biological Invaders in Inland Waters: Profiles, Distribution and Threats.* Dordrecht: Springer.

González-Torres, L.R., R. Rankin & A. Palmarola, eds. 2012. Invasive plants in Cuba. (Plantas invasoras en Cuba.) *Bissea: Boletin sobre Conservacion de Plantad del Jardin Botanico Nacional*, 6, 1–140.

Greca, M.D., A. Molinaro. P. Monaco & L. Previtera. 1991. Acylglycosyl sterols from *Pistia stratiotes*. *Phytochemistry*, 30(7), 2422–2424.

Holm, L.G. 1991. *The World's Worst Weeds: Distribution and Biology.* Malabar, FL: Krieger Publishing Co.

Hossain, J., A. Khan & A. Md. Uddin. 2018. Antimicrobial efficacy and phytochemical analysis of three aquatic plant species in Bangladesh. *Bangladesh Journal of Microbiology*, 35(1), 7–11.

Hussain, Md. S., Md. Ibrahim, Md. M. Hasan, M.T. Aziz, S.A. Suchi & Md. G. Uddin. 2018. An in vivo study of the pharmacological activities of a methanolic acetate fraction of *Pistia stratiotes* L.: an ethnomedicinal plant used in Bangladesh. *Animal Models and Experimental Medicine*, 1(3), 221–227.

Hussner, A., P. Heidbuechel & S. Heiligtag. 2014. Vegetative overwintering and viable seed production explain the establishment of invasive *Pistia stratiotes* in the thermally abnormal Erft River (North Rhine-Westphalia, Germany). *Aquatic Botany*, 119, 28–32.

Jain, A., S. Roshnibala. P. Kanjilal, R. Singh & H.B. Singh. 2007. Aquatic/semi-aquatic plants used in herbal remedies in the wetlands of Manipur, Northeastern India. *Indian Journal of Traditional Knowledge*, 6(2), 346–351.

Jha, M., N. Ganesh & V. Sharma. 2010. In vitro evaluation of free radical scavenging activity of *Pistia stratiotes*. *International Journal of ChemTech Research*, 2(2), 180–184.

Jha, M., V. Sharma, V. & N. Ganesh. 2012a. In vitro antioxidant and cytotoxicity assay of *Pistia Stratiotes* L. against B16F1 and B16F10 melanoma cell lines. In: L.D. Khemani, M.M. Srivastava & S. Srivastava, eds. *Chemistry of Phytopotentials: Health, Energy and Environmental Perspectives.* Berlin: Springer, 19–23.

Jha, M., V. Sharma & N. Ganesh. 2012b. Antioxidant and wound healing potential of *Pistia stratiotes* L. *Asian Pacific Journal of Tropical Disease*, 2, S579–S584.

Karim, M.F.B., H. Imam, M.R. Sarker, N. Uddin, N. Hasan, N. Paul & T. Haque. 2015. Free radical scavenging, antidiarrheal and anthelmintic activity of *Pistia stratiotes* L. extracts and its phytochemical analysis. *Pakistan Journal of Pharmaceutical Sciences*, 28(3), 915–920.

Khan, M.A., K.B. Marwat, B. Gul, F. Wahid, H. Khan & S. Hashim. 2014. *Pistia stratiotes* L. (Araceae): phytochemistry, use in medicines, phytoremediation, biogas and management options. *Pakistan Journal of Botany*, 46(3), 851–860.

Koffuor, G.A., S. Kyei, E. Woode, E. Ekuadzi & I.O. Ben. 2012. Possible mechanism of anti-inflammatory activity and safety profile of aqueous and ethanolic leaf extracts of *Pistia stratiotes* Linn (Araceae). *Journal of Ghanna Science Association*, 14(1), 69–81.

Kumar, H.K.S., A. Bose, A. Raut, S.K. Sahu & M.B.V. Raju. 2010. Evaluation of anthelmintic activity of *Pistia stratiotes* Linn. *Journal of Basic and Clinical Pharmacy*, 1(2), 103–105.

Kumar, H.K.S., M.B.V. Raju, S.C. Dinda, S.K. Sahu & M. Banerjee. 2011. Analgesic, anti-inflammatory and antipyretic activity of *Pistia stratiotes* L. *Rasayan Journal of Chemistry*, 4(3), 506–511.

Kyei, S., G.A. Koffuor & J.N. Boampong. 2012. Antiarthritic effect of aqueous and ethanolic leaf extracts of *Pistia stratiotes* in adjuvant-induced arthritis in Sprague-Dawley rats. *Journal of Experimental Pharmacology*, 5, 41–51.

Langeland, K.A. & K.C. Burks, eds. 1998. *Identification and Biology of Non-Native Plants Found in Florida's Natural Areas*, 1st ed. Gainesville, FL: Institute of Food and Agricultural Sciences, University of Florida.

Langeland, K.A. & B.E. Smith. 1993. Evaluation of triclopyr and diquat for managing mixed populations of water hyacinth (*Eichhornia crassipes*) and water lettuce (*Pistia stratiotes*). In: *Proceedings Southern Weed Science Society: weed science in harmony with the environment, 46th annual meeting*, 18–20 January. Charlotte, North Carolina, USA, 250–254.

Lawal, M., A. Suleiman, N. Matazu, F.A. Dawud, A. Mohammed & I.A. Umar. 2019. Antidiabetic activity of *Pistia strateotes* L. aqueous extract in alloxan-induced diabetic rats. *Tropical Journal of Natural Product Research*, 3(3), 91–94.

Lawal, M., A.K. Verma, I.A. Umar, M. Gadanya, B. Umar, N. Yahaya, B. Auwal, Z. Usman & M. Adam. 2020. Analysis of new potent anti-diabetic molecules from phytochemicals of *Pistia strateotes* with Sglt1 and G6pc proteins of *Homo Sapiens* for treatment of diabetes mellitus. An in silico approach. *IOSR Journal of Pharmacy and Biological Sciences*, 5(4), 59–73.

Lestari, H., S.D. Bastian & S. Sudirman. 2021. Antioxidant activity of the fractions from water lettuce (*Pistia stratiotes*) extract. *Food Research*, 5(2), 451–455.

Ling, Y., F. Wan, B. He, Y. Zhang & J. Zheng. 1999. Chemical constituents of *Pistia stratiotes* L. Zhongguo Zhong Yao Za Zhi = Zhongguo Zhongyao Zazhi = *China Journal of Chinese Materia Medica*, 24(5), 289–290, 318.

Liu, H.-W. 2007. *Chemical constituents of Pistia stratiotes L.* Masters degree. Northwest A and F University, Yangling, Shanxi, China.

Liu, H.-W., L.-Y. He, J.-M. Gao, Y.-B. Ma, X.-M. Zhang, H. Peng & J.-J. Chen. 2008. Chemical constituents from the aquatic weed *Pistia stratiotes*. *Chemistry of Natural Compounds*, 44(2), 236–238.

Lu, Q., Z.L. He, D.A. Graetz, P.J. Stoffella & X. Yang. 2010. Phytoremediation to remove nutrients and improve eutrophic stormwaters using water lettuce (*Pistia stratiotes* L.). *Environmental Science and Pollution Research*, 17(1), 84–96.

Mann, A., J.O. Amupitan, A.O. Oyewale, J.I. Okogun & K. Ibrahim. 2008. An ethnobotanical survey of indigenous flora for treating tuberculosis and other respiratory diseases in Niger State, Nigeria. *Journal of Phytomedicine and Therapeutics*, 12(1), 1–21.

Mishima, D., M. Kuniki, K. Sei, S. Soda, M. Ike & M. Fujita. 2008. Ethanol production from candidate energy crops: water hyacinth (*Eichhornia crassipes*) and water lettuce (*Pistia stratiotes* L.). *Bioresource Technology*, 99(7), 2495–2500.

Monaco, P. & L. Previtera. 1991. A steroid from *Pistia stratiotes*. *Phytochemistry*, 30(7), 2420–2422.

Mukhtar, M.D. & A.Tukur. 2001. Antibacterial activities of aqueous and ethanolic extracts of *Pistia stratiotes* L. *NISEB Journal*, 1(1), 51–59.

Nisha, S.N., B.A. Jothi, B. Geetha & S. Hanira. 2018. Phytochemical analysis of *Pistia stratiotes* by GC-MS analysis. *National Journal of Multidisciplinary Research and Development*, 3(1), 4–6.

Parsons, W.T. & E.G. Cuthbertson. 2001. *Noxious Weeds of Australia*, 2nd ed. Collingwood, Australia: CSIRO Publishing.

PIER. 2013. *Pacific Islands Ecosystems at Risk*. Honolulu, HI, USA: HEAR, University of Hawaii.

Premkumar, V.G. & D. Shyamsundar. 2005. Antidermatophytic activity of *Pistia stratiotes*. *Indian Journal of Pharmacology*, 37(2), 126–128.

Rahman, M.A., M. Hasanuzzaman, S.R. Muhuri & A.A. Rahaman. 2011. Evaluation of antinociceptive and antidiarrhoeal properties of *Pistia stratiotes* (Araceae) leaves. *Journal of Pharmacology and Toxicology*, 6, 596–601.

Ramesh, K., B. Rajani & T. Chinna. 2019. Ethnomedicinal uses of some medicinal plants collected from Amarabad Tiger Reserve Forest. *Journal of Emerging Technologies and Innovative Research* 6(2), 469–481.

Rivers, L. 2002. *Water Lettuce (Pistia stratiotes)*. Gainesville, FL: University of Florida and Sea Grant.

Rixon, C.A.M., I.C. Duggan, N.M.N. Bergeron, A. Ricciardi & H.J. Macisaac. 2005. Invasion risks posed by the aquarium trade and live fish markets on the Laurentian Great Lakes. *Biodiversity & Conservation*, 14(6), 1365–1381.

Sahu, R.K., A. Roy, A.K. Jha & U. Sharma. 2009. Diuretic activity of ethanolic extract of *Pistia stratiotes* in rats. *Biomedical and Pharmacology Journal*, 2(1), 149–152.

Šajna, N., M. Haler, S. Škornik & M. Kaligarič. 2007. Survival and expansion of *Pistia stratiotes* L. in a thermal stream in Slovenia. *Aquatic Botany*, 87(1), 75–79.

Salihu, T., J.O. Olukunle, O.T. Adenubi, C. Mbaoji & M.H. Zarma. 2018. Ethnomedicinal plant species commonly used to manage arthritis in North-West Nigeria. *South African Journal of Botany*, 118, 33–43.

Senthilkumar, M.S.S., D. Vaidyanathan, D. Svikumar & M.G. Basha. 2014. Diversity of ethnomedicinal plants used by Malayali tribals in Yelagiri hills of Eastern ghats, Tamilnadu, India. *Asian Journal of Plant Science and Research*, 4(1), 69–80.

Sharaibi, O.J., O.T. Ogundipe, A.J. Afolayan & D.O. Aworinde. 2014. Ethnobotanical survey and phytochemical analysis of medicinal plants used for the treatment of hyperprolactinemia in Lagos State, Nigeria. *Journal of Medicinal Plant Research*, 8(43), 1284–1288.

Singh, A., M.K. Singh, D.K. Singh & R. Singh. 2013. Ethnomedicinal studies on wetland plant diversity of district Buxar (Bihar, India). *Unique Journal of Pharmaceutical and Biological Sciences*, 1(2), 18–20.

Singh, K., B.K. Dubey, A.C. Tripathi, A.P. Singh & S.K. Saraf. 2014. Natural male contraceptive: phytochemical investigation and anti-spermatogenic activity of *Pistia stratiotes* Linn. *Natural Product Research*, 28(16), 1313–1317.

Somasundaram, G., R.J.I. Israel, S. Sengottuvelu & S. Jaikumar. 2017. Effect of *Pistia stratiotes* leaf extract on hepatic functions against paracetamol induced liver damage in rats. *Scholars Academic Journal of Pharmacy*, 6(1), 1–3.

Tantiado, R.G. 2012. Survey on ethnopharmacology of medicinal plants in Iloilo, Philippines. *International Journal of Bio-Science and Bio-Technology*, 4(4), 11–26.

Tiwari, S., M. Siwakoti, B. Adhikari & K. Subedi. 2005. *An Inventory and Assessment of Invasive Alien Plant Species of Nepal*. Nepal: IUCN, The World Conservation Union.

Tripathi, P., S. Arora, R. Gupta & P.R. Mali. 2011. Diuretic activity of *Pistia stratiotes* leaf extract in rats. *International Research Journal of Pharmacy*, 2(3), 249–251.

Tripathi, P., R. Kumar, A. Sharma, A. Mishra & R. Gupta. 2010. *Pistia stratiotes* (Jalkumbhi). *Pharmacognosy Reviews*, 4(8), 153–160.

Tyagi, T. & M. Agarwal. 2017a. GC-MS analysis of invasive aquatic weed, *Pistia stratiotes* L. and *Eichhornia crassipes* (Mart.) Solms. *International Journal of Current Pharmaceutical Research*, 9(3), 111–117.

Tyagi, T. & M. Agarwal. 2017b. Gas chromatography-mass spectrometry analysis of bioactive constituents in the ethanolic extract of *Pistia stratiotes* L. *International Journal of Basic and Applied Medical Sciences*, 7(1), 14–21.

Tyagi, T. & P. Parashar. 2017. Antimicrobial and antioxidant activity of *Pistia stratiotes* (L.). *International Journal of Pharma and Bio Sciences*, 8(3), 391–399.

Tyagi, T., P. Parashar & M. Agarwal. 2017. Qualitative phytochemical analysis and antioxidant activity of methanolic extract of *Eichhornia crassipes* (Mart.) Solms and *Pistia stratiotes* L. *International Journal of Pharmacognosy and Phytochemical Research*, 9(5), 632–636.

Uleh, M., C. Ikyese & D. Ekhuemelo. 2017. Ethnomedical study of plants used by indigenous people of Nyiev and Mbawa Districts, Makurdi, Benue State, Nigeria. *GSC Biological and Pharmaceutical Sciences*, 1(3), 1–11.

Umair, M., M. Altaf, R.W. Bussmann & A.M. Abbasi. 2019. Ethnomedicinal uses of the local flora in Chenab riverine area, Punjab province Pakistan. *Journal of Ethnobiology and Ethnomedicine*, 15(1), 7.

Uniyal, A.K., C. Singh, B. Singh, M. Kumar & J.A.T. da Silva. 2011. Ethnomedicinal use of wild plants in Bundelkhand Region, Uttar Pradesh, India. *Medicinal and Aromatic Plant Science and Biotechnology*, 5(1), 81–86.

USDA-NRCS. 2012. *The PLANTS Database*. Baton Rouge, FL: National Plant Data Center.

Vanila, D., S. Ghanthikumar & V.S. Manickam. 2008. Ethnomedicinal uses of plants in the plains area of the Tirunelveli-District, Tamilnanu, India. *Ethnobotanical Leaflets*, 12, 1198–1205.

Ward, M., ed. 2017. *Pistia stratiotes* L. *Bulletin OEPP/EPPO Bulletin*, 47(3), 537–543.

Wasagu, I.Z., F.U. Bunza, M.K. Dallatu, A.A. Ngaski, A.S. Mainasara & S.U. Dandare. 2014. Cytotoxic effect of crude aqueous extract of *Pistia stratiotes* leaves (water lettuce) on body weight and renal cells of albino rats. *Bayero Journal of Pure and Applied Sciences*, 7(1), 14–18.

Wasagu, R.S., Lawal, M., Shehu, S., Alfa, H.H. & Muhammad, C. 2013. Nutritive values, mineral and antioxidant properties of *Pistia stratiotes* (water lettuce). *Nigerian Journal of Basic and Applied Sciences*, 21(4), 253–257.

Waterhouse, F.D. 1993. *The Major Arthropod Pests and Weeds of Agriculture in Southeast Asia. ACIAR Monograph No. 21*. Canberra, Australia: Australian Centre for International Agricultural Research.

Witt, A. & Q. Luke. 2017. *Guide to the Naturalized and Invasive Plants of Eastern Africa*. Wallingford, UK: CABI.

Zennie, T.M. and McClure, J.W. 1977. The flavonoid chemistry of *Pistia stratiotes* L. and the origin of the Lemnaceae. *Aquatic Botany*, 3, 49–54.

Zhengyi, W., P.H. Raven & H. Deyuan, eds. 2010. *Flora of China (Acoraceae through Cyperaceae)*. Beijing: Science Press.

10 Biological Activities and Medicinal Uses of *Arundo Donax* L.

*Pranchal Rajput and Kundan Kumar Chaubey**
Division of Research and Innovation, School of Applied and Life Sciences, Uttaranchal University, Arcadia Grant, P.O. Chandanwari, Premnagar, Dehradun, Uttarakhand, India
* Corresponding author (Kundan Kumar Chaubey): chaubeykk1984@gmail.com

10.1 INTRODUCTION

Arundo donax L. is a potent invasive weed of marine and humid areas that belongs to the family Poaceae, and is usually known as giant reed (Figure 10.1). The plant's root structure is strong and has plump, almost bulbous, rhizomes that form fibrous roots that can penetrate deep inside the soil.

The rhizomes of this plant grow horizontally, giving rise to hollow, many- stemmed, bamboo-like clusters, allowing it to form large colonies. It is a destructive species that can reproduce quickly. *A. donax* is one of the world's top hundred worst invasive alien species. It is a high-yielding non-food crop used as paper pulp, bioenergy, and crude wooden material for making buildings. It contains various bioactive metabolites, carbohydrates, lignins, proteins, cellulose, hemicelluloses, alkaloids, indols, fatty acids, and trace elements. It also shows antibacterial, antifungal, antihelminthic, anticancerous, and other pharmacological effects. The *A. donax* L. plant has a variety of medicinal uses, and diverse parts of the plant, like root, stem, leaves, and rhizomes, are used in conventional and traditional formulations to treat various types of disease. It has high ecological worth, i.e. in heavy metal, red mud, and wastewater treatment, as well as high economic worth, i.e. in the production of medicine, paper, ethanol, construction board, and electricity.

10.1.1 Distribution

A. donax is native to humid and moderate regions and all similar climates in the world (Hickman 1993). It has been cultivated for thousands of years in Asia, the Middle East, northern Africa, and southern Europe. In the early 1800s, it was purposely brought from the Mediterranean to the Los Angeles area of North America (Al-Snafi 2015a). It has now broadly spread into many subtropical and warm temperate regions worldwide.

10.1.2 Habitat

A. donax is a hydrophytic plant, but it can survive in a wide variety of situations, i.e. humid, well-drained soil where the water bodies are present near the land as shown in Figure 10.2. The favourable soil for this plant is also widespread. Plants occur on heavy clay, crude sands, harsh soil, and river sediments from clean water and the waterfront.

DOI: 10.1201/9781003256830-10

FIGURE 10.1 Whole *A. donax* L. plant.

FIGURE 10.2 Habitat *of A. donax* L.

10.2 BIOLOGICAL ACTIVITIES OF *ARUNDO DONAX* L.

This plant has a robust root system with plump, almost spherical rhizomes that develop into fibrous roots that can reach deep within the soil. The root, stem, leaf, and rhizome of the *A. donax* plant, among other sections, exhibit various biological activities. i.e. phytochemical effects, pharmacological effects, ethnomedicinal effects, effect on milk production and fattening performance, effect on central nervous system, and effects on smooth muscles.

10.2.1 Phytochemical Effects

The main categories of phytochemicals in *A. donax* include triterpenoids, alkaloids, steroids, phenols, and lignin (Pelaez et al. 1998). Table 10.1 contains a list of the various phytochemical effects that *A. donax* L. has demonstrated (Pelaez et al. 1998).

A. donax produces many different types of alkaloids. Figure 10.3 depicts the chemical structures of certain alkaloids.

A. donax manufactures a wide variety of steroids. Figure 10.4 depicts the chemical structures of certain steroids.

A. donax produces a wide variety of phenolics, lignins, and carotenoids. Figure 10.5 depicts their chemical structures.

TABLE 10.1
Phytochemicals Produced by *Arundo donax* L.

Sl. no.	Category	Source part	Phytoconstituents	References
1.	Alkaloids	Leaves	Phenyl-beta-naphthylamine, donaxarine, donine, donaxine, arundine, gramine, deoxyvasicinone, donaxiridine, donaxanine	Ghoshal et al. 1969; Khuzhaev 2004
		Rhizome	Arundaline, tuberosine B, arundalcohol, *N*-acetyltryptamine, *trans*-*N*-(*p*-coumaroyl) serotonine, *trans*-*N*-feruloylserotonin, donasine, *N*,*N*-dimethyl-tryptamine, 5-methoxy-*N*-methyltryptamine, bufotenine, bufotenidine	Ghoshal et al. 1969; Jia et al. 2008; Liu et al. 2021
		Roots and rhizome	Arundaphine	Khuzhaev 2004
		Epigeal and roots	Arundinine	Zhalolov et al. 1998
		Roots	Arundamine, arundaline, arundanine, arundacine, arundarine, arundavine, arundafine, *N*-methyltetrahydro-beta-carboline	Khuzhaev 2004
		Leaves	Donaxine	Khuzhaev 2004
		Flowers	Gramine, bufotenine, eleganine, *N*,*N*-dimethyltryptamine, 5-methoxy-*N*-methyl-tetramine	Ghoshal et al. 1971
2.	Sterol and triterpenes	Leaves	Beta-sitosterol, beta-amyrin acetate, campesterol, stigmastenol, stigmasterol, dihydrobrassicasterol, stigmasta-3,5-dien-7-one, stigmasta-4-en-3-one, alpha-amyrenone, beta-amyrenone, stigmasta-3,6-dione, cycloartenone	Chaudhuri and Ghoshal 1970
		Rhizome	Squalene	Pansuksan et al. 2020
3.	Phenolics	Rhizome	Yanthone, xanthene guaiacol, *o*-cresol, 4-methyl guaiacol, *m*-cresol, methylguaiacol, 4-ethyl guaiacol, 4-ethylcatechol, 4-vinyl guaiacol, 4-methylsyringol, eugenol, *E*-isoeugenol, 4-vinylsyringol, *Z*-isoeugenol, *E*-propenylsyringol, 4-ethylsyringol, *Z*-syringypropanol	Pansuksan et al. 2020
4.	Lignins	Stem and leaves	Syringyl propane, guaiacyl, *p*-hydroxyphenylpropane	You et al. 2013
5.	Carotenoids		Xanthophyll, beta-carotene	

10.2.2 Pharmacological Effects

Various pharmacological activities are shown by the *A. donax* plant, i.e. antibacterial activity, antifungal activity, antialgal activity, antihelminthic/antiparasitic activity, antioxidant activity, anticancerous activity, antiproliferative activity, antibiofilm activity, antifeedant activity, and toxicological activity.

1. *Antibacterial activity. A. donax* has demonstrated antimicrobial activity, according to several research reports (Pelaez et al. 1998). In addition to methicillin-resistant *Staphylococcus aureus* (MRSA), aqueous and methanolic extracts of *A. donax* nodes and stem exhibit antibacterial activity of 128 µg/ml (Al-Snafi 2015b). The precise bioactive substance responsible for *A. donax*'s antibacterial action has yet to be studied in general research. Prospective investigations must therefore concentrate on determining the molecule's bioactive functioning mode (Pelaez et al. 1998). *A. donax*'s methanolic extract exhibits the most significant impact on *Escherichia coli* and *Pseudomonas aeruginosa* (Al-Snafi 2015a).

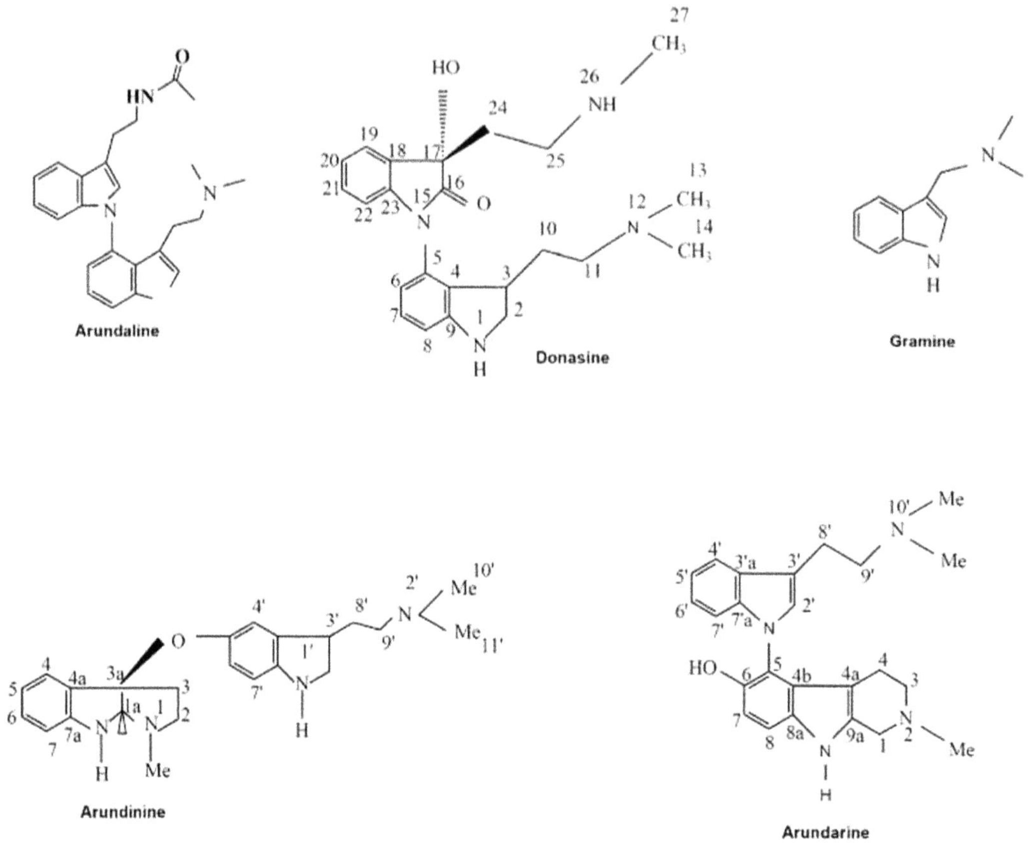

FIGURE 10.3 Chemical structure of alkaloids produced by *A. donax* L.

2. *Antifungal activity*. The four different types of basidiomycetes – *rametes versicolor* CTB, *Coniophora puteana* BAM, *Gloeophyllum trabeum* BAM, and *Postia placenta* FPRL – are stated to exist in opposition to numerous studies (Shatalov and Pereira 2001). Fungal activity can be seen in the plant *A. donax*'s aqueous extract.
3. *Antialgal activity*. According to a study, the methanolic extract of *A. donax* plant shoots inhibits the growth of the cyanobacterium *Microcystis aeruginosa*, which produces algal blooms. It claims that gramine, an allelochemical, exists. This is one of the most potent antialgal chemicals found in marine plants (Pelaez et al. 1998). Other allelochemicals such as skatone, phytone, and phytol discovered in the extract of *A. donax* plant shoots are thought to possess antialgal effects (Hong et al. 2010). The bio-oil and pyrolytic extracted from *A. donax* prevent *Karenina brevis*, a red tide dinoflagellate that can poison aquatic organisms and produce neurotoxin, from growing (Zheng et al. 2018). A study discovered that a neutral portion of the *A. donax* plant's methanolic extract might stop the growth of the potentially harmful alga *Prymmesium parvium* (Patino et al. 2018).
4. *Antihelminthic/antiparasitic activity*. Many researchers have investigated the *A. donax* basic extract of the plant's leaves for its antihelminthic and antiparasitic properties. They found that its ethyl acetate, chloroform, and petroleum spirit fractions were effective against *Haemonchus contortus* (Al-Snafi 2015a). In sheep naturally infected with gastrointestinal nematodes, this plant's crude powder reduced the faecal egg count by 50.5% (Singh and Kaur 2018). In cattle, *A. donax* also exhibits antihelminthic properties against *Ascaris* spp., *Oesophagostomum* spp., and *Paramphistomum* spp. (Pelaez et al. 1998). In addition, it relieves intestinal worms and cough in horses (Pelaez et al. 1998).

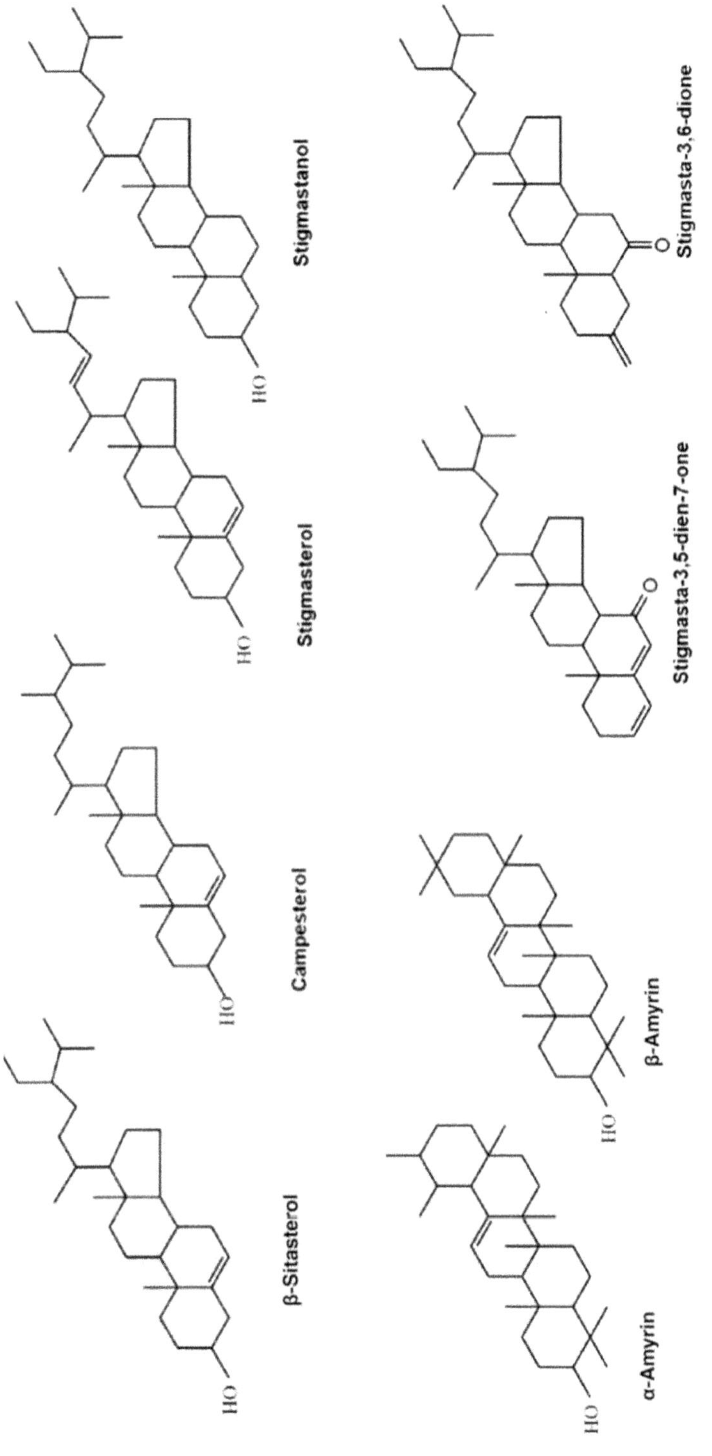

FIGURE 10.4 Chemical structure of steroid produced by *A. donax* L.

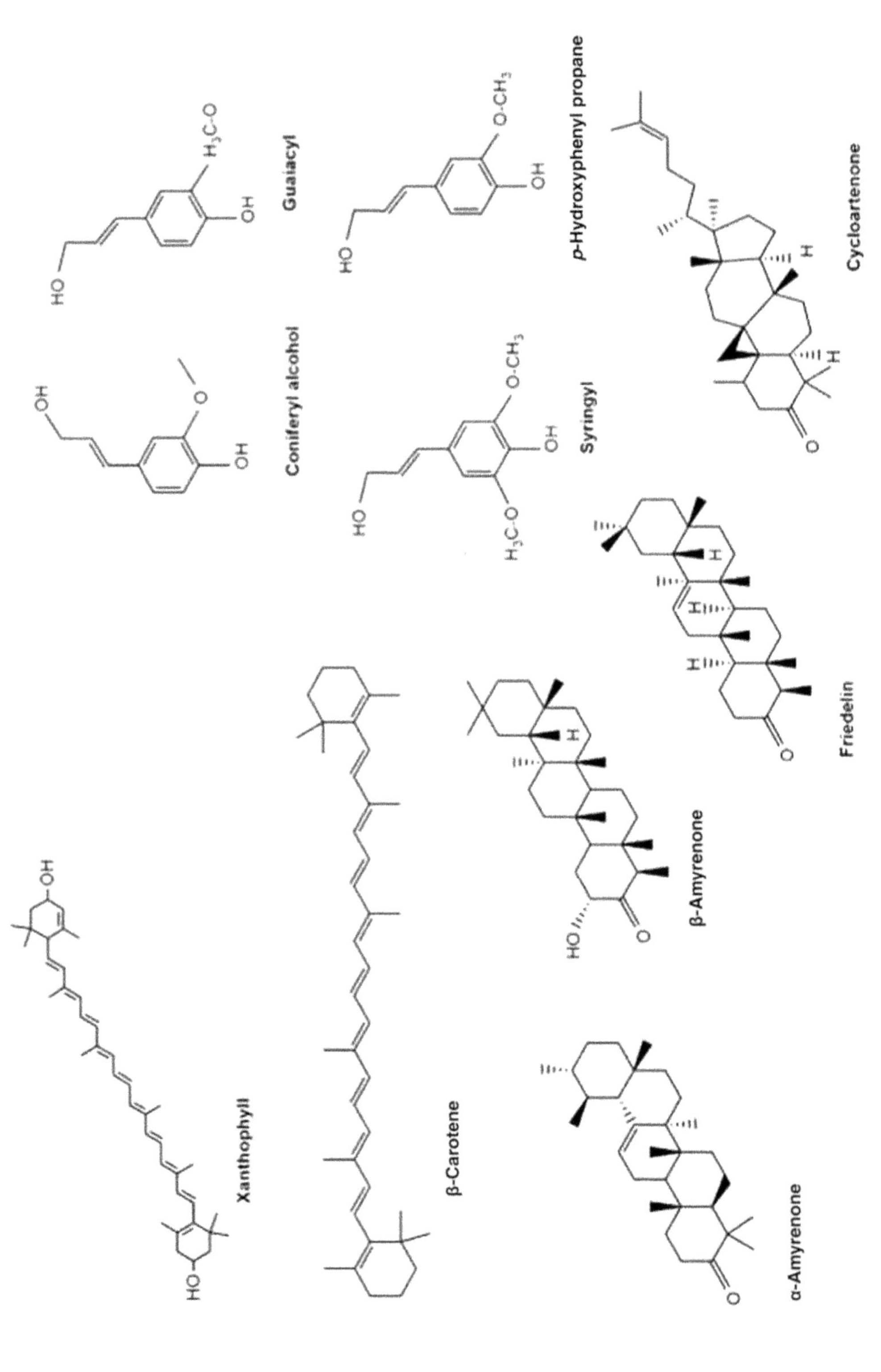

FIGURE 10.5 Chemical structure of phenolics, lignins, and carotenoids produced by *Arundo donax* L.

5. *Antioxidant activity*. Scientists employed the leaf extract from the *A. donax* plant in a phytochemical examination and found that it exhibits total antioxidant activity (Sagir et al. 2018). This plant is a rich source of secondary metabolites with antioxidant characteristics, non-toxic phenol derivatives, tannins, flavonoids, and other phenolic compounds (Piluzza and Bullitta 2011).
6. *Anti-cancerous activity*. All research findings on the anti-cancerous usage of *A. donax* are constrained by ethical standards. The anticancer qualities of this plant have yet to be supported by any scientific evidence (Kaur et al. 2005). However, in Calabria and Southern Italy, *A. donax* is used with *Spartium junceum* L. and *Cynodon daclyton* L. to treat tumours (Leporatti and Impieri 2007). In Pakistan's Bahawalpur district, the leaves and roots of this plant are used against cancer (Nisar et al. 2014). In Muzaffarabad, Pakistan, powder from the stem and root of this plant is administered while adding honey (Ahmed and Akhtar 2016).
7. *Antiproliferative activity*. According to research, *N*-acetyl-D-glucosamine, a lectin derived from the rhizome of *A. donax*, has shown antiproliferative characteristics against human cancer cell lines and mitogenic properties towards human peripheral blood monolayer cells (Kaur et al. 2005). *Dysdercus peruvianus* cells were cytotoxically affected by the lactin isolated from *A. donax*, and *Meloidogyne incognita* also exhibited nematicidal activity (Pelaez et al. 1998).
8. *Antibiofilm activity*. The traditional use of this plant is suggested to be beneficial for biofilm-associated disease. *A. donax* is regarded as one of the medicinal plants; in a study, it was reported that aqueous extract, which is extracted from the nodes of this plant and contains white hemicelluloses membrane and confirmed antibiofilm activity, avoids the formation of biofilm of MRSA and interrupts the establishment of biofilm (Quave et al. 2008).
9. *Antifeedent activity*. *A. donax* displays antifeedant action next to the *Anthonomous grandis* or boll weevil. Numerous bioassays show that, notwithstanding sterol's modest antifeedant action, the main compounds, tetramethyl-*N*,*N*-bis-2,6-dimethyl-phenyl cyclobutane-1,3-diaamine, tricontarol, and tricin, all have antifeedant activity (Al-Snafi 2015a).
10. *Toxicological activity*. The rhizome extract of the *A. donax* plant was noted to be harmful to rats and albino mice in the first study concerning the toxicological activity displayed by this plant, which was published in 1969 (Ghoshal et al. 1969). The alkaloid bufotenidine, a toad toxin, has qualities as a neuromuscular blocking agent. The primary toxicity-causing alkaloid in A. *donax* is bufotenine. When the rats and albino mice received a dose of bufotenidine, i.e. 10 mg/kg, ataxia was evident in all ten rats and death resulted in all five mice. The *A. donax* plant, according to certain research, does not exhibit any toxicological activity towards humans (Pelaez et al. 1998).

10.2.3 Effect on Milk Production and Fattening Performance

The leaves of the *A. donax* plant are poorly absorbed because it has a high fibre and low protein content. A commercially produced powder mixture from various plants, including the *A. donax* plant, increased milk production in cows which was attributed to the presence of a component later determined to be galactogogues (Behera et al. 2013). Compared to *A. donax* hay, lambs of the same age gave better fattening performance with fresh *A. donax* forage (Dekorne 1995).

10.2.4 Effect on Central Nervous System

A. donax plant rhizome extract contains five different tryptamines, including *N*,*N*-DMT, 5-MeO-DMT, and bufotenidine, which have no psychedelic effects, according to certain studies (Christian 1998). For example, 50 mg of rhizome extract had no psychotropic effects (Ana et al. 2003).

However, it can result in minor allergic reactions that continue for a while, such as bloodshot eyes, conjunctivitis, hives, and swollen, watery eyes (Christian 1998).

10.2.5 Effects on Smooth Muscles

The defatted rhizome of the *A. donax* plant's ethanolic extract has been shown to have antispasmodic and hypotensive properties in addition to those of acetylcholine, serotonin, and histamine – including spasms. In an investigation of the effects of bufotenidine on smooth muscles, such as an intestinal loop of guinea pigs, the intestine of dogs, the uterus of guinea pigs and albino rats, and the tracheal chain of dogs, a number of characteristic effects were present (Ghoshal et al. 1969).

10.3 MEDICINAL USES OF *ARUNDO DONAX* L.

Different parts of the *A. donax* plant have various medicinal uses; the root, stem, leaves, and rhizome of this plant are used as traditional and conventional formulations which help to treat a variety of diseases, as shown in Table 10.2. *A. donax* is widely used in communities to cure skin diseases. In Sardinia, Italy, it was used for wound healing, promoting the formation of lactate and haemostasis, and had anti-inflammatory and antiseptic uses (Palmese et al. 2001). The Sushruta (the earliest Indian text) states that the decoction of *A. donax* with *Salic alba* L. is given to people to treat chronic fever (Khare 2007).

In treating rabies, erysipelas, and the retention of urine and fever, the trin panchmula kwath of the *A. donax* plant is used in the Ayurvedic formulation (Singh and Kaur 2018). In addition, an oral dose of about 100–250 g of fresh leaves of this plant treats worm infections in cattle; this method is used by the people of Jhang, Pakistan (Babar 2011). It also treats reproductive disorders, and dermatological and inflammatory infections. Some other ethnomedicinal uses of the different parts of the *A. donax* plant are listed in Table 10.2.

TABLE 10.2
Medicinal Uses of *A. donax* L.

Sl. no.	Category	Condition	Part used	Mode of use	References
1.	Circulatory disorders	Haemostasis	Stem (clum)	For the prevention of infections, to induce cicatrization and wound healing, the internal membrane of stem and dividing membrane of nodes are used	Palmese et al 2001; Pieroni et al. 2002, 2004; Bullitta et al. 2007
		Dropsy	Rhizomes and roots		Egamberdiena et al. 2017; Al-Snafi 2018
2.	Dermatological disorders	Leprosy	Leaves and shoots		Basumalary et al. 2014
		Skin problems	Tender shoot	The affected area is cured by applying pounded new shoot	Kom et al. 2018
		Eczema	Roots	For the treatment of affected area about 20 g of root paste is applied externally	Shekhar et al. 2011

TABLE 10.2 (Continued)
Medicinal Uses of *A. donax* L.

Sl. no.	Category	Condition	Part used	Mode of use	References
		Condyloma	Roots and leaves		Nisar et al. 2014; Al-Snafi 2018
		Astringent	Whole plant		Vanila et al. 2008
3.	Digestive	Diarrhoea and dysentery	Whole plant		Vanila et al. 2008
		Intestinal worms and typhoid	Stem	A mixture of honey and fresh green shoot is used	Jain et al. 2007
		Persistent haemorrhoids	Leaves		Ningthoujan et al. 2013
4.	Excretory problems	Diaphoresis	Roots and stem (clum)		Upadhyay 2016, Bullitta et al. 2007
5.	Urinary disorders	Diuretic	Rhizomes		Kamatchi and Parvathi 2017
6.	Inflammatory	Fever	Leaves and shoot		Jia et al. 2008; Hussain et al. 2010; Basumalary et al. 2014; Singh and Kaur 2018
		Headache	Roots and leaves		Nisar et al. 2014
		Otitis	Leaves	Use vapours	Ballero et al. 2001; Dall'Acqua et al. 2008
7.	Respiratory	Asthma	Stem	Orally given the mixture of honey and tender shoot	Khan and Yadav 2010
		Pertussis	Root sap		Passalacqua et al. 2007
8.	Skeletal	Bone dislocation and leg pain	Root and stem	Powder is taken with honey	Basumalary et al. 2014; Ahmed and Akhtar 2016
9.	Reproductive disorders	Reproductive disorders	Rhizome	Decoction of rhizome can be used	Ghoshal et al. 1969; Sarmah 2010; Kumar and Bhagat 2012; Kamatchi and Parvathi 2017
		Menolipsis (amenorrhoea) in humans	Roots	Mixture of aqueous decoction of fresh root and long pepper (*Piper longum*) paste is given on an empty stomach early in the morning to women	Sahal et al. 2014
		Abortifacient	Root	Orally take the mixture of root decoction and jiggery as abortifacient	Bhatia et al. 2014
10.	Cancer		Rhizome and rootstock	The mixture of honey and decoction of rhizome is taken in wine	Al-Snafi 2018

10.4 CONCLUSION

The plant *A. donax* L. exhibits a wide range of phytochemical, biological, and ethnomedicinal effects, as well as ecological and economic uses. Different bioactive metabolites, carbohydrates, lignins, proteins, cellulose, hemicelluloses, alkaloids, indoles, fatty acids, and trace elements are present. Additional pharmacological actions include antibacterial, antifungal, antialgal, antihelminthic, anti-cancerous, antiproliferative, antioxidant, antifeedant, toxicological, and other properties. It is used to treat a wide range of illnesses, including circulatory difficulties, skin problems, digestive issues, excretory issues, urinary problems, inflammatory problems, respiratory problems, skeletal problems, and cancer. Additionally, it has high ecological value in the treatment of wastewater, heavy metals, red mud, and other pollutants, as well as high economic value in the production of medicines, paper, ethanol, artificial wood, and electricity, all of which are beneficial for the welfare of both humans and animals.

REFERENCES

Ahmed, M.J. and T. Akhtar. 2016. Indigenous knowledge of the use of medicinal plants in Bheri, Muzaffarabad, Azad Kashmir, Pakistan. *European Journal of Integrative Medicine* 8(4):560–569.

Al-Snafi, A.E. 2015a. The constituents and biological effect of *Arundo donax* – a review. *International Journal of Phytopharmacy Research* 6(1):34–40.

Al-Snafi, A.E. 2015b. Therapeutic properties of medicinal plants: a review of their antibacterial activity (part 1). *International Journal of Pharmacy and Technology* 6(3):137–158.

Al-Snafi, A.E. 2018. Traditional uses of Iraqi medicinal plants. *International Organisation of Scientific Research Journal of Pharmacy* 8(8):32–96.

Ana, M.L., A.S. Seca Jose, F.M.J. Cavaleiro, A.J.D. Domingues, D.E. Silvestre and P.N. Carlos. 2003. Structural characterization of the lignins from the nodes and internodes of *Arundo donax* L. *Journal of Agricultural and Food Chemistry* 48:817–824.

Babar, N. 2011. *Documentation of Indigenous Antiparasitic Practices and Scientific Evaluation of Some Ethnobotanicals for their Antihelmenthic Activity*. Dissertation. University of Agriculture, Faisalabad, Pakistan.

Ballero, M., F. Poli, G. Sacchetti, et al. 2001. Ethnobotanical research in the territory of Fluminimaggiore (South-western Sardinia). *Fitoterapia* 72(7):788–801.

Basumalary, N., R. Teron and M. Saikia. 2014. Ethnomedicinal practices of the bodo-kachari tribe of KarbiAnglong district of Assam. *International Journal of Life Sciences Biotechnology and Pharma Research* 3(1):161–167.

Behera, P.C., D.P. Tripathy and S.C. Parija. 2013. Shatavari: potentials for galactogogue in dairy cows. *Indian Journal of Traditional Knowledge* 12(1):9–17.

Bhatia, H., Y.P. Sharma, R.K. Manhas and K. Kumar. 2014. Ethnomedicinal plants used by the villagers of district Udhampur J&K, India. *Journal of Ethnopharmacology* 151(2):1005–1018.

Bullitta, S., G. Peluzza and L. Viegi. 2007. Plant resources used for traditional ethnoveterinaryphytotherapy in Sardinia Italy. *Genetic Resources and Crop Evolution* 54(7):1447–1464.

Chaudhuri, R.K. and S. Ghoshal. 1970. Triterpenes and sterol of the leaves of *Arundo donax*. *Phytochemistry* 9:1895–1896.

Christian, R. 1998. *The Encyclopedia of Psychoactive Plants: Ethnopharmacology and its Applications*. Rochester, VT: Park Street Press.

Dall'Acqua, S., R. Cervellati, M.C. Loi and G. Innocenti. 2008. Evaluation of in vitro antioxidant properties of some traditional Sardinian medicinal plants: investigation of the high antioxidant capacity of *Rubus ulmifolius*. *Food Chemistry* 106(2):745–749.

Dekorne, J.. 1995. *Arundo donax* L. *Entheogene* 4:27–28.

Duke, J.A. 1975. Ethnobotanical observations on the Cuna Indians. *Economic Botany* 29:278–293.

Egamberdiena, D., N. Mamedov, E. Ovidi, A. Tiezze and L. Craker. 2017. Phytochemical and pharmacological properties of medicinal plants from Uzbekistan: a review. *Journal of Medicinally Active Plants* 5(2):59–75.

Ghoshal, S., R.K. Chaudhuri and S.K. Dutta. 1971. Alkaloids of the flowers of *Arundo donax*. *Phytochemistry* 10:2852–2853.

Ghoshal, S., S.K. Datta, A.K. Sanyal and S.K. Bhattacharya. 1969. *Arundo donax* L. (Graminae) phytochemical and pharmacological evaluation. *Journal of Medicinal Chemistry* 12(3):380–383.

Girotto, L., A.C. Franco, C.V. Nunz, S.C.C. Oliveria, M.C. Scheffler de Souza, M.T. Fachin-Esipnar and C.S. Ferreira. 2021. Phytotoxicity and allelopathic potential of extracts from rhizomes and leaves of *Arundo donax* L. an invasive grass in neotropical savannas. *Notulae Botanicae Horti Agrobotanici Cluj-Napoca* 49(3):12440.

Hickman, J.C. 1993. *The Jepson Manual: Higher Plants of California*. Berkeley, CA: University of California Press.

Hong, Y., H.Y. Hu, A. Skoda and M. Sagehasi. 2010. Isolation and characterization of antialgal allelochemicals from *Arundo donax* L. *Journal of Allelopathy* 25(2):357–368.

Hussain, K., M.F. Nisar, A. Majeed, K. Nawaz and K.H. Bhatti. 2010. Ethnomedicinal survey for important plants of Jalalpurjattan district, Gujrat, Punjab, Pakistan. *Ethnobotanical Leaflets* (7):11.

Jain, A., S. Roshnibala, P.B. Kanjilal, R.S. Singh and H.B. Singh. 2007. Aquatic/semi aquatic plants used in herbal remedies in the wetland of Manipur, northeastern India. *Indian Journal of Traditional Knowledge* 6(8):346–351.

Jia, A.L., X.Q. Ding, D.L. Chen, Z.Z. Chao, Z.Y. Liu and R.B. Chao. 2008. A new indole alkaloid from *Arundo donax* L. *Journal of Asian National Products Research* 10(1–2):105–109.

Kamatchi, A. and A.S Parvathi. 2017. Ethnomedico and ethnoeconomic studies of grasses utilization by paliyar's tribal in Sadhuragiri hills, a part of Western Ghats, Tamil Nadu, India. *World Journal of Pharmacy and Pharmaceutical Sciences* 6(8):2478–2491.

Kaur, A., J. Singh, S.S. Kamboj, A.K. Sexana, R.M. Pandita and M. Shamnugavel. 2005. Isolation of an *N*-acetyl-D-glucosamine specific lectin from the rhizome of *Arundo donax* with antiproliferative activity. *Phytochemistry* 66(16):1933–1940.

Khan, M. and P.S. Yadav. 2010. Herbal remedies of asthma in Thoubal district of Manipur in north east India. *Indian Journal of Natural Products and Resources* 1(1):180–184.

Khare, C.P. 2007. *Indian Medicinal Plants: An Illustrated Dictionary*. Bertin: Springer-Verlag.

Khuzhaev, V.U. 2004. Alkaloid of the flora of Uzbekistan, *Arundo donax*. *Chemistry of Natural Compounds* 40(2):160–162.

Khuzhaev, V.U., I. Zhalolov, K.K. Turgunov, B. Tashkhodzhaev, M.G. Levkovich, S.F. Aripova, et al. 2004. Alkaloids from *Arundo donax*. XVII. Structure of the dimeric indole alkaloid arundaphine. *Chemistry of Natural Compounds* 40(3):269–272.

Khuzhaev, V.U., I. Zhalolov, M.G. Levkovich, S.F. Aripova and A.S. Shashkov. 2004. Alkaloids of *Arundo donax* L. *Russian Chemical Bulletin* 53(8):1765–1767.

Kom, L.E., K. Tilotama, T.D. Singh, A.K. Rawat and D.S. Thokchom. 2018. Ethnomedicinal plant used by the kom community of Thayong village, Manipur. *Journal of Ayurvedic and Herbal Medicine* 4(4):171–179.

Kumar, P., S. Singh, A. Sharma, G. Kaur, R. Kaur and A.N. Singh. 2021. *Arundo donax* L. an overview on its traditional and ethnomedical importance, phytochemistry and pharmacological aspects. *Journal of Herbmed Pharmacology* 10(3):269–280.

Kumar, R. and N. Bhagat. 2012. Ethnomedicinal plants of district Kathua (J&K). *International Journal of Medicine and Aromatic Plants* 2(4):603–611.

Leporatti, M.L. and M. Impieri. 2007. Ethnobotanical notes about some uses of medicinal plants in Alto Tirreno Cosentino area (Calabria, southern Italy). *Journal of Ethnobiology and Ethnomedicine* 3:34.

Liu, Q.R., J. Li, X.F. Zhao, B. Xu, X.H. Xiao, J. Ren, et al. 2021. Alkaloids and phenylpropanoid from rhizome of *Arundo donax* L. *Natural Product Research* 35(3):465–470.

Miguez, C., A. Cavcela, A. Sanchez and X. Alvarez. 2022. Possibilities for exploitation of invasive species, Arundo donax L., as a source of phenol compound, waste and biomass valorization. *Waste and Biomass Valorization*, 13(10):4253–4265.

Neto, C.P., A. Seca, A.M. Nunes, M.A. Coimbra, F. Domingues, D. Evtuguin, et al. 1997. Variations in chemical composition and structure of macromolecular components in different morphological regions and maturity stages of *Arundo donax*. *Industrial Crops and Products* 6(1):51–58.

Ningthoujan, S.S., A. Das Talukdar, K.S. Potsangbam and M.D. Chaudhury. 2013. Traditional uses of herbal vapour therapy in Manipur, north-east India: an ethnobotanical survey. *Journal of Ethnopharmacology* 147(1):163–147.

Nisar, M.F., F. Jaleel, M. Waseem, S. Tsmail, Y. Toor, S.M. Haider, et al. 2014. Ethnomedicinal uses of plants from district Bahawalpur, Pakistan. *Current Research Journal of Biological Science* 6:183–190.

Palmese, M.T., R.E. Uncini Manganelli and P.E. Tomie. 2001. An ethnopharmacobotanical survey in the Sarrabus district (south-east Sardinia). *Fitoterapia* 72(6):619–643.

Pansuksan, K., S. Sukprasert, N. Karaket. 2020. Phytochemical compounds in *Arundo donax* L. rhizome and antimicrobial activity. *Pharmacognosy Journal* 12(2):287–292.

Passalacqua, N.G., P.M. Guarrera and G. De Fine. 2007. Contribution to the knowledge of the folk plant medicine in Calabria region (southern Italy). *Fitoterapia* 78(1):52–68.

Patino, R., R.H. Reshel, A. Rubio and S. Longing. 2018. Growth suppressing and algicidal properties of an extract from *Arundo donax*, an invasive riparian plant, against *Prymnesium parvum*, an invasive alga. *Harmful Alga* 71:1–9.

Pelaez, F., J. Collado, F. Arenal, A. Basilio, A. Cabello, et al. 1998. Endophytic fungi from plants living on gypsum soil as a source of secondary metabolites with antimicrobial activity. *Mycological Research* 102(6):755–761.

Pieroni, A., C. Quave and R.F. Santoro. 2004. Pharmaceutical knowledge in the territory of the Dolomiti Lucane, inland southern Italy. *Journal of Ethnopharmacology* 95(2–3):373–384.

Pieroni, A., C. Quave, S. Nebel and M. Heinrich. 2002. Ethnopharmacy of the ethnic Albanians (Arbereshe) of northern Basilicata, Italy. *Fitoterapia* 73(3):217–241.

Piluzza, G. and S. Bullitta. 2011. Correlation between phenolic content and antioxidant properties in twenty four plant species of traditional ethnoveterinary use in the Mediterranean area. *Pharmaceutical Biology* 49(3):240–247.

Quave, C.L., L.R.W. Plane and B.C. Bennett. 2008. Effects of extracts from Italian medicinal plants on planktonic growth, biofilm formation and adherence of methicillin-resistant *Staphylococcus aureus*. *Journal of Ethnopharmacology* 118:418–428.

Sagir, C., Z.A. Everest and Y. Keles. 2018. The comparative investigation of the antioxidant activities of some species belonging to the Lamiaceae and Poaceae families. *Anatolian Journal of Botany* 2(2):52–59.

Sahal, A., A. Hoque, S.K. Mallick and S. Panda. 2014. Medicinal uses of grasses by the tribal people in West Bengal: an overview. *International Journal of Basic and Applied Research* 3:63–70.

Sarmah, R. 2010. Commonly used non-timber forest products (NTFPs) by the Lisu tribe in Changlang district of Arunachal Pradesh, *Journal of Frontiers research* 5:68–77.

Sharma, A., S.J. Singh, P. Kumar and A.N. Singh. 2020. Evaluation of antibacterial potential of an invasive grass (*Arundo donax* L.) in a lower Shiwalik region, India. *Plant Archives* 20(2):3580–3584.

Shatalov, A.A. and H. Pereira. 2001. *Arundo donax* L. reed: new perspective for pulping and bleaching, raw material characterization. *Tappi Journal Review Edpeper* 84(1):1–12.

Shekhar, J., G.P. Pratap, G. Sudharshanam and G.P. Prasad. 2011. Data on herbal remedies from the tribe of Chittoor district, Andhra Pradesh in India. *Life Sciences Leaflets* 17:621–630.

Singh, B. and H. Kaur. 2018. In silico documentation of medicinal plants in Lacchiwala range, Dehradun forest division, Uttarakhand (India). *Journal of Phytopharmacology* 7(1):92–102.

Upadhyay, D. 2016. Ethnobotanically useful plants of Katarniaghat Wildlife Sanctuary, Uttar Pradesh, India. *International Journal of Applied Research* 2(2):24–29.

Vanila, D., S. Ghanthikumar and V.S. Manickam. 2008. Ethnomedicinal uses of plant in the plain area of the Tirunelveli district, Tamil Nadu, India. *Ethnobotanical Leaflets* 12:1198–1205.

You, T.T., J.Z. Mao, T.Q. Yuan, J.L. Wen and F. Xu. 2013. Structural elucidation of the lignins from stem and foliage of *Arundo donax* L. *Journal of Agricultural and Food Chemistry* 61(22):5361–5370.

Zhalolov, I., V.U. Khuzhaev, B. Tashkhodzhaev, M.G. Levkovich, S.F. Aripova and N.D. Abdullaev. 1998. Alkaloids of *Arundo donax* VII a spectroscopic and X-ray structural investigation arundinine – a new dimeric alkaloid from the epigeal part of *Arundo donax*. *Chemistry of Natural Compounds* 34(6):706–1710.

Zhang, D., Q.W. Jiang, D.Y. Liang, S. Huang and J. Liao. 2021. The potential application of giant reed (*Arundo donax*) in ecological remediation. *Frontiers in Environmental Science* 9:652367.

Zheng, H., C. Sun, X. Hau, M. Wu, Y. Yao and F. Li. 2018. Pyrolysis of *Arundo donax* L. to produce pyrolytic vinegar and its effects on the growth of dinoflagellate *Karenia Brevis*. *Bioresource Technology* 247:273–281.

11 *Sagittaria sagittifolia*
An Important Aquatic Medicinal Plant with Numerous Applications

Pranchal Rajput, Ambika Saxena, Kundan Kumar Chaubey, and Sujata Hariharan*
Division of Research and Innovation, School of Applied and Life Sciences, Uttaranchal University, Arcadia Grant, P.O. Chandanwari, Premnagar, Dehradun, Uttarakhand, India
* Corresponding author (Kundan Kumar Chaubey): kundan2006chaubey@gmail.com

11.1 INTRODUCTION

An annual watery herb known as *Sagittaria sagittifolia* can grow in the mud of ponds and slow-moving streams in various temperate regions (CISEH 2018; Holm et al. 1997). The arrow-shaped emerging leaves have petioles reaching a length of 45 cm (PFAF 2020). The upper flowers are male and lower flowers are female. Fruits have wings that measure 4–6 mm. Additionally, plants can reproduce vegetatively by producing tubers (Hroudova et al. 1988). See Figure 11.1.

The US federal government recognizes that *S. sagittifolia* tubers are significant cultural foods in Asia (United States Department of Agriculture 2020) and has investigated its *geographic potential*: according to how certain the climatic match is, anywhere from 22% to 87% of the USA is appropriate for establishing *S. sagittifolia*. The largest area where it believes the environment is potentially suitable is represented by the distribution of the following Köppen–Geiger climate classes: steppe, Mediterranean, humid subtropical, marine west coast, humid continental warm summers, humid continental cool summers, subarctic, and tundra (United States Department of Agriculture 2020). This area also includes regions with precipitation ranging from 10 to 100 inches annually. Only these three climatic characteristics were used to pinpoint the region of the USA that is climatically acceptable. The locations where this species is likely to establish itself may change due to other factors like soil, hydrology, disturbance regime, and species interactions. As *S. sagittifolia* is a rooted aquatic plant that thrives in the mud of ponds, slow-moving streams, ditches, and canals (Stace, 2010; PFAF 2020). as well as in rice fields and wetlands (Zhirong 1990) it would be limited to wetlands and marine ecosystems, permitting the importation of tubers in California, Hawaii, and New York (Figure 11.2).

The environment is composed of different elements which help to enhance the interaction of humans with other species, which are vital to survival, and plants are one of the main environmental elements (Jain et al. 2011). Wetlands are rich in plant species, which are very useful for humanity as they have innumerable benefits. One of the rich plants is *S. sagittifolia*, commonly known as 'arrowhead'. The *S. sagittifolia* belongs to the Alismataceae family, which includes flowering plants, and it is an edible, leading to an increased presence in the human diet due to its benefits (Jain et al. 2011). *S. sagittifolia* is an herbaceous aquatic perennial plant and the large leaves are arrow-shaped, with a glossy appearance (Hroudová et al. 1988). *S. sagittifolia* is a mid-summer plant where seeds ripen up to autumn and are monoecious, leading to insect pollination. *S. sagittifolia* has an arrangement of

DOI: 10.1201/9781003256830-11

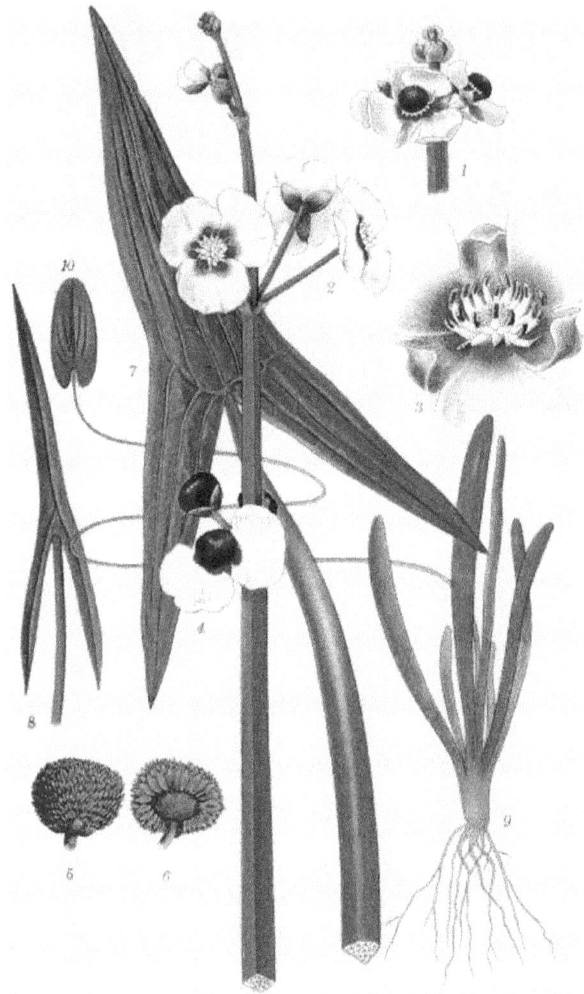

FIGURE 11.1 Drawing of the parts of *S. sagittifolia*. (From Wikipedia: (https://commons.m.wikimedia.org/wiki/File:485_Sagittaria_sagittifolia.jpg.)

leaves in a spiral manner which has long triangular petioles and reaches a height of about 1 m. The first leaves of *S. sagittifolia* form a rosette at the stem base, which is ribbon-shaped, followed by the next leaves, which are transitory and spathulate. Further leaves are heart-shaped followed by the final leaves, which are arrow-shaped (Hroudová et al. 1988) (Figure 11.3).

S. sagittifolia L. has a long history of usage as a nutritious food and in traditional Chinese medicine. Despite this, there still needs to be more data on the bioactivity and chemical characterization of *S. sagittifolia* L. polysaccharides (J. Gu et al. 2020). Plant species present in wetlands are rich in components that can effectively be used to address some health issues due to their high mineral content (Table 11.1) (Abemsana et al. 2021).

One such plant with high food and medicinal value is *S. sagittifolia*, as its different parts have rich content with high medicinal use leading to increased benefits in human welfare. *S. sagittifolia* is native to Europe, followed by China, Turkey, Russia, and Lebanon; it is considered a weed in the agricultural system but does not negatively impact plants. The fruits in *S. sagittifolia* are 4–6 mm winged achenes, and reproduction in the plant occurs vegetatively through the production of tubers (Abemsana et al. 2021).

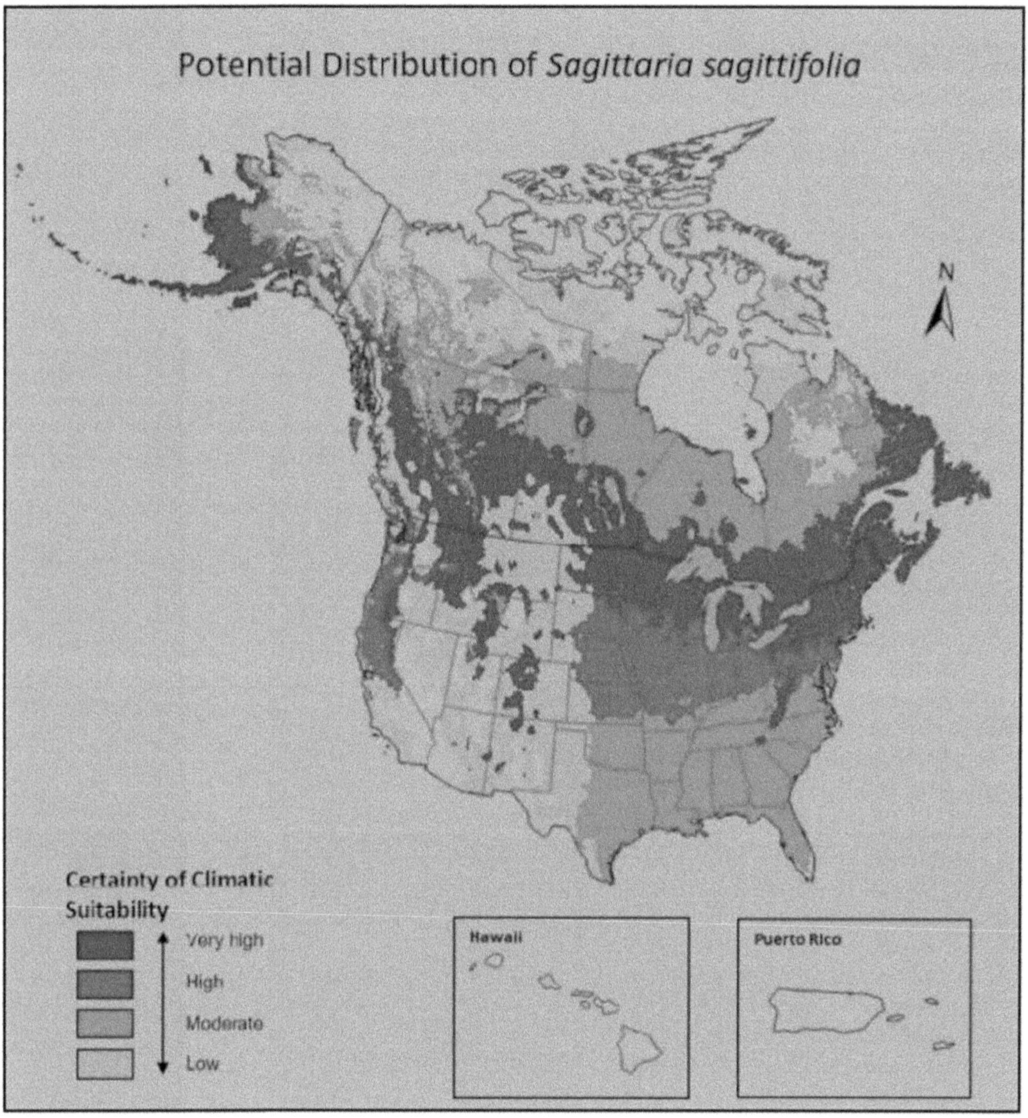

FIGURE 11.2 Distribution of *S. sagittifolia* in the world. (From United States Department of Agriculture. 2020. Weed Risk Assessment for *Sagittaria sagittifolia* (Alismataceae) – Arrowhead. United States Department of Agriculture, *Animal and Plant Health Inspection Service*.)

11.2 SOME APPLICATIONS OF *SAGITTARIA SAGITTIFOLIA*

- *Used as an ornamental plant*: *S. sagittifolia* is primarily used as an ornamental plant in gardens as they are very attractive with a beautiful flower type, and they are present in slow-moving water like reservoirs, swamps, or ponds (Gao et al. 2022) (Figure 11.4). *S. sagittifolia* has a long flowering time that lasts about half a year, leading to its unisexual and monoecious state in which the female flower appears earlier than the male flower over the third whorl of inflorescence (Gao et al. 2022).
- *Used as an edible plant*: cooked roots (Figure 11.5). The tubers have a particular flavour and are starchy. The flavour is similar to potatoes and they are delicious if roasted. The tubers

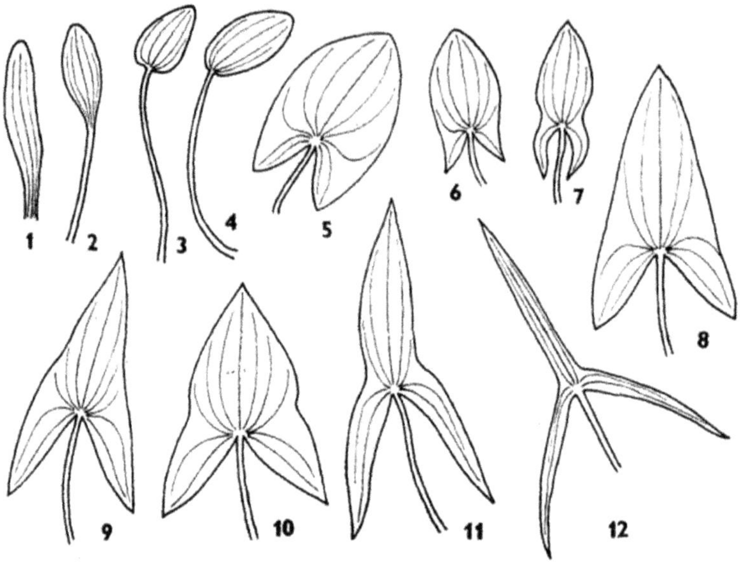

FIGURE 11.3 Leaf forms of *S. sagittifolia*. (From Hroudová et al. (1988).)

TABLE 11.1
Mineral Content in Selected Aquatic Plants

Sl. no.	Botanical name	Part of the plant used	Magnesium (ppm)	Zinc (ppm)	Calcium (ppm)	Iron (ppm)
1.	*Neptunia oleraceae*	Tender shoot	18.01 ± 0.07	1.01 ± 0.04	51.45 ± 0.79	8.07 ± 0.005
2.	*Trapannatas*	Tender shoot	21.20 ± 0.018	0.80 ± 0.003	54.01 ± 0.21	12.89 ± 0.01
3.	*Enhydra fluctuans*	Tender shoot	7.11 ± 0.06	1.44 ± 0.09	16.11 ± 0.19	12.78 ± 0.002
4	*Ipomea aquatica*	Tender shoot	4.64 ± 0.13	1.70 ± 0.002	9.45 ± 0.29	8.44 ± 0.27
5.	*Sagittaria sagittifolia*	Tuber	0.98 ± 0.001	1.47 ± 0.04	2.13 ± 0.03	15.21 ± 0.05
6.	*Jussiaea repens*	Tender shoot	7.10 ± 0.02	1.17 ± 0.20	14.10 ± 0.002	22.11 ± 0.41
7.	*Marsilea quadrifolia*	Whole plant	3.82 ± 0.05	1.03 ± 0.07	7.70 ± 0.19	16.44 ± 0.10
8.	*Eleocharis dulcis*	Rhizome	1.03 ± 0.07	0.61 ± 0.02	2.01 ± 0.16	11.18 ± 0.04
9.	*Zizania latifolia*	Rhizome	0.62 ± 0.01	1.61 ± 0.01	1.50 ± 0.01	3.13 ± 0.04

should be consumed only after being cooked. It is necessary to remove the peel just after the tubers have been cooked because it is quite bitter. Additionally, tubers can be dried and processed into a powder that can be added to cereal flours to make bread or used as a gruel or other dish.

- *Used as antiscorbutic*: the leaves and root part of *S. sagittifolia* are used as a diuretic and an antiscorbutic. Paste of leaf is externally applied for different types of skin problems. A decoction of the tuber is used by tribal people as a means of birth control as it induces premature terminations (Shankar and Mishra 2012).
- There has been a long history of treating livestock with different medicinal plants, as traditional veterinary medicine has been a trusted approach since prehistoric times (Mahmud and Shah 2009). One such medicinal plant which is effective in addressing the health issue of

FIGURE 11.4 Ornamental use of *S. sagittifolia*. (From https://pondtrends.com/plants/sagittaria-sagittifolia/).

FIGURE 11.5 Edible part of *S. sagittifolia* (tubers). (From https://pixen.wordpress.com/tag/sagittaria-sagittifolia/).

livestock is *S. sagittifolia*, as the leaf extract has been essential to addressing skin issues which include warts, scabies, or lesions when given with buttermilk. *S. sagittifolia* not only addresses the skin issues, but there are pieces of evidence that highlight the medicinal use of the medicinal plant in poor milk production, dyspepsia, cuts or wounds, and diarrhoea (Mahmud and Shah 2009). Used as a medicinal plant, *S. sagittifolia* proves its medicinal value to humanity, and animals benefit from its positive effect on livestock diseases (Figure 11.6).

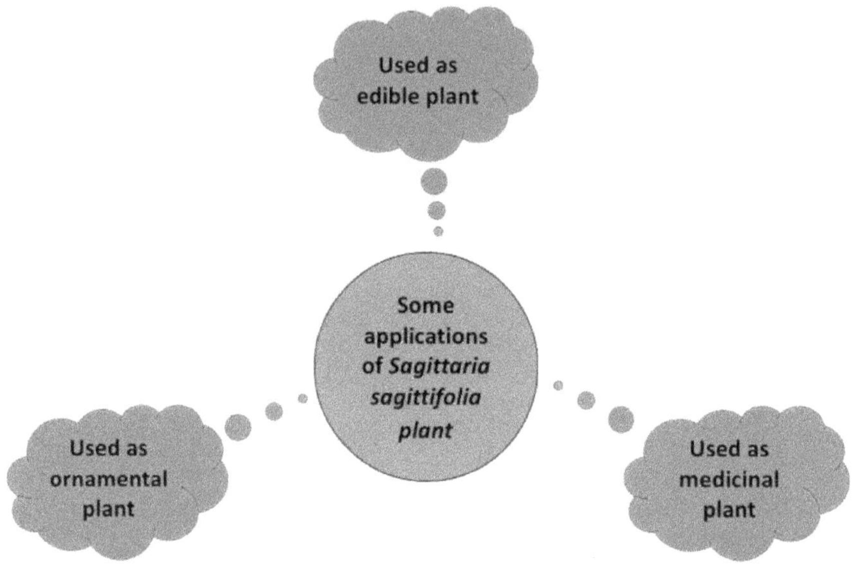

FIGURE 11.6 Some applications of *S. sagittifolia*.

11.3 MEDICINAL USES OF *SAGITTARIA SAGITTIFOLIA*

One of the medicinal uses of *S. sagittifolia* is that it can help to suppress bladder cancer, leading to an increased need to research its use in the healthcare domain (Gong et al. 2019). *S. sagittifolia* can help suppress bladder cancer in combination with other medicinal plants like *Astragalus*, *Poria*, *Paeonia*, or *Polyporus*, positively affecting the cancer state. Water extract of *S. sagittifolia* also produces a hepatoprotective effect on the injured liver, reducing the complications produced by isoniazid and rifampin (Gong et al. 2019). Moreover, another study presented by Palombo (2011) proposed that one of the medicinal uses of *S. sagittifolia* is in addressing oral diseases, which are caused by about 750 species of bacteria leading to major health problems. Both Gram-positive and other bacteria increase individual complications by causing acidogenic dental caries and periodontal diseases, respectively (Palombo 2011). An alternative intervention to address dental caries issues is a plant extract of *S. sagittifolia*, one of the medicinal plants that can inhibit the growth of oral pathogens, leading to a healthier oral cavity. *S. sagittifolia* is expected to have antibacterial activity against *S. mutans* or *Actinomyces naeslundii*, leading to activity against Gram-positive bacteria involved in oral diseases (Palombo 2011).

There has been a long history of treating livestock with different medicinal plants, as traditional veterinary medicine has been a trusted approach since prehistoric times (Mahmud and Shah 2009). One such medicinal plant effective in addressing livestock health issues is *S. sagittifolia*, as the leaf extract has been essential to addressing skin issues, including warts, scabies, or lesions, when given with buttermilk. *S. sagittifolia* not only addresses the skin issue, but there are pieces of evidence that highlight the medicinal use of the plant in poor milk production, dyspepsia, cuts or wound, and diarrhoea (Mahmud and Shah 2009). The article by Mahmud and Shah (2009) added that *S. sagittifolia* not only has medicinal value to humanity but animals also benefit due to its positive effect on livestock diseases.

Moreover, the study presented by Mishra (2012) added that each ethnic community has its own healthcare system leading to increased utilization of different medicinal plants in the treatment process. Ethnic therapeutics has been using different parts of the plant, like leaves, roots, stems, or fruits, to address some of the health issues which require urgent attention leading to a reduction in the side effects (Mishra 2012). One such medicinal use of *S. sagittifolia*, which has been effectively used by one section of the community, is that its tubers can induce premature termination. Thus, *S. sagittifolia* has been used as a means of birth control by different tribes leading to the termination of pregnancy at an early stage (Mishra 2012).

Tang et al. (2020) described how *S. sagittifolia* is one of the famous perennial aquatic plants with high medicinal value and it is edible, also of great importance for humanity. The nature of *S. sagittifolia* is that it acts as a detoxifying agent and it is used for different clinical conditions such as gonorrhoea, retention of placenta, haemoptysis, and soreness (Tang et al. 2020). *S. sagittifolia* has also been effective in addressing different issues through its antioxidative activity, leading to a role in a clinical setting along with high phytochemical demand (Tang et al. 2020). Gu et al. (2020) stated that *S. sagittifolia* has been effectively used as a healthy food source and in Chinese medicine due to its high number of medicinal uses, leading to its popularity in the population. *S. sagittifolia* has effectively delivered adjuvant therapy for night blindness, tuberculosis, urinary tract infection, pancreatitis, tracheitis, and diabetes. *S. sagittifolia* has different medicinal benefits due to different bioactive agents like proteins, polysaccharides, fibres, or vitamins that support its clinical efficacy (Gu et al. 2020). See Table 11.2.

TABLE 11.2
Some Applications of *Sagittaria sagittifolia* in Different Studies

Sl. no.	Application of plant	Uses	Reference
1.	Used as an ornamental plant	*S. sagittifolia* is primarily used as an ornamental plant in gardens as they are attractive	Gao et al. 2022
2.	Used as an edible plant	Cooked roots. The flavour is similar to potatoes and is delicious if roasted	www.naturalmedicinalherbs.net
3.	Used as a medicinal plant	*S. sagittifolia* has various medicinal uses	Mahmud and Shah 2009
4.	Used in bladder cancer	*S. sagittifolia* can help to suppress bladder cancer	Gong et al. 2019
5.	Used in oral diseases	*S. sagittifolia* is used to address oral diseases which are caused by about 750 species of bacteria and lead to major health problems	Palombo 2011
6.	Used for treatment of livestock	*S. sagittifolia* not only has medicinal value to humans, but animals also benefit due to its positive effect on livestock diseases	Mahmud and Shah 2009
7.	Used to induced premature birth	*S. sagittifolia* has been used by different tribes leading to the termination of pregnancy at an early stage	Mishra 2012
8.	Used as a detoxifying agent	*S. sagittifolia* acts as a detoxifying agent which is used clinically in treating gonorrhoea, retention of placenta, haemoptysis, and soreness	Tang et al. 2020
9.	Used as an antioxidant	*S. sagittifolia* has been effective in addressing different issues through its antioxidative activity	Tang et al. 2020
10.	Used in delivering adjuvant therapy	*S. sagittifolia* has been effective in delivering adjuvant therapy for night blindness, tuberculosis, urinary tract infection, pancreatitis, tracheitis, and diabetes	Gu et al. 202
11.	Used as bioactive agents	*S. sagittifolia* has different bioactive agents like proteins, polysaccharides, fibres, or vitamins, which support its clinical efficacy	Gu et al. 202
12.	Used as antiscorbutic	Paste of leaf is externally applied in different types of skin problems. Decoction of tuber is used by tribal people as mean of birth control as it induces premature termination	Shankar and Mishra 2012

11.4 DISCUSSION

S. sagittifolia is one of the wetland plants known for its effective consideration in dietary and medicinal utilization, leading to increased demand. The various characteristics of *S. sagittifolia* have been discussed in the introduction section, which can be used to understand its nature, being a member of the Alismataceae family. Moreover, in the second section, the different medicinal values of *S. sagittifolia* were discussed, leading to a better understanding of the medicinal importance of the plant.

11.5 CONCLUSION

S. sagittifolia is one of the aquatic plants famous in China for its multiple uses which has led to increased demand by the population. It has been effective in addressing both dietary as well as medicinal requirements of the population. Thus, it should be further investigated to understand its potential role and beneficial results in different health issues in humans and livestock.

REFERENCES

Abemsana, O., M.B. Saikia and S.S. Ningthoujam. 2021. Traditional usage of underutilized aquatic plants found in the Imphal valley of Manipur. *Research Square*, 39(1):449–456.

CISEH. 2018. *Invasive Plant Atlas of the United States*. Athens, GA: University of Georgia, Center for Invasive Species and Ecosystem Health.

Gao, M., W. Jiang, Z. Lin, Q. Lin, Q. Ye, W. Wang, Q. Xie, X. He, C. Luo and Q. Chen. 2022. SMRT and Illumina RNA-Seq Identify potential candidate genes related to the double flower phenotype and unveil SsAP2 as a key regulator of the double-flower trait in *Sagittaria sagittifolia. International Journal of Molecular Science* 23(2240):1–11.

Gong, H., W. Chen, L. Mi, D. Wang, Y. Zhao, C. Yuan and A. Zhao. 2019. Qi Sailing decoction suppresses bladder cancer growth by inhibiting the Wnt/B-catenin pathway. *Pharmaceutical Biology* 57(1):507–513.

Gu, J., H. Zhang, H. Yao, J. Zhou, Y. Duan and H. Ma. 2020. Comparison of characterization, antioxidant and immunological activities of three polysaccharides from *Sagittaria sagittifolia* L. *Carbohydrate Polymers* 235:1–11.

Holm, L., J. Doll, E. Holm, J. Pancho and J. Herberger. 1997. Sagittaria sagittifolia L. Alismataceae, water plantain family. In *World Weeds: Natural Histories and Distribution*. New York: John Wiley and Sons, Inc, 699–707.

Hroudová, Z., L. Hrouda, P. Zákravskyand and I. Ostry. 1988. Ecology and distribution of *Sagittaria sagittifolia* L. in Czechoslovakia. *Folia Geobotanica et Phytotaxonomica* 23(4):337–373.

Jain, A., M. Sundriyal, S. Roshnibala, R. Kotoky, P.B. Kanjilal, H.B. Singhand and R.C. Sundriyal. 2011. Dietary use and conservation concern of edible wetland plants at Indo-Burma hotspot: a case study from Northeast India. *Journal of Ethnobiology and Ethnomedicine* 7(29):1–17.

Mahmud, S. and N.H. Shah. 2009. Use of aquatic and marshy plants in ethnoveterinary practices by tribals and rural people of Jammu province, (J&K), India. *International Journal of Plant Sciences Muzaffarnagar* 4:471–474.

Mishra, P.K. 2012. Study of aquatic medicinal plants of Hazaribag district of Jharkhand, India. *International Research Journal of Pharmacy* 3(4):405–409.

Palombo, E.A. 2011. Traditional medicinal plant extracts and natural products with activity against oral bacteria: Potential application in the prevention and treatment of oral diseases. *Evidence Based Complementary and Alternative Medicine* 2011:680354.

PFAF. 2020. Sagittaria sagittifolia L. In: *Plants For a Future database*. Dawlish, UK: Plants For A Future. Available from: www.pfaf.org/USER/Default.aspx

Shankar, L.H. and P.K. Mishra. 2012. Study of aquatic medicinal plants of Hazaribag District of Jharkhand, India. *International Research Journal of Pharmacy* 3(4):405–409.

Stace, C. 2010. *New Flora of the British Isles*, 3rd edn. Cambridge: Cambridge University Press.

Tang, Y., J. Lv, J. Wan, B. Li, Q. Zhang, Z. Liu, X. Ke, W. Luo, Y. Linand and Y. Liao. 2020. *Sagittaria sagittifolia* polysaccharide, a Chinese herbal extract, protects against isoniazid- and rifampicin-induced hepatotoxicity in the in-vitro models. *Pharmacognosy Magazine* 16:637–643.

United States Department of Agriculture. 2020. *Weed Risk Assessment for Sagittaria sagittifolia (Alismataceae) – Arrowhead.* United States Department of Agriculture, Animal and Plant Health Inspection Service. Available from: www.aphis.usda.gov/plant_health/plant_pest_info/weeds/downloads/wra/sagittaria-sagittifolia.pdf

Zhirong, W. 1990. *Farmland Weeds in China: A Collection of Coloured Illustrative Plates*. Beijing: Agricultural Publishing House.

12 Pharmacological and Phytochemical Potential of *Laurencia obtusa* (Hudson) J.V. Lamouroux

A Red Marine Alga

*Varsha Nigam**
Department of Botany, Guru Nanak Khalsa College, Yamuna Nagar, Haryana, India
* Corresponding author (Varsha Nigam): varshanigam29@gmail.com

12.1 INTRODUCTION

Traditional natural treatments are being promoted for the treatment of chronic diseases, as synthetic pharmaceuticals might have unknown side effects in such circumstances. The recent rise in bioactive compounds obtained from terrestrial plants has opened doors to the underutilized marine algae, which appears to be a promising natural resource opportunity (Baker 1984). Marine macroalgae (sustainable resources) are used as food, feed, and fertilizer globally (Chapman 1998, pp.299–300; Bhatia et al. 2022). In addition to their ecological qualities, marine algae also have interesting nutritive food value. Results revealed that the abundant algae in this ecosystem have a large potential for use as sources of carbohydrates, amino acids, proteins, phenols, and lipids in the manufacture of nutrient supplements, drugs, and fine chemical synthesis in the food and pharmaceutical industries (Sethi et al. 2020). Algae and their preparations contain biologically active compounds. Benefits to humans, animals, and plants have been found and acknowledged in the development of novel biotechnological products in the past and today (Chojnacka et al. 2012). Bioactive compounds found in marine algae have different effects on a number of ailments than those algae found in terrestrial habitats, which could lead to the discovery of new secondary metabolites with different actions and mechanisms (Li et al. 1998; Alarif et al. 2012). A wide number of marine bioactive chemicals are being researched and are used to treat a variety of ailments (Robert 2004).

In Southeast Asian countries, many species of *Laurencia* are used as human food in the form of puree, raw salad, curry, and so on. Minerals, trace elements, fibrous material, proteins, and vitamins can all be extracted for a reasonable cost from them (Kolanjinathan et at. 2014). *Laurencia* spp. are known for producing a vast variety of remarkable phytochemicals with a wide range of structural diversification and bioactivity (Kladi et al. 2007; Alarif et al. 2016).

Laurencia obtusa (Hudson) J.V. Lamouroux, is a marine alga belonging to the class Rhodophyceae of order Ceramiales (Figure 12.1). It is also called rounded brittle fern weed, and is native to southern England, but it may now be found in warm and temperate waters all over the world, including the Mediterranean Sea, and the Atlantic Ocean. They are also found on the intertidal rocks of warm seas all around the world (Masuda et al. 1996). The AlgaeBase database contains 421 species names, 146 of which have been flagged as currently accepted taxonomically (Guiry and Guiry 2022).

DOI: 10.1201/9781003256830-12

FIGURE 12.1 *L. obtuse*, a marine alga.

Laurencia spp. are always distinguished by the presence of new phytochemicals, particularly sesquiterpenes, di- and tri-terpenes, sterols, alkaloids, and halogenated C15 acetogenins, among the marine algal flora. Generally, only one halogen atom can distinguish these phytochemicals. *L. obtusa* is a rich source of polyhalogenated sesquiterpenes with different types of skeletons among other different natural products such as hurgadenyne, β-snyderyl acetate, hurgadol, *n*-heptadecane, obtusane, 3-β-bromo-8-epicaparrapi oxide, (*E*)-2-tridecyl-2-heptadecenal, β-synderol, palmitic acid, and cholesterol (Martin et al. 1989; Deyab 2011). A somewhat high level of halogenation characterizes the bulk of these phytochemicals. *L. obtusa* has been demonstrated to be a rich source of new secondary metabolites with antioxidant, anticancer, anti-inflammatory, and anti-bacterial activity (Figure 12.2). *L. obtusa* was evaluated as a biofertilizer for *Zea mays* (L.) plants. The results of the study showed that the parameters of plant production increased when this alga was used as a biofertilizer.

12.2 PHYTOCHEMICALS OF *LAURENCIA OBTUSA*

Secondary metabolites are abundant in marine organisms, and some of them have important biological functions. *Laurencia* spp., in particular, are known to process over 700 chemicals, including a range of frameworks and halogenated compounds (Gribble 2015).

12.2.1 Brominated Labdane Diterpenoid

Laurencia, a red alga, produces a dizzying array of structurally unique secondary metabolites that have sparked intense inquiry since the first experiments on a marine organism. Halogenated labdanes 1 and 2 were obtained from *L. obtusa*, which contain an 8- or 7-membered ether ring embedded with a decalin core, a labile carbon–bromine bond at the third carbon atom, and an epoxide ring or a OH group enclosing even asymmetric centres, which was collected from the Mitikas Bay in the Ionian Sea, Greece, as shown in Figure 12.3 (Iliopoulou et al. 2003a; Hagiwara et al. 2020).

12.2.2 C15 Acetogenins

Two principal phytochemicals have been derived from the Saudi Red Sea red alga *L. obtusa.* One has an unusual substituted allene side chain and the other inserts furan as a new class of acetogenins, namely laurentusenin (1) and laurenfuresenin (2), respectively. These two metabolites displayed chemotaxonomic value to the genus *Laurencia.* A DNA fragmentation assay and microscopic examination were used to show that these two chemicals cause apoptosis. These findings imply that compounds 1 and 2 are involved in the onset and propagation of inflammatory responses via

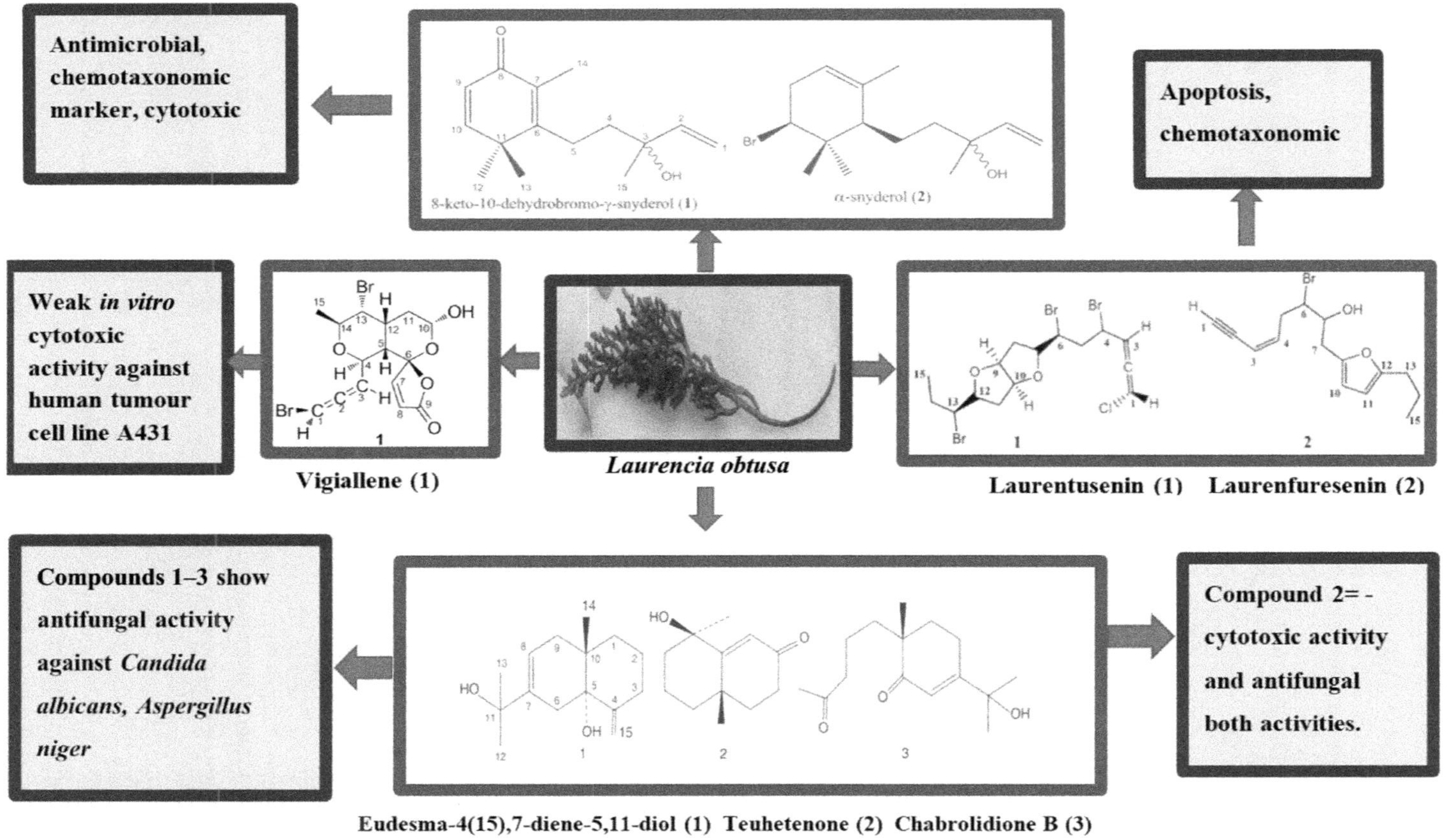

FIGURE 12.2 Phytochemical uses of *L. obtuse*.

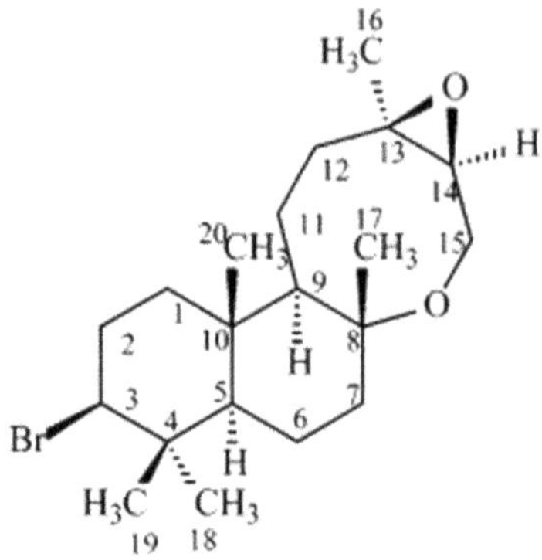

1

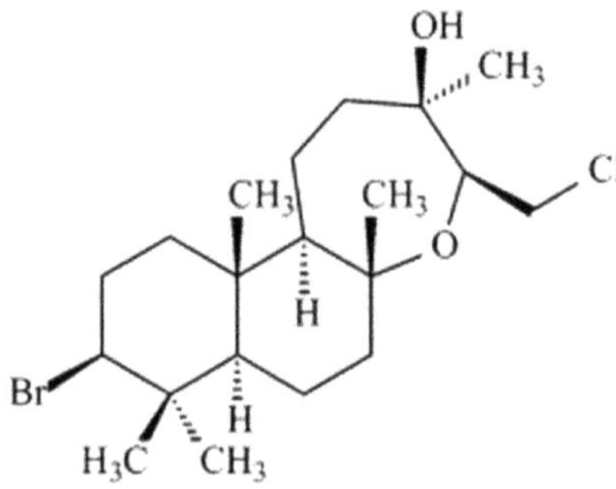

2

FIGURE 12.3 Structure of halogenated labdanes 1 and 2. (From Iliopoulou et al. (2003a).)

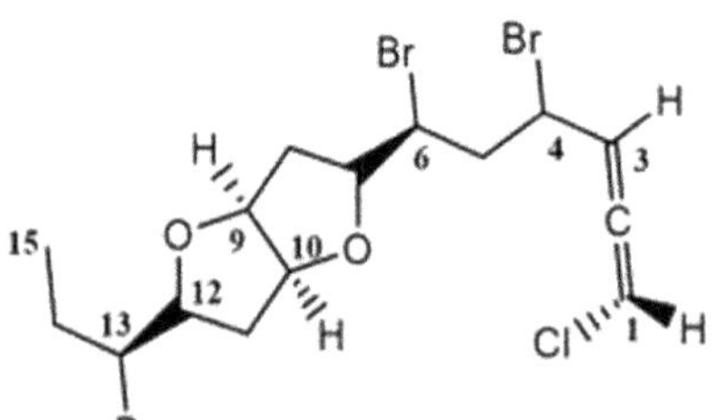

1

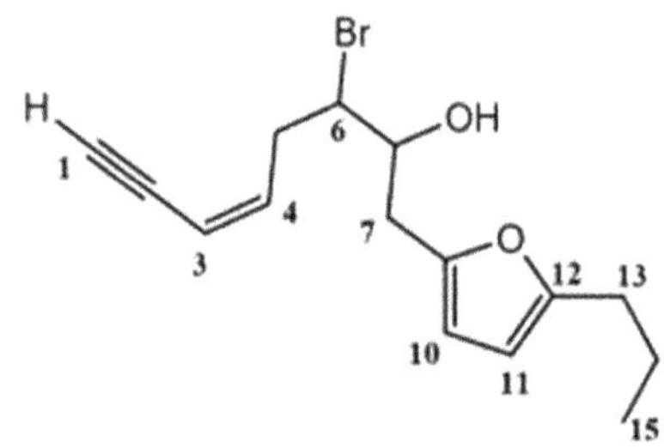

2

FIGURE 12.4 Structure of laurentusenin (1) and laurenfuresenin (2). (From Ghandourah et al. (2019).)

regulating programmed death. The isolated metabolite 1 had a unique substituted allene side chain, whereas metabolite 2 had a novel acetogenin nucleus with an inserted furan ring (as shown in Figure 12.4). Both chemicals could be involved in the activation of apoptosis as well as the beginning and propagation of inflammatory responses (Ghandourah et al. 2019).

Vagiallene (1), a reorganized C15 acetogenin with the chemical formula $C_{15}H_{16}Br_2O_5$ (Figure 12.5) and a carbon skeleton found nowhere else in nature, was discovered in a minute ingredient from the organic extract of the red alga *L. obtusa* from the Ionian Sea's Lefkada island. Vagiallene (1) was obtained as a colourless oil (Perdikaris et al. 2019).

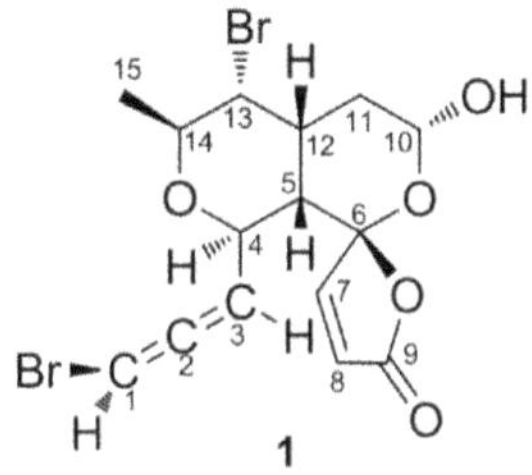

1

FIGURE 12.5 Structure of vagiallene (1). (From Perdikaris et al. (2019).)

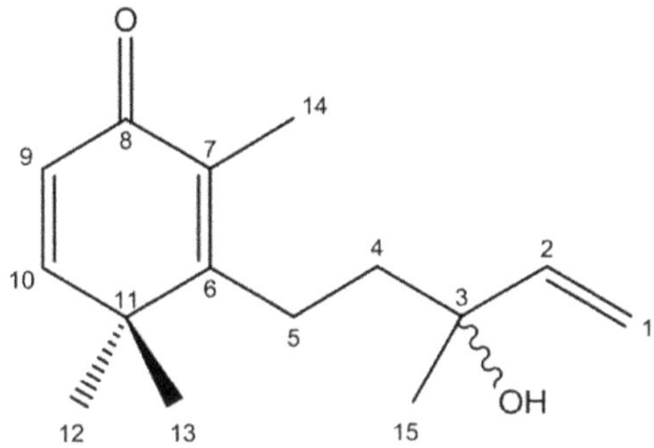

1

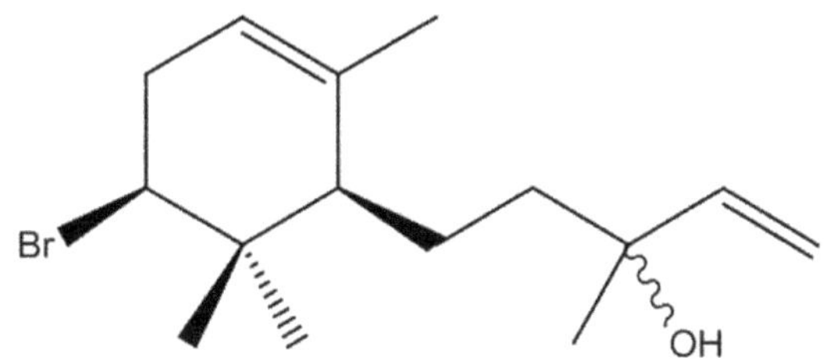

2

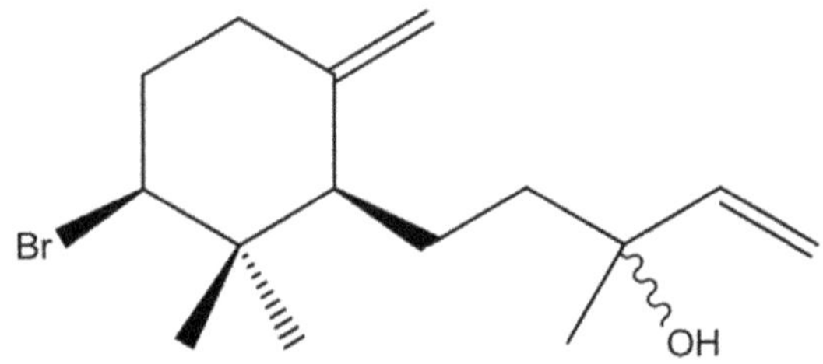

3

FIGURE 12.6 Structure of 8-keto-10-dehydrobromo-γ-snyderol (1), α-snyderol (2), and β-snyderol (3). (From Esselin et al. (2019).)

12.2.3 Sesquiterpenes

The major group of natural compounds generated by the *Laurencia* genus is sesquiterpenes.

A significant variety of structure, such as laurane, snyderane, and cuparane, contribute to the exceptional structural diversity of these compounds. With 57 members discovered to date, snyderane is the third bulk main structure of sesquiterpenes produced in the *Laurencia* spp. (Harizani et al. 2016). Previously, *L. obtusa* yielded ten compounds with a snyderane skeleton and snyderols derivatives as shown in Figure 12.6 (Esselin et al. 2019).

The $C_4H_8O_2$ extract of *L. obtuse*, collected in Corsica, France, yielded 8-keto-10-dehydrobromo-γ-snyderol (1), with the unique γ-snyderane structure (Figure 12.6). Spectroscopic analysis, was used to identify the first compound.

The Red Sea red alga *L. obtusa* has yielded a novel eudesmane sesquiterpenoid, eudesma-4(15),7-diene-5,11-diol (1), as well as the known teuhetenone (2), and chabrolidione B (3) as shown

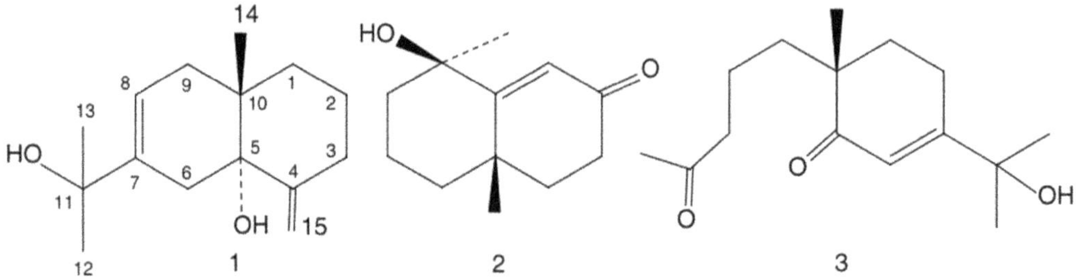

FIGURE 12.7 Structure of eudesma-4(15),7-diene-5,11-diol (1), teuhetenone (2), and chabrolidione B (3). (From Alarif et al. (2016).)

in Figure 12.7. Compounds 1 and 3 showed antifungal activity and compound 2 showed anticancer activity against cancer cells (Alarif et al. 2016).

12.3 RECENT PROGRESS IN PHARMACEUTICAL USE OF *LAURENCIA OBTUSA*

12.3.1 Anti-inflammatory Activity

The physiological reaction of vascular tissues to adverse stimuli such as damage, disease, or hypersensitivity is inflammation, which is characterized by erythema, abscess, soreness, heat, and tissue and organ dysfunction (Lawrence et al. 2002). Inflammation is a multi-step process that is frequently associated with soreness and discomfort. *In vivo*, the raw extract and its polar fractions were found to have significant anti-inflammatory properties, blocking the pro-inflammatory cytokine tumour necrosis factor alpha and decreasing inflammatory swelling. They also demonstrated strong antinociceptive and gastrointestinal protection. *L. obtusa* possesses anti-inflammatory properties as well as the capacity to guard against gastrointestinal issues. These extracts, unlike non-steroidal anti-inflammatory drugs and corticoids, did not cause mucosal injury in the stomach but did show a marked suppression of ulcer formation. More research is being done on the separation and specification of phytochemicals from *L. obtusa* in an attempt to identify the bioactive components responsible for these critical functions (Lajili et al. 2016).

12.3.2 Antioxidant Activity

An antioxidant is a chemical that can slow or stop other molecules from oxidizing. Oxidation is a chemical reaction in which an electron is shifted from a material to an oxidizer. Free radicals are produced by oxidation, which start chemical reactions that injure cells. Antioxidants are chemicals that suppress oxidation and can mitigate the negative effects of oxidation in body tissue (Nigam and Sodhi 2014). During regular cellular metabolism, reactive oxygen species are continually generated in the body, creating oxidative stress and lipid peroxidation, which leads to damage to the cell membrane, proteins, and nucleic acids, and ultimately cell death. At different proportions, the *L. obtusa* extracts and essential oils showed free radical scavenging activity against the stable free radical DPPH (1,1-diphenyl-2-picryl hydrozyl), decolourizing activity of ABTS (2,2'-azino-bis 3-ethyl benzothiazoline-6-sulphuric acid) radical cation, and inhibition activity of β-carotene bleaching. The chloroform extracts had a strong DPPH radical scavenging activity, but the methanol and hexane extracts had very little. In the ABTS radical cation decolourization experiment, however, the percentage activity of the chloroform and hexane extracts was higher than that of the methanol extracts. Phenolic chemicals are thought to shield algal thalli from the harmful effects of ultraviolet radiation while also acting as free radical scavengers (Zubia et al. 2008). Furthermore, high phenolic levels are linked to antibacterial and antioxidant properties. The antibacterial activity of the *Laurencia* extracts was low, and the antioxidant activity varied according to the test techniques and the chemical makeup of the solvent extracts and essential oils. In order to develop novel useful compounds, the active components responsible for antioxidant and antibacterial actions must be studied. As a result, additional research into the separation and identification of antioxidant and antibacterial components in *Laurencia* spp. for industrial and pharmaceutical uses is recommended (Demirel et al. 2011).

12.3.3 Antimicrobial Activity

Three compounds have been isolated by Alarif et al. 2012, named 12-hydroxy isolaurene (1), 8,11-dihydro-12-hydroxy isolaurene (2), and isolauraldehyde (3). The antibacterial activity of compounds 1–3 was tested using a disc-diffusion technique against two Gram-positive bacteria (*Bacillus subtilis* and *Staphylococcus aureus*) and one Gram-negative bacteria (*Pseudomonas aeruginosa*) at a concentration of 1 mg/ml. Furacin was utilized as a control substance. The minimum inhibitory concentrations (MICs) of all compounds were also determined. The results revealed that compounds 1–3 had strong activity against Gram-positive bacteria *B. subtilis*, with MICs of 46, 39, and 35 mg/ml, respectively, and against *S. aureus*, with MICs of 52, 31, and 27 μg/ml, respectively. All of the chemicals, on the other hand, had no effect on the Gram-negative bacterium *P. aeruginosa.* The antibacterial activity of the red marine alga *L. obtusa* extract and essential oil was examined against ten bacterial strains (*Salmonella typhimurium*, *Enterococcus faecalis*, *S. aureus*, *Staphylococcus epidermidis*, *Enterobacter aerogenes*, *Enterobacter cloacae*, *Escherichia coli*, *P. aeruginosa*, *Proteus vulgaris*, and *B. subtilis*), two specific pathogenic strains (methicillin-oxacillin resistant *S. aureus* and haemorrhagic *E. coli*), and one yeast strain (*Candida albicans*). Antimicrobial properties of methanol, chloroform, hexane extract, and essential oil of *L. obtusa* were examined against Gram-positive and Gram-negative bacteria as well as one yeast strain through the paper disc diffusion assay. The extracts from *L. obtusa* showed weak activity against Gram-positive bacteria, only one Gram-negative bacterium, *E. cloacae*, and yeast strain *C. albicans.* The essential oil of *L. obtusa* was the most active, inhibiting all Gram-positive bacteria examined, including methicillin-oxacillin-resistant *S. aureus* (MRSA), Gram-negative bacteria, *E. cloaceae*, and yeast strains. As a result, all three extracts of *L. obtusa* were discovered to have high antibacterial activity against the pathogens indicated above (Demirel et al. 2011).

12.3.4 Antifungal Activity

From *L. obtusa* three new eudesmane sesquiterpenoid compounds were separated and named as eudesma-4(15),7-diene-5a,11-diol (1), teuhetenone (2), and chabrolidione B (3). Compounds 1 and 3 showed good antifungal activity against *C. albicans*, *Candida tropicalis*, *Aspergillus flavus*, and *Aspergillus niger* in qualitative and quantitative antifungal experiments, while compound 2 showed less fungal activity and greater cytotoxic action (Alarif et al. 2016; Barot et al. 2016). According to Alarif et al. (2012), a compound from *Laurencia* named isolauraldehyde (compound 3) showed antifungal activity. For yeast susceptibility testing, the microtitre broth dilution method is used. At a MIC of 70 mg/ml, compound 3 showed significant activity against *C. albicans*, and moderate activity against *A. fumigatus* and *A. flavus* at MICs of 100 and 1000 mg/ml, respectively.

12.3.5 Antitumour Activity

Cancer is a global problem that has emerged as the most awful disease, affecting people from all walks of life. Cancer patients in India are expected to number 679,421 males and 712,758 females by 2020. Lung cancer affects 1 in 68 men, breast cancer affects 1 in 29 women, and cancer strikes 1 in 9 Indians at some point in their lives (Mathur et al. 2020). Seaweeds have antioxidant systems that produce bioactive compounds and enzymes to help them resist environmental stress. The antioxidant capabilities of *L. obtusa* extracts were investigated, and the findings of the DPPH scavenging assay demonstrated increased antioxidant activity in *L. obtusa* protein extracts (Al Monla et al. 2021). Numerous studies have found a link between antioxidants and the cytotoxic potential of substances (Grigalius and Petrikaite 2017). The antioxidative protein extracts also demonstrated a time- and dose-dependent manner retardation of cell viability in the cancer cell lines examined (HCT-116 and HeLa). Other investigations have found that *L. obtusa* and *Ulva lactuca* gathered from various places had high cytotoxic action against colon cancer (Dellai et al. 2013). A common tumour is Ehrlich ascites carcinoma (EAC). This is a type of undifferentiated carcinoma with a high transplantable capability, no recurrence, rapid proliferation, a shorter life span, and 100% malignancy. It also lacks tumour-specific transplantation antigen (TSTA). Three novel laurene-type sesquiterpenes were discovered in an organic extract of the red alga *L. obtusa*: 12-hydroxy isolaurene (1), 8,11-dihydro-12-hydroxy isolaurene (2), and isolauraldehyde (3). Compound 3 had the highest cytotoxic efficacy (83.1%) in an *in vitro* model of EAC, followed by compound 2 (79.9%) (Alarif et al. 2012)

Alarif et al. (2016) detected three new eudesmane sesquiterpenoid compounds from red marine alga *L. obtusa* named as eudesma-4(15),7-diene-5a,11-diol (compound 1), teuhetenone (compound 2) and chabrolidione B (compound 3). The extracted compounds were given to the non-aggressive human mammary epithelial breast cancer cell line MCF-7 at concentrations ranging from 0 to 250 mg/ml. At 1, 24, 48, and 72 hours, the activities of lactate dehydrogenase (LDH) were measured. The amount of LDH released into the extracellular medium by an injured cell is a reliable cytotoxicity marker.

In addition, the LDH assay was utilized to determine the integrity of the plasma membrane. The release of LDH into the medium was used to assess cytotoxicity. Each cell culture well had 50 μL of supernatant removed and the absorbance measured at 340 nm to determine LDH activity (Hilf et al. 1976). The result showed that in the LDH assay, compound 2 had a strong apoptosis activity against the breast cancer type MCF-7, while compounds 1 and 3 had a good-to-moderate apoptosis activity.

The resazurin assay was used to determine the *L. obtusa* ethyl acetate extract's potential anticancer activity against cancer cells. To broaden the range of this study, we used three human tumour cell lines from different origins a monocytic leukaemia cell line (THP-1), a second tumour cell taken from an osteosarcoma (MNNG-HOS), and a third tumour cell derived from an epithelial cell line from lung adenocarcinoma (A549). In the result, the THP-1 and MNNG-HOS were exhibited to be more sensitive to the *L. obtusa* extract, while A549 was least susceptible to the extract (Esselin et al. 2017).

12.3.6 Antiviral (Anti-Hepatitis C Virus) Activity

Hepatitis C virus (HCV) has infected more than 170 million individuals globally (Pacini et al. 2004). Infection with HCV causes chronic liver disease, which can proceed to hepatocellular carcinoma in the majority of cases. Interferon plus ribavirin, the currently available HCV treatment, has limited advantages due to significant side effects. The ability of *L. obtusa* polysaccharide extracts to scavenge free radicals was used to assess the prevention of lipid peroxidation. The hepatocellular carcinoma HepG2 cells were infected with HCV and then treated with *L. obtusa* polysaccharide. The reverse transcription quantitative polymerase chain reaction (RT-qPCR) was used to detect viral replication inhibition. *L. obtuse* extract polysaccharides are the most effective HCV inhibitors (82.36% at co-adsorption mode). The antiviral effects of *L. obtusa* polysaccharides have very good antiviral qualities in terms of a broad spectrum of activity as well as a complex mode of action. As a result, this marine red alga is a promising possibility for further medication development (Gheda et al. 2016). The World Health Organization has approved the Draft Global Health Sector Initiatives for Viral Hepatitis (2016–2021), which identifies actionable areas and strategies. The National Viral

Hepatitis Control Program (NVHCP) was created by the Indian government in 2018 with the goal of preventing and treating viral hepatitis (hepatitis A, B, C, and E) and providing free screening, diagnosis, treatment, and counselling to all (Anand et al. 2021).

12.3.7 Use of *Laurentia obtusa* in the Development of Probiotic *Saccharomyces cerevisiae*

A significant study on the dietary modification of gut microflora has recently been undertaken, and the results have revealed that a few diets, edibles, or biochemical molecules have an effect on gut microflora. Dietary prebiotics (indigestible carbohydrates) have been studied as potentially beneficial to gastrointestinal health. Polysaccharides have the potential to be utilized as beneficial prebiotic components in human and animal health applications (O'Sullivan et al. 2010). *Saccharomyces cerevisiae* is a human probiotic yeast that has been clinically verified. In the presence of *Laurencia*, the biochemical content of *S. cerevisiae* alters, according to Caf et al. (2017). *L. obtusa* is a high-protein, oligosaccharide, vitamin, vital mineral, and fatty acid-rich plant that is also a good source of tryptophan and methionine and trace elements and *S. cerevisiae* can grow with *L. obtusa* as substrate. *S. cerevisiae* was kept in an incubator with *L. obtusa*, then dry weight, fatty acid contents, vitamins, and phytosterol were measured. After converting fatty acids in the lipid extract into methyl esters, gas chromatography was used to analyse them, and high-performance liquid chromatography (HPLC) was used to quantify vitamin concentrations. When *S. cerevisiae* is treated with *L. obtusa* extract the total fatty acids levels and vitamin contents developed at different values. When comparing the *S. cerevisiae* group incubated with algal extracts to the control group, the levels of myristic acid (14:0) palmitoleic acid (16:1), stearic acid (18:0), oleic acid (18:1), and linoleic acid (18:2) increased Caf et al. (2017). The increased phenolic content and DPPH and ABTS radical scavenging activities of *S. cerevisiae* may be connected to its higher fatty acid biosynthetic activity. The phenolic content of *L. obtusa* was investigated in water and methanol, with the water environment yielding better results. Their total phenolic component content was found to be 33.41 ± 0.11 GAE mg/1 g. After the flavonoid test, it was found that rutin, myricetin, morin, quercetin, kamferol, and naringenin were found in *L. obtusa* extract. Arabinose was the most abundant monosaccharide in *L. obtusa*, followed by maltose. In *S. cerevisiae*, *L. obtusa* has caused a favourable effect in increasing the amount of linoleic and oleic acid. Arabinose is a calorie-free sugar that has been shown to impact glucose and lipid metabolism, as well as reduce obesity and body weight (Hao et al. 2015). The development of *S. cerevisiae*, a beneficial probiotic yeast, was positively influenced by *L. obtusa* extract. The need for chemical protectors can be reduced by adding algae or algal extracts to food products. Finally, when *L. obtusa* extract was utilized in *S. cerevisiae* growth medium, the quantities of fatty acids and vitamins were changed (Caf et al. 2017). See Figure 12.8 and Table 12.1.

12.4 DIETARY *LAURENCIA OBTUSA* IN ENHANCING GROWTH AND BLOOD INDICES

Red tilapia is the second most popular farmed freshwater fish globally. Fast development, more resistance to climate change conditions, appetizing quality, and resistance to strain and sicknesses, are just a few of the characteristics that make it an excellent option for aquaculture (El-Sayed 2019). Tilapia is typically farmed by intensive culture methods, which put the fish yield at risk owing to stress and sickness brought on by many environmental factors like low dissolved oxygen levels or where oxygen is not available in sufficient amounts at the tissue level (Wu et al. 2013). Consequently, diets contain functional qualities that can naturally boost the ability to fight off various stresses. Salem et al. (2021) observed the impact of *L. obtusa* in boosting red tilapia development rate, life span, and food utilizing capability. The diets were provided to fish (4.49 g/fish) for 70 days. According to the

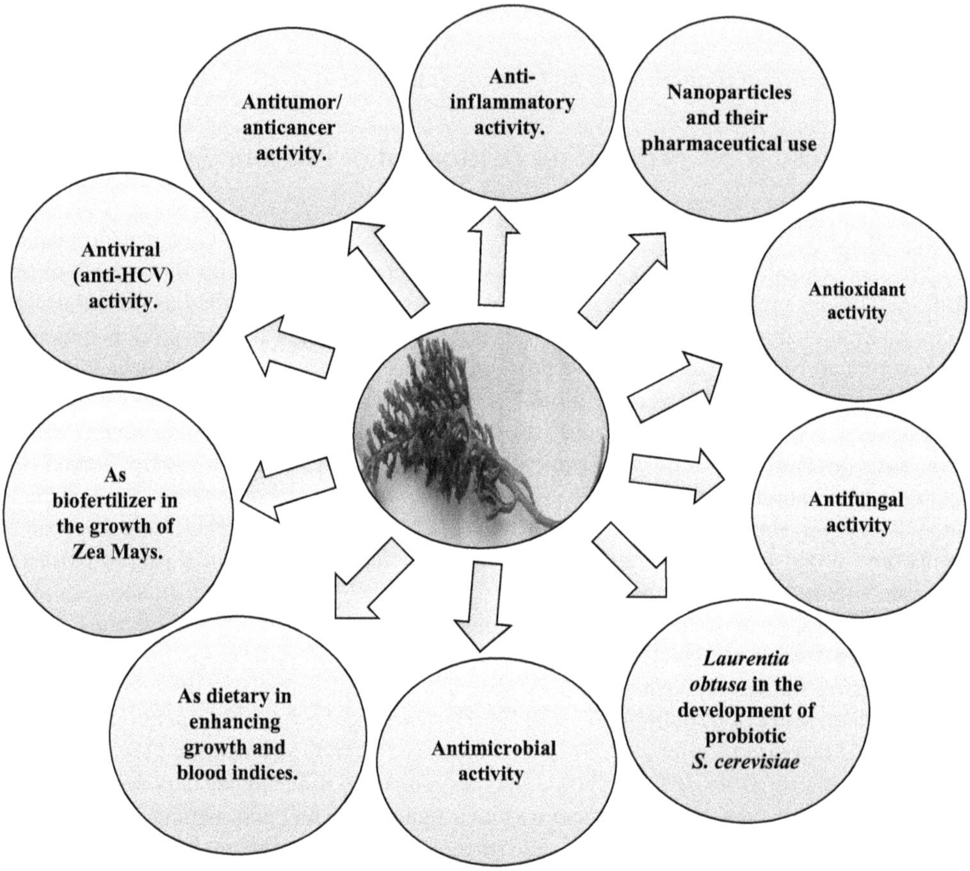

FIGURE 12.8 Pharmaceutical uses of *Laurencia obtuse*.

TABLE 12.1
Phytochemicals of *L. obtusa* and Their Pharmaceutical Uses

Sl. no.	Phytochemicals	Pharmaceutical use	Technique used	References
1.	**C15 Acetogenins**			
	Laurentusenin (1) Laurenfuresenin (2)	Apoptosis Chemotaxonomic value	DNA fragmentation assay, microscopic observation.	Ghandourah et al. 2019
2.	**Sesquiterpenes**			
	8-Keto-10-dehydrobromo-γ-snyderol (1) α-Snyderol β-Snyderol	Antimicrobial Chemotaxonomic marker Cytotoxic	Spectroscopic analysis	Esselin et al. 2017, 2019
3.	**Crude polysaccharides**			
	(sulphate groups of agarose derivatives)	Inhibitor of HCV replication in HepG2 cells	RT-qPCR	Gheda et al. 2016
4	Nitrogen, phosphorous, potassium, protein content, carbohydrate, ash contents, phenol contents	As biofertilizer for the growth and high productivity of *Z. mays*	Agriculture technique	Patel et al. 2017

TABLE 12.1 (Continued)
Phytochemicals of *L. obtusa* and Their Pharmaceutical Uses

Sl. no.	Phytochemicals	Pharmaceutical use	Technique used	References
5	**Eudesmane sesquiterpenoids** Eudesma-4-(15),7-diene-5,11-diol (1) Trinor-sesquiterpene, teuhetenone (2) Seco-eudesmanesesquiterpene, chabrolidione B (3)	Compounds 1–3: antifungal against *C. albicans*, *A. niger*. Compound 2: cytotoxic activity and antifungal		Alarif et al. 2012, 2016
6	Protein extract of *L. obtusa*	Antioxidant activity	DPPH scavenger assay	Al Monla et al. 2021
7.	Protein oligosaccharides, carbohydrate	As substrate for the production of probiotic *S. cerevisiae*	Gas chromatography, HPLC	Caf et al. 2017
8	**AgNPs (silver nanoparticle of *L. obtusa*)** from polysaccharides	Antitumour activity against EAC cell	Fourier transform infrared spectroscopy (FT-IR), X-ray diffraction (XRD), and transmission electron microscopy (TEM)	Durán et al. 2010; Khalifa et al. 2016.
9.	Vigiallene (1) ($C_{15}H_{16}Br_2O_5$)	Vigiallene showed weak anticancer activity in case of human tumour cell line A431	Chromatographic separation	Perdikaris et al. 2019
10	**Sesquiterpentines** 12-Hydroxy isolaurene 8,11-Dihydro-12-hydroxy isolaurene Isolaureldehyde	Antibacterial, antifungal for some yeast Antitumour activity	Disc diffusion technique	Alarif et al. 2012
11	**Brominated diterpenes** Prevezol A, B (2) Prevezol B (6)	Both most effective in antitumour activity	Vacuum column chromatography (VCC) High-pressure liquid chromatography (HPLC)	Iliopoulou et al. 2003b
12	**Maneonene based structure** Jeddahenyne-A (1) Jeddahenyne-A (2) **Isomaneonene based structure:** 12-Debromo-12-methoxy isomaneonene A (3)	Apoptosis induced In the regulation of programmed death In the initiation and propagation of inflammatory responses	Single crystal X-ray analysis DNA fragmentation assay Microscopic observation	Bawakid et al. 2017
13	β-Chemigrane type sesquiterpenes	Antihelminthic	–	Davyt et al. 2001; Demirel et al. 2011

regression test, including *L. obtusa* in a 4.1 g/kg diet is the best way to maximize fish weight gain. According to this study, the fish group which is fed with the maximum algae meal (LO8: 8 g/kg) had minimum development whereas minimum feed intake with LO2 (2 g/kg) showed high growth in the fish group. As a result, including *L. obtusa* alga in red tilapia fish diets increased development and resilience to hypoxic stress.

12.5 *LAURENTIA OBTUSA* AS BIOFERTILIZER IN THE GROWTH OF *ZEA MAYS*

Biofertilizers are a new environmentally benign technology that can help farmers overcome the drawbacks of traditional chemical-based farming. Both soil sustainability and plant development were improved by using biofertilizers. Biofertilizers are a natural alternative to chemical fertilizers that can also boost soil nutrients and crop yields. Marine algae can be used as bio-nutrients in both dried and fresh form as fertilizer in the soil, and this supports the enhancement of organic nutrients such as carbohydrates, lipids, proteins, fibres, ash, phenol, dietary fibre, and others in plants. Marine algae are also a good source of elements that plants require for nutrition (Patel et al. 2017).

Corn is one of the world's three most significant grain crops. Corn is a highly productive crop that is also simple to manage, digest, and is less expensive than other crops. *L. obtusa* along with other algae enhance the whole plant length, dry mass, total hydroxybenzene content, and nitrogen content. Plants treated with seaweed extracts had more protein bands in their protein profiles than control plants (Al-Shakankery et al. 2014).

Safinaz and Ragaa (2013) tested the effect of three species of red marine algae as biofertilizers on maize plants and found it improved the vegetative growth of *Z. mays* plants. They concluded that using algae as biofertilizers improved the growth, yield, and grain quality of maize plants.

12.6 NANOPARTICLES AND THEIR PHARMACEUTICAL USE

One of the key aspects of nanotechnology is green nanotechnology, which strives to produce eco-friendly nanoproducts while lowering environmental risks and enhancing the sustainability of processes. The nanoparticles produced by green nanotechnical methods are either based on plant extracts or on the recycling of rubbish, and they have advantages in terms of ease of use, accessibility, and processing time. The use of nanotechnology in various fields, including environmental preservation (Handojo et al. 2020), nano-adsorbents (Kaliannan et al. 2019), development of drafting super capacitors (Bhat et al. 2020), and nano-silica (Singh and Endley 2020) has been reported.

Marine nanobiotechnology is a new and fascinating field of study. In the near future, industry and biomedicine will be able to use clean, non-toxic, and environmentally favourable nanostructured materials. Because it does not employ toxic chemicals, the biosynthesis of nanoparticles using seaweed is considered a green technique. The development of finer techniques and the applicability of biosynthesized nanostructures, on the other hand, are now being investigated. Because of their potential application in medical therapy, silver and gold are among the most studied metals. Silver nanoparticles (AgNPs) have been tested as antibacterial agents and been found to be effective against a variety of fungi and bacteria.

Metal nanoparticle applications are rapidly developing in the fields of catalysis, biosensors, electronics, and optics. Silver, platinum, gold, and palladium are some of the precious metals that can be used to make nanoparticles.

12.6.1 Antitumour Activity of Silver Nanoparticles Biosynthesized by *Laurencia obtusa* and Other Algae

The effect of different doses of AgNPs made by *L. obtusa* was investigated on EAC cells, and was found to be 86% with 98 g/ml AgNPs. In general, increasing AgNP concentration resulted in a decrease in EAC cell viability. The proliferation of EAC cells was decreased at different concentrations of green-produced AgNPs (42, 56, 69, 85, and 98 g/ml). AgNPs are considered to cause cytotoxicity by interfering with the proper functioning of cellular proteins and creating changes in cellular chemistry. AgNPs produced biologically by *L. obtusa* and other macroalgae are characterized by a consistent shape and tiny size, resulting in AgNPs with significant antitumour activity (Khalifa et al. 2016).

12.6.2 Silver Nanoparticles from Polysaccharide of *Laurencia obtusa*

Red alga was collected from the beach of the Red Sea, Alexandria, Egypt. The cleaned samples were shade dried and homogenized to a fine powder and kept for further treatment in which polysaccharides were extracted from the *Laurentia* powder. Water-soluble polysaccharide is used to make AgNPs, which act as a reducing agent for silver ions as well as a stabilizing agent for the AgNPs. Surface plasmon resonance (SPR) at 350–550 mm, Fourier transform infrared spectroscopy (FT-IR), X-ray diffraction (XRD), and transmission electron microscopy (TEM) were used to confirm the production of AgNPs. The versatility of noble metal nanoparticles in general, and AgNPs in particular, is well known especially in textile industries (Duran et al. 2010), medicine (Chaw 2005), in consumer goods (Jiang et al. 2004; Rai et al. 2009), and in food processing (Tankhiwale and Bajpai 2010) and in terms of antimicrobial, antifungal, and anticancer capabilities (Karthic et al. 2021).

12.7 CONCLUSION AND FUTURE PERSPECTIVES

This chapter seeks to give an overview of the pharmaceutical significance of the red macroalga *L. obtusa*. This alga could be a rich source of critical bioactive substances that are both environmentally favourable and safe. The halogenated phytochemicals discovered in *L. obtusa* are plentiful. Many of these metabolites have been found to have properties such as inhibiting normal feeding behaviour which is caused due to diterpenes, a vermicidal impact which is due to chamigrane-type sesquiterpenes, an antimalarial effect due to brominated sesquiterpenes, a biocidal effect due to sesquiterpenes, antibacterial and antifungal effects, and anticancer effect caused by allolaurinterol of *L. obtusa*.

Polyketides, terpenes, nitrogen-containing compounds, and polysaccharides are four structural kinds of marine chemicals that have cytotoxic characteristics. Kladi et al. (2007) (looking at sesquiterpenes) discovered that *L. obtusa* has antitumour effects in CHO cells taken during ovary biopsies from Chinese hamster and from a human tumour cell line removed from chronic myelogenous leukaemia (K562), a cell line taken from a mammary adenocarcinoma (MCF7), and a cell line taken from an epidermoid prostate adenocarcinoma (PC3). Some cell lines were derived from a cervical adenocarcinoma (HeLa) and from an epidermoid carcinoma (A431). The ability to partition chemicals into living specialized depository structures, such as the glandular part and the corps en cerise (CC), has been linked to the generation of secondary metabolites in seaweed. Optical tweezers and time-lapse video microscopy were used to examine the dynamics of CC and the mechanisms relating to the intracellular transport of halogenated chemicals. The presence of halogenated chemicals in CC, as well as vesicles scattered together with the cytoplasm and within the chloroplasts, is confirmed as one of the primary discoveries. Following the exudation of these halogenated chemicals to the thallus surface, which serves as a protective mechanism against herbivores and inhibits normal feeding behaviour, CC breakdown or cell death occurs (Salgado et al. 2008). It is believed that *L. obtusa* has the potential to be a pharmaceutical for inflammatory illnesses and to protect against gastrointestinal adverse effects based on the findings. Based on the data, it is thought that *L. obtusa* has the potential to provide a pharmaceutical approach to inflammatory disorders and to protect against digestive system harmful effects. More study on the isolation and characterization of bioactive compounds from this red alga is being done in order to establish the bioactive ingredients responsible for these important functions.

Green chemistry is seen to be the best approach for the time ahead of nanomaterials. Marine plants, algae, and bio-nano factories are used in green chemistry to generate nanoparticles (NPs) and bimetallic NPs for biomedical and other uses. *L. obtusa* extract from the Rhodophyceae class contains phytochemicals, which leads to the creation of a shell around metal NPs to increase bioaffinity, and has been investigated for the amalgamation of metallic NPs. Necessary metabolites like alkaloids, flavonoids, and terpenoids in *L. obtusa* have the potential for metal absorption and to reduce metallic ions, creating optimal candidates for generation by functioning as capping sources, and promoting particle formation. In terms of antibacterial, antifungal, and anti-cytotoxic capabilities, this chapter has focused on extensive research on algae-based NP production.

The molecular characterization of potential bioactive chemicals present in *L. obtusa* must be evaluated and explored in order to identify less hazardous and more effective medications that will benefit the world's population. *L. obtusa* is also used as a biofertilizer to improve the growth, yield, and grain quality of maize plants.

ACKNOWLEDGEMENTS

The author wishes to express her gratitude to all of the authors whose work has been used in the preparation of this chapter, as well as to Dr (Maj) Harinder Singh Kang, Principal, Guru Nanak Khalsa College, Yamun Nagar, Haryana, India, for providing overall facilities, and Dr Amarjit Singh, Dr Prabhjot Kaur, and Dr Daljeet Kaur for their help and support.

REFERENCES

Anand, A. and Shalimar, D.M. 2021. Hepatitis C virus in India: challenges and successes. *Clinical Liver Disease*, 18(3):150.

Al Monla, R., Salma, Y., Kouzayha, A., Gali-Muhtasib, H., Dassouki, Z. and Mawlawi, H. 2021. Antioxidative, cytotoxic, and anti-metastatic potentials of *Laurencia obtusa* and *Ulva lactuca* seaweeds. *Asian Pacific Journal of Tropical Biomedicine*, 11(7):308–316.

Alarif, W.M., Al-Footy, K.O., Zubair, M.S., Halid P.H.M., Ghandourah, M.A., Basaif, S.A., Al-Lihaibi, S.S., Ayyad, S.E.N. and Badria, F.A. 2016. The role of new eudesmane-type sesquiterpenoid and known eudesmane derivatives from the red alga *Laurencia obtusa* as potential antifungal–antitumour agents. *Natural Product Research*, 30(10):1150–1155.

Alarif, W.M., Al-Lihaibi, S.S., Ayyad, S.E.N., Abdel-Rhman, M.H. and Badria, F.A. 2012. Laurene-type sesquiterpenes from the Red Sea red alga *Laurencia obtusa* as potential antitumor–antimicrobial agents. *European Journal of Medicinal Chemistry*, 55:462–466.

Al-Shakankery, F.M., Hamouda, R.A. and Ammar, M.M. 2014. The promotive effect of different concentrations of marine algae as biofertilizers on growth and yield of maize (*Zea mays* L.) plants. *Journal of Chemical, Biological and Physical Sciences*, 4:43201–43211.

Baker, J.T. 1984. Seaweeds in pharmaceutical studies and applications. In: *Proceedings of the Eleventh International Seaweed Symposium*. Dordrecht: Springer, 29–40.

Barot, M., Kumar, N.J.I. and Kumar, R.N. 2016. Bioactive compounds and antifungal activity of three different seaweed *species Ulva lactuca, Sargassum tenerrimum* and *Laurencia obtusa* collected from Okha coast, Western India. *Journal of Coastal Life Medicine*, 4(4):284–289.

Bawakid, N.O., Alarif, W.M., Ismail, A.I., El-Hefnawy, M.E., Al-Footy, K.O. and Al-Lihaibi, S.S. 2017. Bioactive maneonenes and isomaneonene from the red alga *Laurencia obtusa*. *Phytochemistry*, 143:180–185.

Bhat, V.S., Jayeoye, T.J., Rujiralai, T., Sirimahachai, U., Chong, K.F. and Hegde, G. 2020.Influence of surface properties on electrochemical supercapacitors utilizing *Callerya atropurpurea* pod derived porous nanocarbons: structure-property relationship between porous structures to energy storage devices. *Nano Select*, 1(2):226–243.

Bhatia, L., Bachheti, R.K., Garlapati, V.K. and Chandel, A.K. 2022. Third-generation biorefineries: a sustainable platform for food, clean energy, and nutraceuticals production. *Biomass Conversion and Biorefinery* 12:4215–4230.

Caf, F. and Åžen í–zdemir, N. 2017. Potential of *Laurencia obtusa* as a substrate for the development of a probiotic *Saccharomyces cerevisiae*. *Cellular and Molecular Biology*, 63(8):71–76.

Chapman, A.P. 1998. *Seaweeds and Their Uses*. London: Camelot Press.

Chaw, K.C., Manimaran, M. and Tay, F.E. 2005. Role of silver ions in destabilization of intermolecular adhesion forces measured by atomic force microscopy in *Staphylococcus epidermidis* biofilms. *Antimicrobial Agents and Chemotherapy*, 49(12):4853–4859.

Chojnacka, K., Saeid, A., Witkowska, Z. and Tuhy, L. 2012. Biologically active compounds in seaweed extracts-the prospects for the application. *The Open Conference Proceedings Journal*, 3(1):20–28.

Davyt, D., Fernandez, R., Suescun, L., Mombrú, A.W., Saldana, J., Domínguez, L., Coll, J., Fujii, M.T. and Manta, E. 2001. New sesquiterpene derivatives from the red alga *Laurencia scoparia.* Isolation, structure determination, and anthelmintic activity. *Journal of Natural Products*, 64(12):1552–1555.

Dellai, A., Laajili, S., Le Morvan, V., Robert, J. and Bouraoui, A. 2013. Antiproliferative activity and phenolics of the Mediterranean seaweed *Laurencia obusta. Industrial Crops and Products*, 47:252–255.

Demirel, Z., Yilmaz-Koz, F.F., Karabay-Yavasoglu, N.U., Ozdemir, G. and Sukatar, A. 2011. Antimicrobial and antioxidant activities of solvent extracts and the essential oil composition of *Laurencia obtusa* and *Laurencia obtusa* var. *pyramidata. Romanian Biotechnological Letters*, 16(1):5927–5936.

Deyab, M.A. 2011. Biochemical constituents and bioactivity of *Laurencia Obtusa* (Ceramiales, Rhodophyta) in relation to seasonal and ecological factors at Red Sea coast – Egypt. *Egyptian Journal of Phycology*, 12(1):33–46.

Durán, N., Marcato, P.D., Alves, O.L., Da Silva, J.P., De Souza, G.I., Rodrigues, F.A. and Esposito, E. 2010. Ecosystem protection by effluent bioremediation: silver nanoparticles impregnation in a textile fabrics process. *Journal of Nanoparticle Research*, 12(1):285–292.

El-Sayed, A.F.M. 2019. *Tilapia Culture*, 2nd edn. London: Elsevier/Academic Press.

Esselin, H., Sutour, S., Bighelli, A. and Tomi, F. 2019. Snyderol derivatives *from Laurencia obtusa* collected in Corsica. *Biochemical Systematics and Ecology*, 82:24–26.

Esselin, H., Sutour, S., Liberal, J., Cruz, M.T., Salgueiro, L., Siegler, B., Freuze, I., Castola, V., Paoli, M., Bighelli, A. and Tomi, F. 2017. Chemical composition of *Laurencia obtusa* extract and isolation of a new C15-acetogenin. *Molecules*, 22(5):779.

Ghandourah, M.A., Alarif, W.M. and Bawakid, N.O. 2019. New bioactive C15 acetogenins from the red alga *Laurencia obtusa. Pharmacognosy Magazine* 15(61):199–203.

Gheda, S.F., El-Adawi, H.I. and El-Deeb, N.M. 2016. Antiviral profile of brown and red seaweed polysaccharides against hepatitis C virus. *Iranian Journal of Pharmaceutical Research: IJPR*, 15(3):483–491.

Gribble, G.W. 2015. Biological activity of recently discovered halogenated marine natural products. *Marine Drugs*, 13(7):4044–4136.

Grigalius, I. and Petrikaite, V. 2017. Relationship between antioxidant and anticancer activity of trihydroxy flavones. *Molecules*, 22(12):2169.

Guiry, M.D. and Guiry, G.M. 2022. *AlgaeBase*. World-Wide Electronic Publication, National University of Ireland, Galway. Available from: www.algaebase.org.

Hagiwara, H., Fujiwara, S., Iibachi, C., Suzuki, T. and Hoshi, T. 2020. A synthetic approach toward a brominated oxocane labdane diterpenoid isolated from *Laurencia obtusa. Natural Product Communications*, 15(3):1–4.

Handojo, L., Pramudita, D., Mangindaan, D. and Indarto, A. 2020. Application of nanoparticles in environmental cleanup: production, potential risks and solutions. In: Bharagava, R. (ed) *Emerging Eco-friendly Green Technologies for Wastewater Treatment*. Singapore: Springer, 45–76.

Hao, L., Lu, X., Sun, M., Li, K., Shen, L. and Wu, T. 2015. Protective effects of L-arabinose in high-carbohydrate, high-fat diet-induced metabolic syndrome in rats. *Food & Nutrition Research*, 59(1):28886.

Harizani, M., Ioannou, E. and Roussis, V. 2016. The *Laurencia paradox*: An endless source of chemodiversity. *Progress in the Chemistry of Organic Natural products* 102:91–252.

Hilf, R., Rector, W.D. and Orlando, R.A. 1976. Multiple molecular forms of lactate dehydrogenase and glucose 6-phosphate dehydrogenase in normal and abnormal human breast tissues. *Cancer*, 37(4):1825–1830.

Iliopoulou, D., Mihopoulos, N., Roussis, V. and Vagias, C. 2003a. New brominated labdane diterpenes from the red alga *Laurencia obtusa. Journal of Natural Products*, 66(9):1225–1228.

Iliopoulou, D., Mihopoulos, N., Vagias, C., Papazafiri, P. and Roussis, V. 2003b. Novel cytotoxic brominated diterpenes from the red alga *Laurencia obtusa. Journal of Organic Chemistry*, 68(20):7667–7674.

Jiang, H., Manolache, S., Wong, A.C.L. and Denes, F.S. 2004. Plasma-enhanced deposition of silver nanoparticles onto polymer and metal surfaces for the generation of antimicrobial characteristics. *Journal of Applied Polymer Science*, 93(3):1411–1422.

Kaliannan, D., Palaninaicker, S., Palanivel, V., Mahadeo, M.A., Ravindra, B.N. and Jae-Jin, S. 2019. A novel approach to preparation of nano-adsorbent from agricultural wastes (*Saccharum officinarum* leaves) and its environmental application. *Environmental Science and Pollution Research*, 26(6):5305–5314.

Karthic, A., Wagh, N.S. and Lakkakula, J.R. 2021. Algae-assisted synthesis of nanoparticles. In: Ghosh, S. and Webster, T.J. (eds) *Nanobiotechnology*. Amsterdam: Elsevier, 145–165

Khalifa, K.S., Hamouda, R.A., Hanafy, D. and Hamza, A. 2016. In vitro antitumor activity of silver nanoparticles biosynthesized by marine algae. *Digest Journal of Nanomaterials and Biostructures*, 11(1):213–221.

Kladi, M., Vagias, C., Papazafiri, P., Furnari, G., Serio, D. and Roussis, V. 2007. New sesquiterpenes from the red alga *Laurencia microcladia*. *Tetrahedron*, 63(32):7606–7611.

Kolanjinathan, K., Ganesh, P. and Saranraj, P. 2014. Pharmacological importance of seaweeds: a review. *World Journal of Fish and Marine Sciences*, 6(1):1–15.

Lajili, S., Azouaou, S.A., Turki, M., Muller, C.D. and Bouraoui, A. 2016. Anti-inflammatory, analgesic activities and gastro-protective effects of the phenolic contents of the red alga, *Laurencia obtusa*. *European Journal of Integrative Medicine*, 8(3):298–306.

Lawrence, T., Willoughby, D.A. and Gilroy, D.W. 2002. Anti-inflammatory lipid mediators and insights into the resolution of inflammation. *Nature Reviews Immunology*, 2(10):787–795.

Li, H.Y., Matsunaga, S. and Fusetani, N. 1998. Antifungal metabolites from marine sponges. *Current Organic Chemistry*, 2(6):649–682.

Martin, J.D., Caballelo, P., Fernandez, J.J., Norte, M., Perez, R. and Rodriguez, M.L. 1989. Metabolites from *Laurencia obtuse*. *Phytochemistry*, 28(12):3365–3367.

Masuda, M., Abe, T., Suzuki, T. and Suzuki, M. 1996. Morphological and chemotaxonomic studies on *Laurencia composita* and *L. okamurae* (Ceramiales, Rhodophyta). *Phycologia*, 35(6):550–562.

Mathur, P., Sathishkumar, K., Chaturvedi, M., Das, P., Sudarshan, K.L., Santhappan, S., Nallasamy, V., John, A., Narasimhan, S., Roselind, F.S. and ICMR-NCDIR-NCRP Investigator Group. 2020. Cancer statistics, 2020: Report from national cancer registry programme, India. *JCO Global Oncology*, 6:1063–1075.

Nigam, V. and Sodhi, J.S. 2014. Some medicinal plants with antioxidant activity: a review. *International Journal of Pharmacy and Biological Sciences*, 4:73–78.

O'Sullivan, L., Murphy, B., McLoughlin, P., Duggan, P., Lawlor, P.G., Hughes, H. and Gardiner, G.E. 2010. Prebiotics from marine macroalgae for human and animal health applications. *Marine Drugs*, 8(7):2038–2064.

Pacini, L., Bartholomew, L., Vitelli, A. and Migliaccio, G. 2004. Reporter substrates for assessing the activity of the hepatitis C virus NS3-4A serine protease in living cells. *Analytical Biochemistry*, 331(1):46–59.

Patel, R.V., Pandya, K.Y., Jasrai, R.T. and Brahmbhatt, N. 2017. A review: scope of utilizing seaweed as a biofertilizer in agriculture. *International Journal of Advanced Research*, 5(7):2046–2054.

Perdikaris, S., Mangoni, A., Grauso, L., Papazafiri, P., Roussis, V. and E. Ioannou. 2019. Vagiallene, a rearranged C15 acetogenin from *Laurencia obtusa*. *Organic Letters*, 21(9):3183–3186.

Rai, M., Yadav, A. and Gade, A. 2009. Silver nanoparticles as a new generation of antimicrobials. *Biotechnology Advances*, 27(1):76–83.

Robert, A.H. 2004. Marine natural products. *Annual Reports on the Progress of Chemistry. Section B*, 100:169–189.

Safinaz, A.F. and Ragaa, A.H. 2013. Effect of some red marine algae as biofertilizers on growth of maize (*Zea mays* L.) plants. *International Food Research Journal*, 20(4):1629.

Salem, M.E.S., Abdel-Ghany, H.M. and Almisherfi, H.M. 2021. Role of dietary *Laurencia obtusa* in enhancing growth, blood indices, and hypoxia resistance of red tilapia (*Oreochromis niloticus* x *O. mossambicus*). *Journal of Applied Phycology*, 33(4):2617–2628.

Salgado, L.T., Viana, N.B., Andrade, L.R., Leal, R.N., Da Gama, B.A.P., Attias, M., Pereira, R.C. and Amado Filho, G.M. 2008. Intra-cellular storage, transport and exocytosis of halogenated compounds in marine red alga *Laurencia obtusa*. *Journal of Structural Biology*, 162(2):345–355.

Sethi, P. and Nandhagopal, P.K. 2020. Anatomical studies of *Laurencia Obtusa* (Hudson) Lamouroux (Ceramiales) of Rhodophyceae. *Chemistry & Material Sciences Research Journal*, 2(2):42–46.

Singh, S.P. and Endley, N. 2020. Fabrication of nano-silica from agricultural residue and their application. In: Husen, A. and Jawaid, M. (eds) *Nanomaterials for Agriculture and Forestry Applications*. Amsterdam: Elsevier, 107–134.

Tankhiwale, R. and Bajpai, S.K. 2010. Silver-nanoparticle-loaded chitosan lactate films with fair antibacterial properties. *Journal of Applied Polymer Science*, 115(3):1894–1900.

Wu, Y.R., Gong, Q.F., Fang, H., Liang, W.W., Chen, M. and He, R.J. 2013. Effect of *Sophora flavescens* on non-specific immune response of tilapia (GIFT *Oreochromis niloticus*) and disease resistance against *Streptococcus agalactiae*. *Fish & Shellfish Immunology*, 34(1):220–227.

Zubia, M., Payri, C. and Deslandes, E. 2008. Alginate, mannitol, phenolic compounds and biological activities of two range-extending brown algae, *Sargassum mangarevense* and *Turbinaria ornata* (Phaeophyta: Fucales), from Tahiti (French Polynesia). *Journal of Applied Phycology*, 20(6):1033–1043.

13 Brown Algae (Phaeophyta)

A Source of Different Phytochemicals and Their Medicinal Applications

Shaza H. Aly,[1] Esraa A. Elhawary,[2] Ahmed M. Elissawy,[2,3] Nada M. Mostafa,[2,3] Omayma A. Eldahshan,[2,3] and Abdel Nasser B. Singab[2,3,]*

[1] Department of Pharmacognosy, Faculty of Pharmacy, Badr University in Cairo, Cairo, Egypt

[2] Department of Pharmacognosy, Faculty of Pharmacy, Ain Shams University, Abbassia, Cairo, Egypt

[3] Center of Drug Discovery Research and Development, Ain Shams University, Cairo, Egypt

[*] Corresponding author (Abdel Nasser B. Singab): dean@pharma.asu.edu.eg

13.1 INTRODUCTION

Seaweeds/marine macroalgae are macroscopic multicellular organisms that are usually an important part of the ecosystem worldwide. According to Fritsch (1935), there are 11 classes of algae, and the main classes of algae include the brown algae (Phaeophyta), green algae (Chlorophyta), red algae (Rhodophyta), blue-green algae (Cyanobacteria), and diatoms (Bacillariophyta) (Figure 13.1). Their colours differ due to their pigment-production ability, nutrition, and chemical composition (Hakim and Patel 2020). They are also classified according to their habitat into aquatic algae, terrestrial algae, parasitic algae, symbiotic algae, and algae with special habitats (Sangha et al. 2014). There are 30,000 to 1 million known algal species with ubiquitous distribution. Brown algae are more abundant in shallow rocky coastal areas with low tides. Brown and red algae are utilized as food by humans due to their richness in minerals, vitamins, protein, amino acids, and other compounds, especially in Chinese cuisine (Lomartire and Gonçalves 2022).

The unique, diverse chemical nature of the brown algae and their fascinating biological activities have put them as one of the primary natural sources for new powerful drug leads for many foods, agricultural and health-related applications. Marine algae are usually found in challenging environmental conditions, which lead to the formation of potent secondary metabolites which help them cope with such harsh conditions to protect them from invaders and which are responsible for their reported biological activities (Hakim and Patel 2020). Chemically, they show diverse secondary metabolites, including different polysaccharides, proteins, and glycoproteins. Kelp, brown algae, and sea cucumber synthesize different forms of mucilage, which differ in their sulphated polysaccharide. Fucoidans, for example, are oligosaccharides with many monomeric sugars such as glucose, mannose, and others together with fucose and have many interesting biological activities (Pooja 2014). Other forms of biogenic molecules have been isolated from marine algae, especially

DOI: 10.1201/9781003256830-13

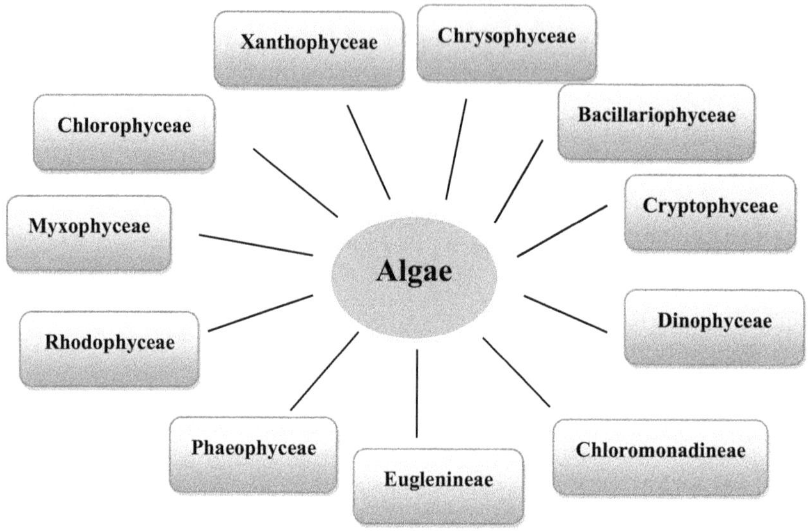

FIGURE 13.1 Schematic diagram showing different classes of algae.

the brown seaweeds, including terpenoids, sterols, fatty acids, hydrocarbons, polyphenolics, tannins, polysaccharides, and pigments. Brown algae represent a diverse class (Phaeophyceae) that varies from small filamentous forms to large/giant complex seaweeds. The fucoxanthin pigment imparts their characteristic colours. Brown algae are also rich in growth hormones like cytokines, auxins, and gibberellins, which make them excellent natural plant fertilizers in terms of their effect on growth and for industrial use (Hakim and Patel 2020).

Class Phaeophyta comprises about 265 genera and 2040 species distributed mainly in cold to temperate waters (Hakim and Patel 2020). A variety of polyphenolic compounds in such algae impart potent antioxidant, anti-inflammatory, and cytotoxic characteristics for use in many health issues such as cancer, osteoarthritis, diabetes, etc. (Nigam et al. 2022). Their sulphated polysaccharide content, especially the alginates and the fucoidans, are necessary anticoagulants from non-animal sources. Secondary metabolites from brown seaweed are of great importance for human health; they may act as strong antimicrobial agents to defend against resistant bacteria, fungi, and viruses. The fucoxanthin pigment works as an antiwrinkle and anti-ageing compound for cosmetic use; phlorotannins are used as skin whiteners for dark spots and fairer skin tone.

13.2 BROWN ALGAE MEDICINAL APPLICATIONS

The brown algae (Phaeophyta), or brown seaweeds, possess a number of exciting phytoconstituents and nutraceuticals which provoke a wide array of biological activities such as antimicrobial, antioxidant, anti-inflammatory, cytotoxic, hepatoprotective activities, and many others (Abdelwahab 2017). The research interest in the biological activity of such macroalgae was usually related to the folk food and medicinal practices of coastal communities (Ghazali et al. 2017). The isolated phytoconstituents and potential biological activities of brown seaweeds have gained much interest in recent years in many review articles (Sugiura et al. 2007; Brownlee et al. 2012; Thomas and Kim 2013; Pati et al. 2016; Khan et al. 2018; Hakim and Patel 2020; Pacheco et al. 2020; Rashad and El-Chaghaby 2020; Rashed et al. 2020; Rushdi et al. 2020; Begum et al. 2021; Okeke et al. 2021; Ślusarczyk et al. 2021; Lomartire and Gonçalves 2022; Nigam et al. 2022), and many of the medicinal applications of the brown algae from different geographical resources can be detailed as follows (Figure 13.2).

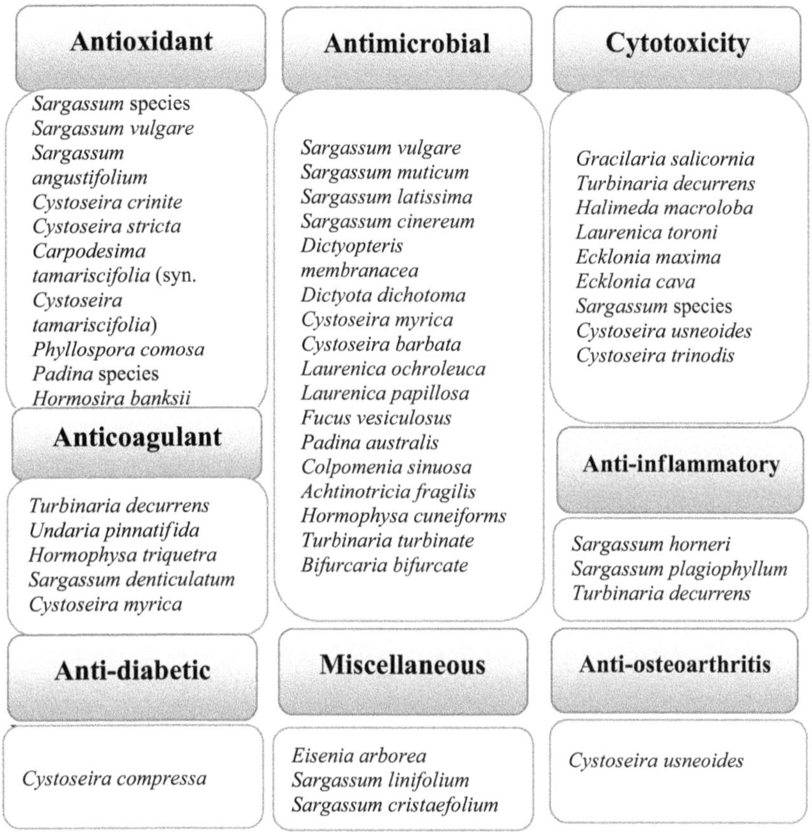

FIGURE 13.2 Schematic diagram showing different brown algal species and their reported biological activities.

13.2.1 Folk Medicinal Uses of Brown Seaweed

For centuries, brown algae, especially *kelp* and *sargassum*, have been utilized by humans as a food source and in folk medicinal practices, especially by coastal communities. The main utilized brown algae genera were *Alaria*, *Himanthalia*, *Bifurcaria*, *Laminaria*, *Saccharina*, *Undaria*, *Ascophyllum*, *Fucus*, and *Sargassum* (Brownlee et al. 2012). These brown seaweeds are a great source of carbohydrates in the form of polysaccharides, so they are used by Chinese people as a vegetable which is full of nutrients and vitamins with low fat and protein. These algae are a great source of vitamin D, iodine, folic acid, calcium, magnesium, iron, and amino acids. The sulphated polysaccharides such as carrageenan provide excellent gelling and thickening agents for food such as ice cream, salad dressings, and others. For medical applications, algae are used to treat goitre, tumour, oedema, testicular pain, scrofula, vermifuge, and swelling (Hoppe 1979; Pooja 2014).

13.2.2 Antioxidant Activity

The potent antioxidant activity of brown algae is usually attributed to its high phenolic and flavonoid content. The antioxidant power depends mainly on the ability to scavenge free radicals from oxidation (Al-Madhagy et al. 2019; Mostafa et al. 2021; El-Zahar et al. 2022). Many *in vitro* assays have been reported to evaluate the antioxidant potential of crude extracts. 2,2-Diphenyl-1-picrylhydrazyl (DPPH), ferric reducing antioxidant power (FRAP), 2,2'-azino-bis(3-ethylbenzothiazoline-6-sulphonic acid

($ABTS^+$), NO^- radical scavenging, inhibition of lipid peroxidation, Trolox equivalent antioxidant capacity (TEAC), and many others (Apak et al. 2016). *Sargassum* species showed 44–48% activity in the DPPH assay (Budhiyanti et al. 2012). Meanwhile, *Cystoseira crinita*, *Cystoseira stricta*, and *Sargassum vulgare* collected from Libya presented superior antioxidant power with half-maximal inhibitory concentration (IC_{50}) of 50–150 µg/ml (Alghazeer et al. 2018). Dang et al. (2018) evaluated the antioxidant potential of three *Sargassum* species and three other brown algae, namely *Phyllospora comosa*, *Padina* sp., and *Hormosira banksii*, through ABTS, DPPH, and FRAP assays. The antioxidant activity of the six algae was dose dependent, where *H. banksii*, *S. vestitum* and *Padina* sp. showed superior activity as revealed by the ABTS assay ($p > 0.05$), while *S. vestitum* was the strongest through both DPPH and FRAP assays. The antioxidant activity of the known brown algae *Carpodesmia tamariscifolia* (syn. *Cystoseira tamariscifolia*) was evaluated through a DPPH assay, where it showed a dose-dependent radical scavenging activity attributed to its richness in polyphenolics (Moussa et al. 2020). The phlorotannins isolated from different ethanol/water extracts, using different ratios from both, from *Sargassum angustifolium* showed a vigorous antioxidant activity by radical scavenging in the DPPH assay and ferric reducing power where the antioxidant activity varied according to the solvent ratio used and thus the phlorotannin content of each extract (Hodhodi et al. 2022).

13.2.3 Antidiabetic Activity

Diabetes mellitus is a metabolic disorder linked to high serum glucose levels and insulin resistance caused by glucose intolerance (El-Nashar et al. 2020a, 2021b). This health condition affects many humans worldwide, with some ethnic groups showing a higher tendency for the disease. It is considered an irreversible but still curable and controllable health issue. Researchers are frequently searching for new, safe, and natural precursors for new drug leads to replace or work in synchrony with the existing chemical entities; thus, plant and marine sources are of much potential in this context (Deutschländer et al. 2009). In a streptozotocin-induced diabetes model in rats, the phlorotannins from *Cystoseira compressa* presented a pronounced activity. The rats in the test groups were treated with 60 mg/kg of the phlorotannin extract intraperitoneally. The phlorotannins decreased the serum glucose level, the level of hepatic malondialdehyde, and the enzymes glucosidase and α-amylase (Gheda et al. 2021).

13.2.4 Cytotoxic Activity

Many natural products have shown potential cytotoxic activities (Ashmawy et al. 2019; El-Nashar et al. 2020b; Moussa et al. 2020). Phlorotannins from the brown algae *Ecklonia cava* showed a synergistic activity when taken with the anticancer drug cisplatin, and decreased its nephrotoxicity in an *in vivo* study. The cytotoxicity activity for these tannins was assayed through MTT and annexin V assays, where the tannin extract helped in inducing apoptosis in cancer cells and increased intracellular reactive oxygen species levels (Yang et al. 2015). The presence of phlorotannins and sterols in the brown algae *Ecklonia maxima* accounted for its superior cytotoxicity against HeLa cancer cell lines with IC_{50} 125 µM (Mwangi et al. 2017). Different brown seaweeds from Sulawesi, *Gracilaria salicornia*, *Turbinaria decurrens*, *Halimeda macroloba*, and *Laurencia tronoi*, were collected and tested against the cancer cell line HeLa. The IC_{50} values for the extracts were 432.63 µg/g for *G. salicornia*, 41.027 µg/g for *T. decurrens*, 137.38 µg/g for *H. macroloba*, and 78.53 µg/g for *L. tronoi* (Sanger et al. 2021). Seaweed polysaccharides could serve as a promising source of anticancer and chemotherapeutic agents. In this context, the polysaccharides extracted from the *Sargassum* species acted as a potent anticancer agent against the human breast cancer cell line (MCF-7) with IC_{50} 39.46 mg/ml (Karthick et al. 2019). Different meroterpenoids isolated from

Cystoseira usneoides were tested against colorectal cancer cell lines. They showed a wide array of activity by selectively inhibiting the growth of HT-29 cancer cells and inducing apoptosis for them and being less toxic to the other non-cancerous cells (Zbakh et al. 2020). *Cystoseira trinodis* is a macroalgal strain collected from Egypt's Red Sea where its hexane fraction provoked a potent cytotoxic agent when tested on HepG-2 (IC_{50} 14.3 mg/ml) and MCF-7 (IC_{50} 19.2 mg/ml) cell lines (Ahmed et al. 2021).

13.2.5 Anti-inflammatory Activity

The anti-inflammatory activity of different brown algal strains was the focus of many research articles, and many of them provoked superior activity. Anti-inflammatory activity can be evaluated via inhibition of inflammatory mediators, such as tumour necrosis factor alpha (TNF-α) and interleukins; inflammation pathways, such as nitric oxide (NO) production; and enzymes, such as cyclooxygenase (COX) and 5-lipoxygenase (LOX) (Edmond et al. 2021; Abdallah et al. 2022; Mostafa et al. 2022). The polysaccharides extracted from *Sargassum horneri* collected from Korea inhibited the NO production (IC_{50} 95.7 µg/ml), TNF-α and interleukin (IL)-1β as measured by Griess assay and ELISA, respectively; thus, this seaweed polysaccharide extract can act as a potential anti-inflammatory precursor for many ailments (Sanjeewa et al. 2017). Hexane extracts from the brown seaweeds *Sargassum plagiophyllum* and *Turbinaria decurrens* powerfully inhibited the proinflammatory enzyme 5-LOX (IC_{50} 0.4–0.6 mg/ml), which made them a novel source of anti-inflammatory drug leads (Thambi and Chakraborty 2022).

13.2.6 Antimicrobial Activity

The brown algae carry a relatively strong antimicrobial potential against many organisms such as bacteria, fungi, and viruses. This was the focus of many research groups aiming to provide more natural and renewable antimicrobial agents. From the Western Libyan coast, *Sargassum vulgare*, *Cystoseira barbata*, *Dictyopteris membranacea*, *Dictyota dichotoma*, and *Colpomenia sinuosa* were collected and extracted separately with solvents of different polarities and then tested against eight fungal strains (*Alternaria alternata*, *Cladosporium cladosporioides*, *Fusarium oxysporum*, *Epicoccum nigrum*, *Aspergillus niger*, *Aspergillus ochraceus*, *Aspergillus flavus*, and *Penicillium citrinum*) where the most potent antifungal activities were traced for the *n*-hexane extracts of all the tested algae compared to the other solvents so the hexane extracts of these brown seaweeds could provide a potential source of antifungal drug leads (Khallil and Daghman 2015). In a recent study, the Australian brown algae (*Padina australis*) ethanol extract provoked a strong antibacterial activity through the agar-well diffusion method by inhibiting the growth of *Staphylococcus aureus*, *Streptococcus mutants* and *Escherichia coli* (11.3–11.8 mm) compared to ciprofloxacin (11–14 mm) as a positive control and this was linked to its richness in alkaloids, flavonoids, tannins, phenols, terpenoids, steroids and saponins (Singkoh et al. 2021). The 80% methanol extracts of six seaweeds (*Actinotrichia fragilis*, *Cystoseira myrica*, *Hormophysa cuneiformis*, *Laurencia papillosa*, *Sargassum cinereum*, and *Turbinaria turbinate*) collected from Hurghada, Red Sea, Egypt showed a potential antimicrobial activity as measured by their inhibition zones created by the agar disc diffusion assay (Osman et al. 2019). Hamrun et al. (2020) investigated the antibacterial activity of the alginate fucoidan against oral cavity bacteria responsible for the dental caries process, namely *Streptococcus mutans*, *Porphyromonas gingivalis*, and *Fusobacterium*. The alginate extracts significantly inhibited the oral cavity bacterial growth, thus serving as natural anti-dental caries agents. Meanwhile, Silva et al. 2021 assayed the antibacterial activity of nine brown algal strains from *Pinensula* against different food-borne bacteria. Each of the nine algae was extracted with solvents with different polarities, namely *n*-hexane, chloroform, ethanol, acetone, and ethyl acetate. The resulting extracts were screened against an array of Gram-positive and Gram-negative bacteria using an agar diffusion assay. The antibacterial activity was affected by the algal strain and the solvent where four algae were potentially active (*B. bifurcata*, *S. latissima*, *L. ochroleuca*, and *S. muticum*), respectively, and *B. bifurcata* extract was the strongest among them. Fucoidans, isolated from *Fucus vesiculosus* cell walls, were considered potential antibacterial agents. They were tested against different Gram-positive and Gram-negative bacterial strains showing minimum inhibitory concentration values of 4–6 mg/ml with bacteriostatic effect, especially against *E. coli* (Ayrapetyan et al. 2021).

13.2.7 Anticoagulant Activity

The coagulation process is a natural and healthy way to stop bleeding. However, in various conditions this process may cause many complications for patients, and anticoagulants are required to reverse the process (Palta et al. 2014). Marine macroalgae represent one of the primary natural sources of sulphated polysaccharides, which act mainly as anticoagulants. To investigate ths, the sulphated polysaccharides extract of *Hormophysa triquetra*, *Sargassum denticulatum*, and *Cystoseira myrica* were collected from the shores of the Red Sea, Egypt, and they showed significantly high anticoagulant activity by prolonging the prothrombin time. *Hormophysa triquetra* showed a value of 26.50 ± 0.10 seconds, while *Cystoseira myrica* showed a value of 1.13 ± 0.06 seconds. The brown algae are naturally rich in alginate polymers with a sulphated polymer called fucoidan. *Turbinaria decurrens's* fucoidan content was extracted with ethanol and loaded into silver nanoparticles, and it showed a powerful anticoagulation activity compared to heparin (Shanthi et al. 2021). Faggio et al. (2015) found that the polysaccharide extract of the Mediterranean brown seaweed *Undaria pinnatifida* significantly prolonged prothrombin time, making it a promising anticoagulant agent.

13.2.8 Anti-osteoarthritis Activity

Osteoarthritis has become a widespread and concerning health issue in recent decades. It is typically classified as a chronic, progressive, and irreversible process that usually affects women more than men, especially after menopause; thus early diagnosis and prevention could delay the disease progression (Hame and Alexander 2013). The process of osteoarthritis is usually linked to inflammatory and catabolic enzymes. Thus the presence of a potent anti-inflammatory could help in halting this process. Amentadione, a diterpenoid isolated from *Cystoseira usneoides*, acted as cartilage preserver in a pre-clinical study where it inhibited COX-II and IL-6, and down-regulated two chondrocyte hypertrophic factors and nuclear factor kappa B expression (Araújo et al. 2020).

13.2.9 Cosmetics Application

Cosmetics from natural sources became the new trend worldwide due to their wide range of safety and acceptance by customers. Nowadays, the discovery of new extracts and products for cosmetics use from marine resources is challenging due to the limited number of organisms, legal prohibition over their collection and use, and the difficulty in collecting and preserving their quality (Michalak 2020). Skin fairness and whitening products are prevalent in many countries, which prompted the need for newer and safer bioresources. Phlorotannins from marine algae play an important role in many skin hypopigmentation products which lighten skin tone by inhibiting melanosis (Azam et al. 2017). The presence of fucoxanthin in many brown algal strains accounted for their new role in the cosmetics industry. Fucoxanthin is a pigment from the xanthophyll class which imparts colour to algae. In cosmetics, it has an anti-ageing activity in the skin by decreasing oxidative stress and lipid peroxidation (Lourenço-Lopes et al. 2021).

13.2.10 Miscellaneous Activities

The brown seaweed *Eisenia arborea* was reported to have an anti-allergic activity through inhibition of *β*-hexosaminidase release in rats which was attributed to its phlorotannin content as described by Sugiura et al. (2007). The nutritive value of brown algae was evaluated by Ismail (2017), where *Sargassum linifolium* was collected from Egypt and was found to be rich in proteins, especially aspartic acid, glutamic acid, alanine, leucine, and proline together with polyunsaturated fatty acids, lipids, phenols, flavonoids, and ascorbic acid. One of the interesting biological activities exhibited by brown algae (*Sargassum cristaefolium*) was shown by its synergistic activity as a fertilizer with urea, which helps in producing tomato crops with higher quality and yield (14–15 fruits/plant compared to 9 fruits/plant for the control) together with a longer shelf life for the crop (Widyastuti et al. 2019).

13.3 PHYTOCONSTITUENTS

Secondary metabolites from natural sources, especially marine plants, show a diverse range of chemical structures and pharmacological effects (Elkhawas et al. 2020; Aly et al. 2019, 2021, 2022; El-Nashar et al. 2021a; Ads et al. 2022). Brown algae is among the well-known marine seaweeds widely distributed across the Red Sea (Rushdi et al. 2022). Brown algae comprise about 265 genera and the most well-known genera are *Cystoseira*, *Dictyopteris*, *Laminaria*, *Sargassum*, *Fucus*, and *Padina* (Balboa et al. 2013; Hakim and Patel 2020). Brown algae have a plethora of secondary metabolites such as steroids, terpenoids, alkaloids, flavonoids, tannins, and phenolic compounds (Ayyad et al. 2003a, 2003b; Navarro et al. 2004; El-Sheekh et al. 2020; Ahmed et al. 2021). Moreover, they are a rich source of minerals, vitamins, proteins, and amino acids that could be used in human nutrition (Hakim and Patel 2020) (Figure 13.3).

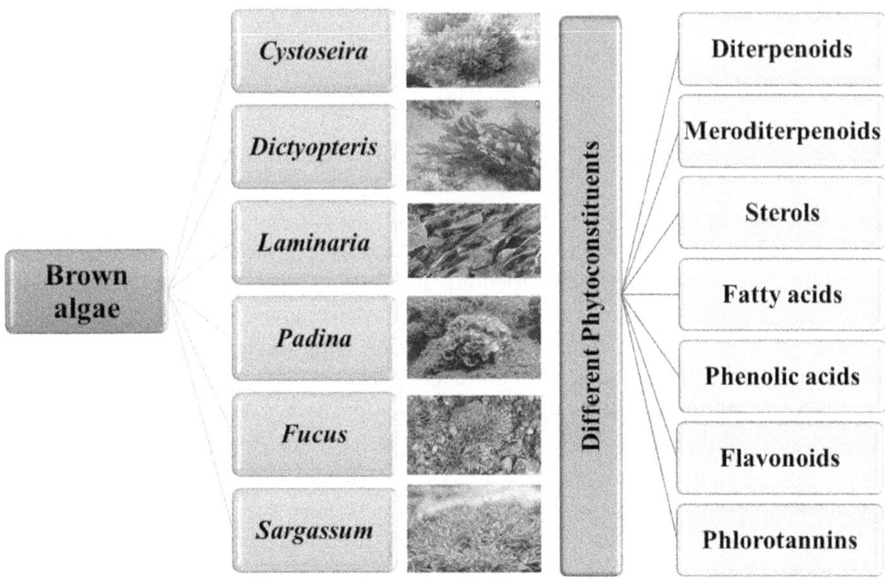

FIGURE 13.3 Schematic diagram showing the different genera of brown algae and their major phytoconstituents.

13.3.1 Terpenoids

Several compounds of a terpenoid nature were isolated from different genera of brown algae, as illustrated in Figure 13.4. An in-depth phytochemical study was carried out on the ethanolic extract of the brown alga (*Cystoseira myrica*) (Ayyad et al. 2003a). It revealed the isolation of six hydroazulene diterpenes, namely, pachydictyol A (**1**) [bold numbers in brackets refer to the compound numbers shown in Figures 13.4–13.8], dictyone (**2**), dictyone acetate (**3**), dictyol F monoacetate (**4**), isodictytriol monoacetate (**5**), and cystoseirol monoacetate (**6**). Dictyone (**2**) was also isolated from the ethanolic extract of *Sargassum asperifolium* (Ayyad et al. 2003b). In another study, three new acyclic diterpenes were been isolated from *C. crinite* and identified as (2*E*,10*E*)-1,6-dihydroxy-7-methylene-13-keto-3,11,15-trimethylhexadeca-2,10,14-triene (**7**), (2*E*,5*E*,10*E*)-1,7-dihydroxy-13-keto-3,7,11,15-tetramethylhexadeca-2,6,10,14-tetraene (**8**), and (2*E*,10*E*)-l-hydroxy-6,13-diketo-7-methylene-3,11,15-trimethylhexadeca-2,10,14-triene (**9**) (Amico et al. 1981). Gouveia et al. (2013) reported that in a cytotoxic study on *C. abies-marina*, two new meronorsesquiterpenes were characterized as cystoazorones A and B (**10** and **11**). In addition, two new merediterpenes were identified as cystoazorols A and B (**12** and **13**). New merediterpenes have also been isolated from specimens of *Cystoseira* spp. collected around the Canary Islands and characterized as amentol (**14**), cystoseirone diacetate (**15**), amentol chromane diacetate (**16**), amentol triacetate (**17**), and 14-methoxyamentol chromane (**18**) (Navarro et al. 2004). Another study on the *C. baccata* collected from the Atlantic coasts of Morocco, revealed the isolation of seven new merediterpenoids and their derivatives. They are characterized and named by their molecular formula as $C_{28}H_{40}O_4$ (**19**), $C_{28}H_{42}O_5$ (**20**), $C_{28}H_{38}O_4$ (**21**), $C_{28}H_{38}O_4$ (**22**), $C_{16}H_{26}O_2$ (**23**), $C_{18}H_{28}O_3$ (**24**), and $C_{18}H_{26}O_2$ (**25**) (Mokrini et al. 2008). El Shoubaky and Salem (2014) reported that the main terpene in the saponifiable matter of *Padina pavonica* and *Hormophysa triquetra* was the hemiterpene 3-furoic acid (**26**) followed by the diterpene phytol (**27**) by using gas chromatography–mass spectrometry (GC-MS) analysis. Zbakh et al. (2020) investigated the active constituents of the methanol extract of *C. usneoides*. Eight algal meroterpenoids were isolated and identified as usneoidone Z (**28**), 11-hydroxy-1′-*O*-methylamentadione (**29**), cystomexicone B (**30**), cystomexicone A (**31**), 6-*cis*-amentadione-1′-methyl ether (**32**), amentadione-1′-methyl ether (**33**), cystodione A (**34**), and cystodione B (**35**). Another antioxidant and anti-inflammatory studies reported the isolation of six new meroterpenoids, namely cystodiones A–F (**34–39**), from the acetone/MeOH extract of *C. usneoides*. In addition, known compounds were identified as 6-*cis*-amentadione-1′-methyl ether (**32**), amentadione-1′-methyl ether (**33**), cystomexicone A (**31**), cystomexicone B (**30**), usneoidone Z (**28**), and 11-hydroxyamentadione-1′-methyl ether (**40**) (De Los Reyes et al. 2013).

A mixture of kjellmanianone (**41**) and loliolide (**42**) were isolated from *C. trinodis* (Ahmed et al. 2021) and isololiolide (**43**) was identified by liquid chromatography (LC)-MS analysis in *C. tamariscifolia* (Moussa et al. 2020). A novel osteoarthritis protective agent was isolated from *C. usneoides* and identified as a merediterpenoid, amantadine (**44**) (Araújo et al. 2020). De Sousa et al. (2017) carried out a study to investigate the antileishmanial activity of merediterpenoids isolated from *C. baccata* and characterized as (3*R*)- and (3*S*)-tetraprenyltoluquinol (1a/1b) (**45, 46**) and (3*R*)-and(3*S*)-tetraprenyltoluquinone (2a/2b) (**47, 48**). From *C. tamariscifolia* extract five new phloroglucinol–merediterpenoid hybrids have been isolated and identified as cystophloroketals A–E (**49–53**) (El Hattab et al. 2015). From the brown seaweed *Sargassum ilicifolium*, a new anti-inflammatory xenicane-type diterpenoid was isolated and characterized as sargilicixenicane (**54**) (Dhara and Chakraborty 2021). Another study conducted on *S. fusiforme* revealed the isolation of three monoterpenes, dehydrololiolide (**55**), speciospongone A (**56**), and (3*S*,5*R*)-dihydroxymegstigma-6,7-dien-9-one (**57**) (Wang et al. 2020). Another phytochemical investigation of *S. siliquastrum* resulted in isolation of seven meroterpenoids identified as (9*S**,10*S**)-13-(3,4-dihydro-6-hydroxy-2,8-dimethyl-2*H*-1-benzopyran-2-yl)-2,6,10-trimethyl-trideca-(2*E*,6*E*)-diene-4,5,10-triol (**58**), (9*S**,10*R**)-13-(3,4-dihydro-6-hydroxy-2,8-dimethyl-2*H*-1-benzo-pyran-2-yl)-2,6,10-trimethyl-trideca-(2*E*,6*E*)-diene-4,5,10-triol (**59**),

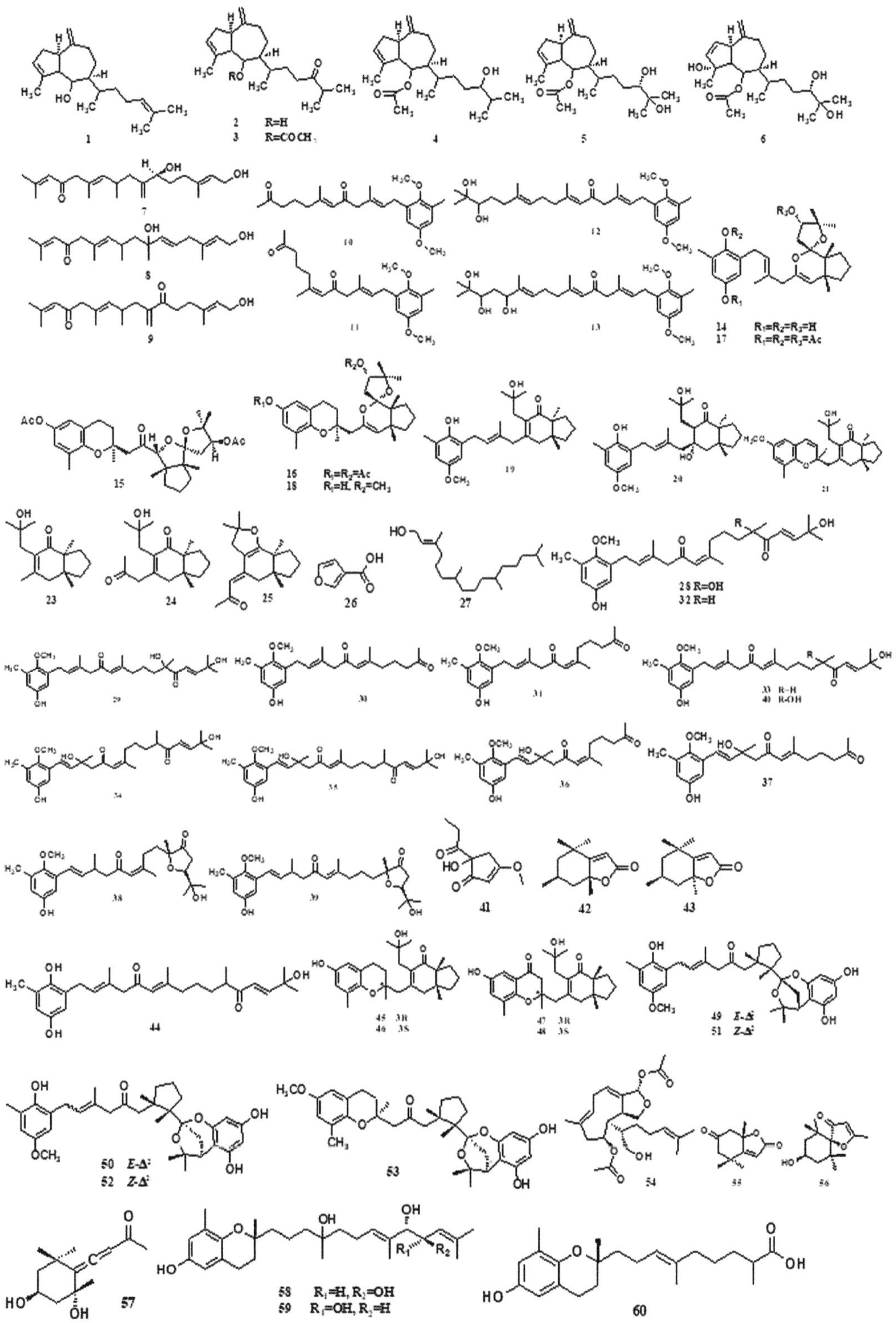

FIGURE 13.4 The structures of terpenoids (**1–131**) isolated from brown algae.

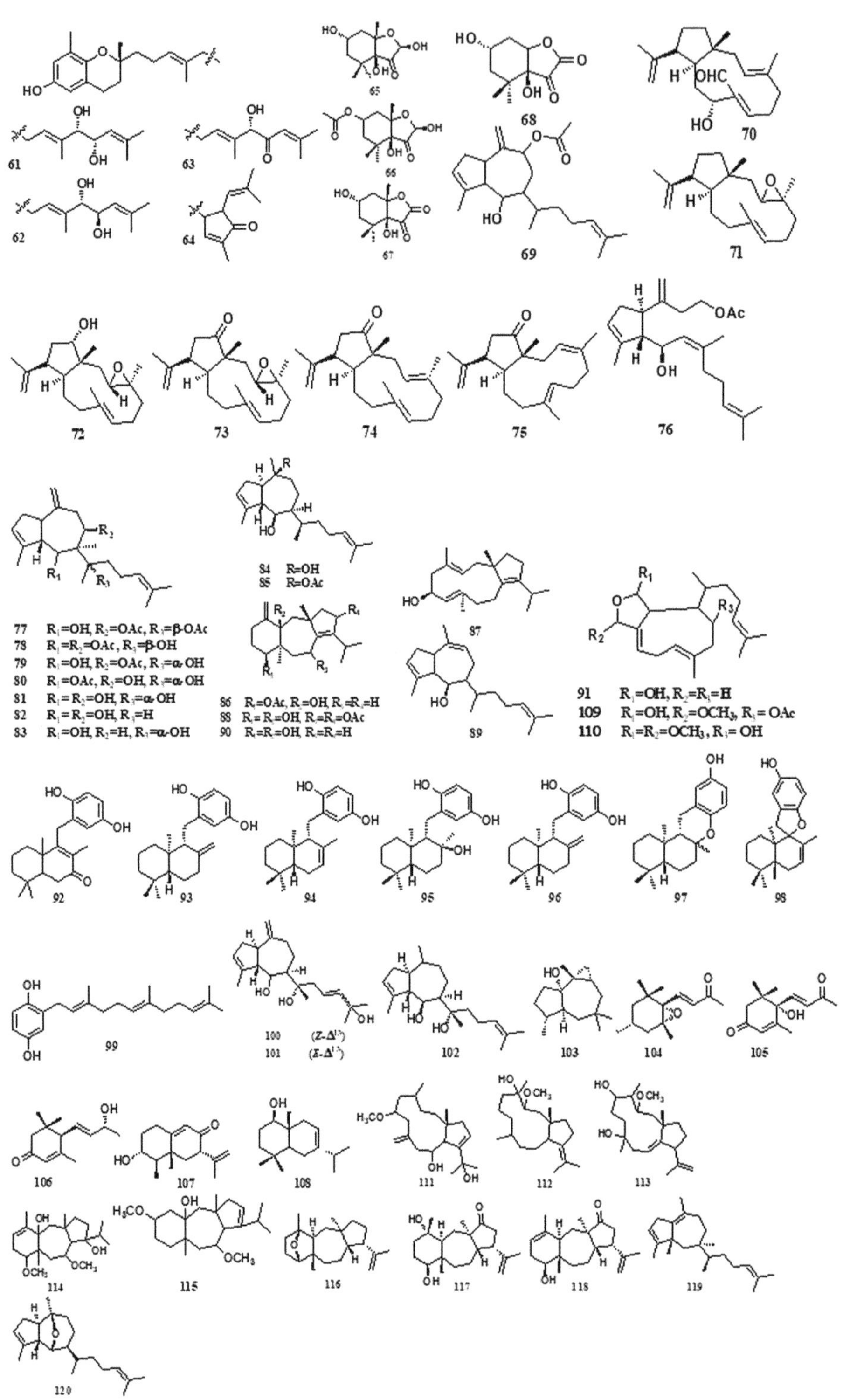

FIGURE 13.4 (Continued)

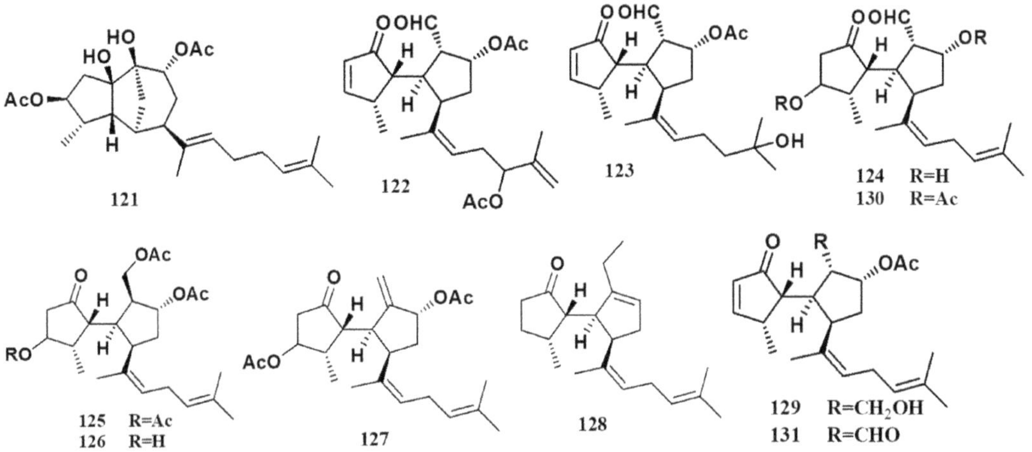

FIGURE 13.4 (Continued)

9-(3,4-dihydro-6-hydroxy-2,8-dimethyl-2*H*-1-benzopyran-2-yl)-2,6-dimethyl-(6*E*)-nonanoic acid (**60**), sargachromanol E (**61**), sargachromanol D (**62**), sargachromanol K (**63**), and sargachromanol P (**64**) (Lee et al. 2013). In another study, a new C11-norisoprenoid derivative was isolated and identified as sargassumone (**65**) from *Sargassum naozhouense*, together with six known norisoprenoids and a highly oxygenated cyclopentene: (2*R*,6*S*,8*S*,9*S*)-hexahydro-2,9-dihydroxy-4,4,8-trimethyl-6-acetyloxy-3(2*H*)-benzofuranone (**66**), (6*S*,8*S*,9*R*)-hexahydro-6,9-dihydroxy-4,4,8-trimethyl-2(2*H*)-benzofuranone (**67**), (6*S*,8*S*,9*R*)-hexahydro-6,9-dihydroxy-4,4,8-trimethyl-2(2*H*)-benzofuranone (**68**), loliolide (**42**), (+)-epiloliolide (**43**), spheciospongones A (**56**), and (+)-kjellmanianone (**41**) (Peng et al. 2018). Two diterpenes were isolated from the hexane extracts of *Dictyota ciliolate*, *Turbinaria tricostata*, and *Padina sanctae-crucis* and characterized as pachydictyol A (**1**) and dictyol B acetate (**69**) (Caamal-Fuentes et al. 2014). Another study reported isolation of a new dolabellane aldehyde spiralyde A (**70**) from *D. spiralis* along with five diterpenes namely 3,4-epoxy-7,18-dolabelladiene (**71**), 3,4-epoxy-14α-hydroxy-7,18-dolabelladiene (**72**), 3,4-epoxy-14-oxo-7,18-dolabelladiene (**73**), 14-oxo-3,7,18-dolabellatriene (**74**), 14-oxo-3,7,18-dolabellatriene (**75**) (Chiboub et al. 2019). Previous studies on *Dictyota* spp. brown alga have shown the isolation of a new secohydroazulene derivative, 7*Z*-7,8-*seco*-7,11-didehydro-8 acetoxypachydictyol A (**76**) and four new hydroazulenes identified as (8*R*,11*R*)-8,11-diacetoxypachydictyol A (**77**), (8*R**,11*R**)-6-*O*-acetyl-8-acetoxy-11-hydroxypachydictyol A (**78**), (8*R**,11*S**)-8-acetoxy-11-hydroxypachydictyol A (**79**), and (8*R**,11*S**)-6-*O*-acetyl-8,11-dihydroxypachydictyol A (**80**), along with known compounds 8β,11-dihydroxypachydictyol A (**81**), 8β-hydroxypachydictyol A (**82**), dictyol E (**83**), dictyol C (**84**), and acetyldictyol C (**85**) (Wu et al. 2021). From *Dictyota dichotoma* var. *implexa* nine diterpenes were isolated, among them three new diterpenes were characterized as amijiol acetate (**86**), dolabellatrienol (**87**) and amijiol-7,10-diacetate (**88**) along with known ones, pachydictyol A (**1**), isopachydictyol A (**89**), 8β-hydroxypachydictyol A (**82**), amijiol (**90**), isodictyohemiacetal (**91**) and dictyol C (**84**) (Ayyad et al. 2011). Another phytochemical investigation on the methanol extract of *D. undulata* afforded a novel sesquiterpene hydroquinone identified as zonarenone (**92**), along with seven known sesquiterpene hydroquinones, control (**93**), isozonarol (**94**), yahazunol (**95**), zonaroic acid (**96**), chromazonarol (**97**), isochromazonarol (**98**), and 2-geranylhydroquinone (**99**) (Ishibashi et al. 2013). Abou-El-Wafa et al. (2013) have reported two new diterpenoids, namely pachydictyol B (**100, 101**) and pachydictyol C (**102**), from the dichloromethane extract of *D. dichotoma*, along with known compounds pachydictyol A (**1**), dictyol E (**83**), and *cis*-africanan-1α-ol (**103**). Phytochemical investigation of *Padina pavonia* resulted in the isolation of six terpenes namely 3α-hydroxy-5,6-epoxy-7-megastigmen-9-one (**104**), (+)-dehydrovomifoliol (**105**), loliolide (**42**), (6*R*,7*E*,9*R*)-9-hydroxy-4,7-megastigmadien-3-one (**106**), petasol (**107**) and oplodiol (**108**) (Germoush et al. 2019). Another cytotoxic study on *P. pavonia* hexane fraction revealed the isolation of xenicane diterpenes namely 18,19-epoxyxenic-19-methoxy-18-hydroxy-4-acetoxy-6,9,13-triene (**109**) and 18,19-epoxyxenic-18,19-dimethoxy-4-hydroxy-6,9,13-triene (**110**) (Awad et al. 2008). A variant class of diterpenoids dolabellane and dolastane were isolated from *P. tetrastromatica* and characterized as new dolabellane 6-methoxy-dolabella-8(17),12-diene-10β,18-diol (**111**), 3-methoxy-dolabella-12(18)-ene-4β-ol (**112**), 3-methoxydolabella-10,18(19)-diene-5α,8β-diol (**113**), and dolastanes 2,7-dimethoxy-14α-hydroxy-dolasta-1(15),9-diene (**114**) and 4,7-dimethoxy-9β,14 α-dihydroxy-dolasta-1-ene (**115**) (Antony et al. 2021). Another study reported the isolation of three new dolastanes from *Dilophus spiralis*, namely 1,4-epoxy-17-dolostone (**116**), 1,4-dihydroxy-11-oxo-17-dolostone (**117**), and 4-hydroxy-11-oxo-1,17-dolastadiene (**118**) along with five known perhydroazulenes characterized as dictytriene B (**119**), dictyoxide (**120**), pachydictyol A (**1**), isopachydictyol A (**89**), and dictyol E (**83**) (Ioannou et al. 2013). A recent study by Cuevas et al. (2021) revealed the isolation of ten new diterpenoids from *Rugulopteryx okamurae*, characterized as rugukadiol A (**121**), rugukamurals A–C (**122–124**), and ruguloptones A–F (**125–129**); besides the major metabolites in the extract are known compounds dilkamural (**130**) and its elimination product (**131**).

13.3.2 Sterols

Sterols were one of the predominant secondary metabolites in the brown algae, as illustrated in Figure 13.5. From the *n*-hexane fraction of *Cystoseira trinodis* (24*R*),(24*S*)-saringosterols (**132, 133**) and β-sitosterol (**134**) were isolated (Ahmed et al. 2021). Yoon et al. (2008) reported the isolation of sterols from *Ecklonia stolonifera n*-hexane fraction and identified them as fucosterol (**135**) and 24-hydroperoxy 24-vinylcholesterol (**136**). Fucosterol is the major sterol in different genera of brown algae (Balboa et al. 2013). It was previously reported in *C. foeniculacea*, *Dictyota ciliolate*, *Sargassum angustifolium*, *S. asperifolium*, and *Undaria pinnatifida* (Sánchez-Machado et al. 2004; El Shoubaky and Salem 2014; Hakim and Patel 2020). Another study reported the isolation of a new steroid from *Sargassum asperifolium* and identified it as saringosterone (**137**) along with known saringosterol (**132**) (Ayyad et al. 2003b). Also, a recent study characterized fucosterol (**135**) from *S. horridum* using the LC-MS technique (Castro-Silva et al. 2021). In a cytotoxic study, saringosterol acetate (**138**) was isolated from *Sargassum fusiformis* (Lee et al. 2019; Li et al. 2022). From the edible seaweed *S. fusiforme*, 24(*S*)-saringosterol (**133**) was isolated and characterized as a good candidate for the prevention of cognitive decline (Martens et al. 2021). In another study, saringosterol (**133**) was isolated from *S. fusiforme* and revealed effective cholesterol-lowering and anti-atherogenic properties (Yan et al. 2021). Another study conducted on *S. fusiforme* revealed the isolation of three sterols, namely, sargassuol A (**139**), sargassuol B (**140**), and (24*S*)-stigmasta-5,28-diene-3β,24-diol-7-one (**141**) (Wang et al. 2020). GC-MS analysis study on the petroleum extract of *S. aquifolium* revealed the identification of eight compounds where the steroidal derivatives are the predominant compounds as stigmasta-5,24(28)-dien-3-ol, (3β,24*Z*) (**142**) and stigmasterol (**143**) (Moni et al. 2021). Another study reported that the GC-MS analysis of the saponifiable matter of *Hormophysa triquetra* revealed the presence of β-sitostanol (**144**), stigmasterol (**143**), β-sitosterol glucoside (**145**), and campesterol (**146**). At the same time, the saponifiable matter of *Padina pavonica* revealed the presence of β-sitosterol (**134**), campesterol (**146**), β-sitosterol glucoside (**145**), and stigmasterol glucoside (**147**) (El Shoubaky and Salem 2014).

Phytochemical investigation on the edible brown seaweed *Laminaria japonica* revealed the presence of five sterols identified as 29-hydroperoxy-stigmasta-5,24(28)-dien-3β-ol (**148**), saringosterol (**132**), 24-methylenecholesterol (**149**), fucosterol (**135**), and 24-hydroperoxy-24-vinyl-cholesterol (**136**) (Lu et al. 2022). Rehman et al. (2019) isolated two sterols from the methanol

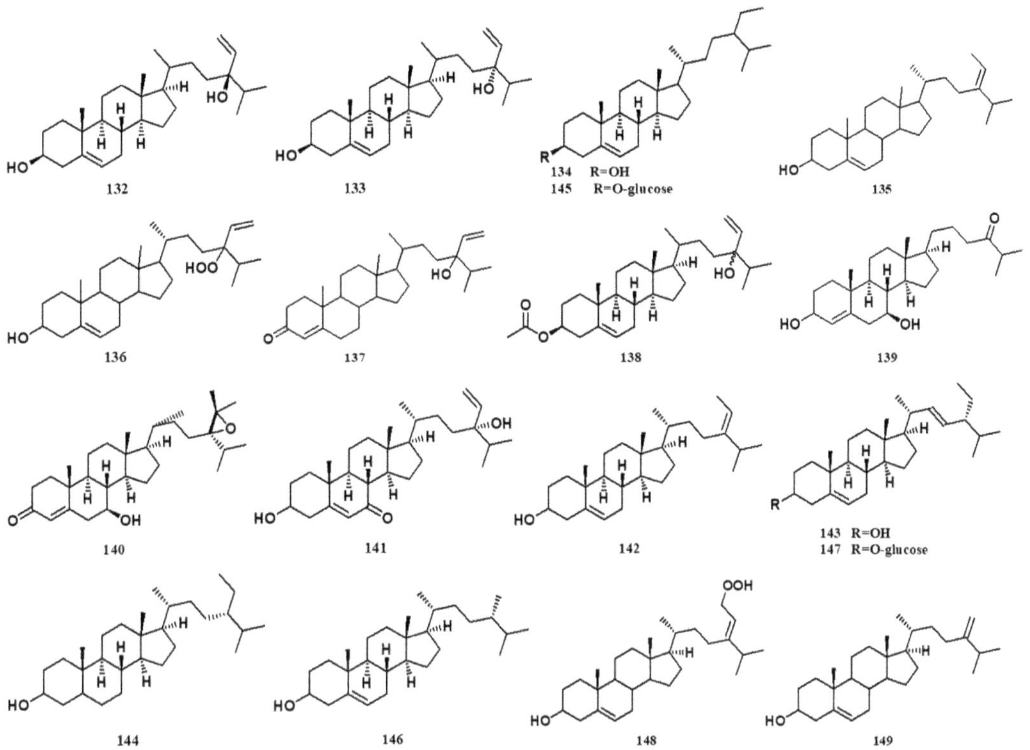

FIGURE 13.5 The structures of sterols (**132–149**) isolated from brown algae.

extract of *Dictyopteris hoytii* and identified them as fucosterol (**135**) and β-sitosterol (**134**). Caamal-Fuentes et al. (2014) reported that in a cytotoxic study on the bioactive constituents of hexane extracts of *Dictyota ciliolate*, *Turbinaria tricostata*, and *Padina sanctae-crucis* they isolated two sterols: fucosterol (**135**) and 24β-hydroperoxy-24-vinylcholesterol (**136**). Another phytochemical investigation on the methanolic extract of *Nizamuddinia zanardinii* resulted in the isolation of hydroperoxy sterol characterized as 24β-hydroproxy-24-vinylcholesterol (**136**) (Moghadam et al. 2013).

13.3.3 Fatty Acids and Hydrocarbons

Different fatty acids and hydrocarbons were identified and isolated from brown algae, as illustrated in Figure 13.6. Three fatty acids isolated from the *n*-hexane fraction of *Cystoseira trinodis* and identified as octacosanoic acid (**150**), glyceryl trilinoleate (**151**), and oleic acid (**152**) (Ahmed et al. 2021). Seasonal variation of *C. compressa* was investigated by GC-MS analysis, that revealed palmitic acid (**153**) as the major component in samples from May, June, and July (Mekinić et al. 2021a).

Through bioassay guided isolation of the methanol extract of *Dictyopteris hoytii*, four fatty acids were isolated and identified as *n*-hexadecanoic acid, methyl ester (**154**), cerotic acid (**155**), *n*-octacos-9-enoic acid (**156**), and 11-eicosenoic acid (**157**) (Rehman et al. 2019). Moreover, four fatty acids were isolated from the methanol extract of *D. hoytii* and identified as pentatetracontanoic acid (**158**), tricosylic acid (**159**), *n*-hexadecanoic acid (**153**), and lacceroic acid (**160**) (Rafiq et al. 2021).

GC-MS analysis of the petroleum ether extract of *Sargassum tenerrimum* revealed the presence of hexadecanoic acid methyl ester (**154**) and 17-pentatriacontene (**161**). In addition, the methanol

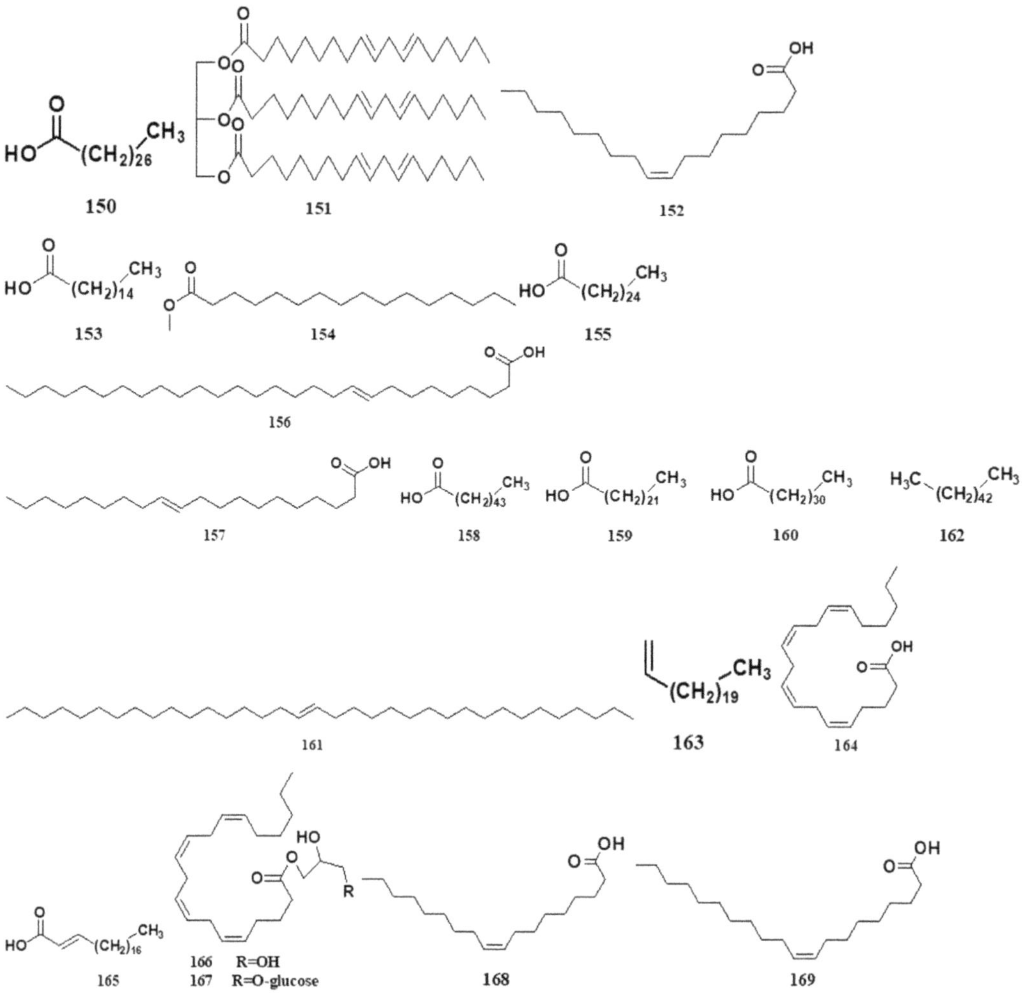

FIGURE 13.6 The structures of fatty acids and hydrocarbons (**150–169**) isolated from brown algae.

extract is rich with tetratetracontane (**162**) and 1-docosene (**163**) along with phthalic and benzoic acid derivatives (Albratty et al. 2021). Moreover, arachidonic acid (**164**), eicosenoic acid (**165**), 1-*O*-arachidonyl-glycerol (**166**), and 1-*O*-arachidonyl-3-*O*-(*a*-D-glucopyranosyl) glycerol (**167**) were isolated and identified from *S. cinereum* (Alzarea et al. 2021). Two fatty acids were isolated from *S. fusiforme* and identified as *cis*-octadec-9-enoic acid (**168**) and (*Z*)-9-eicosenoic acid (**169**) (Wang et al. 2020).

13.3.4 Phenolic Compounds

Phenolic compounds, including phenolic acids and flavonoids isolated from brown algae, are illustrated in Figure 13.7. Using high-pressure liquid chromatography (HPLC) analysis, five phenolics, namely, kaempferol (**170**), ellagic acid (**171**), delphinidin-3-*O*-glucoside (**172**), naringenin (**173**), and ferulic acid (**174**), were identified from the seaweed *Padina pavonica* (Sudha and Balasundaram 2018). A bioassay-guided fractionation of *P. boergesenii* revealed the cytotoxicity of the polyphenol-rich fraction. It was characterized by HPLC analysis as gallic acid (**175**), caffeic acid (**176**), rutin (**177**), quercetin (**178**), and ferulic acid (**174**) (Rajamani and

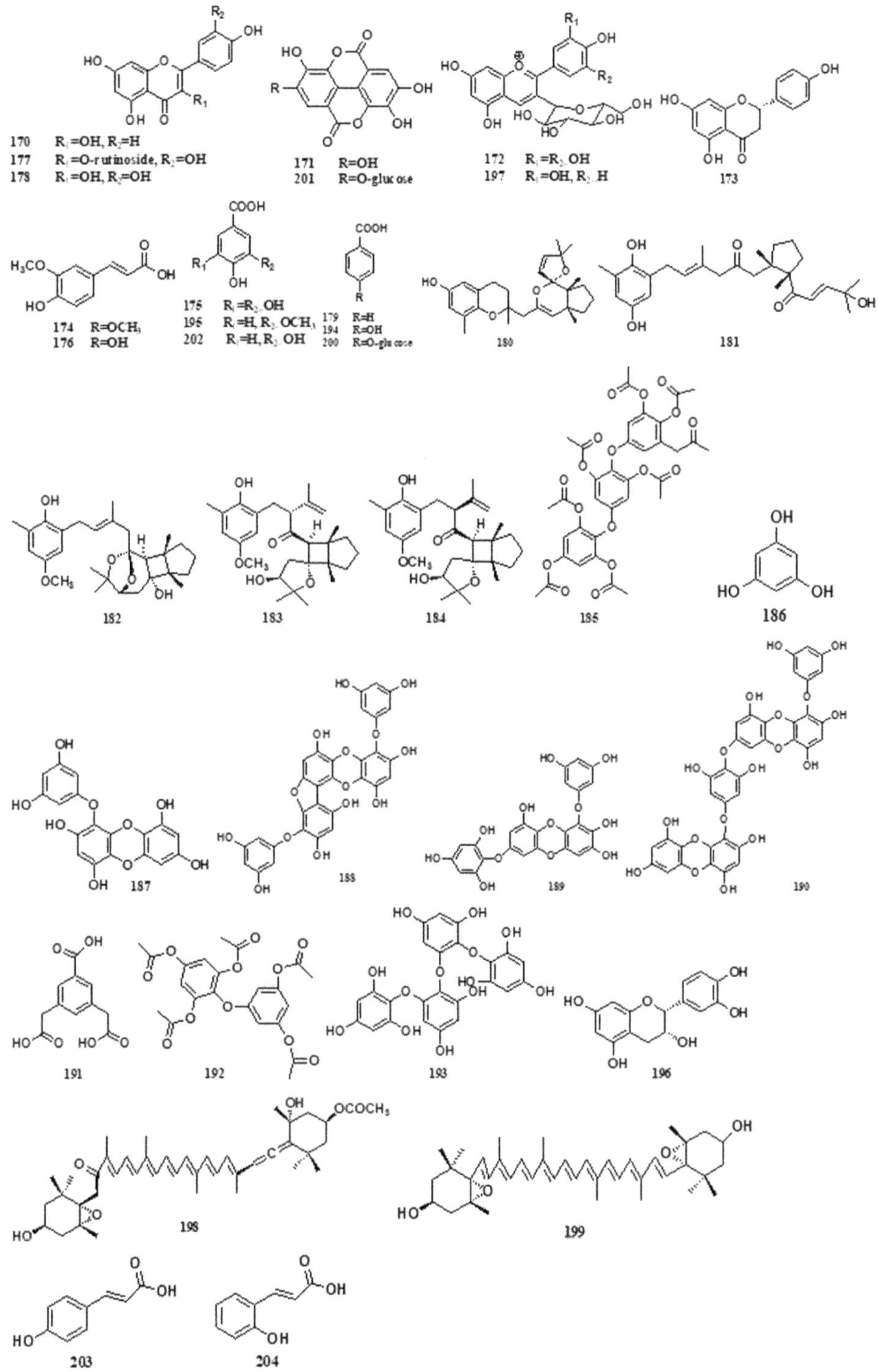

FIGURE 13.7 The structures of phenolic compounds (**170–204**) isolated from brown algae.

Thirugnanasambandan 2018). Also, benzoic acid (**179**) was isolated from *Cystoseira abies-marina* (Gouveia et al. 2013).

Phenolic compounds in the brown algae are represented as hydroquinone and prenylated hydroquinone; LC-MS analysis of different fractions of *C. tamariscifolia* revealed the presence of dimethoxy cystoketal chromane (**180**), methoxybifurcarenone (**181**), strictaketal (**182**), neobalearone (**183**), and epineobalearone (**184**) (Moussa et al. 2020). Gheda et al. (2021) characterized the phlorotannins of *C. compressa* by the ultra performance liquid chromatography–tandem mass spectrometry (UPLC-MS/MS) technique. The main components were identified as trifuhalol (**185**), phlorotannins tetramer, hydroxyl tetrafuhalol, and trihydroxy hexafuhalol (Gheda et al. 2021). Phlorotannins are polyphenolic compounds and result from polymerization of phloroglucinol, and they have been isolated from *Ecklonia cava* and identified as phloroglucinol (**186**), eckol (**187**), phlorofucofuroeckol A (**188**), 7-phloroeckol (**189**), and dieckol (**190**) (Yoon et al. 2008). Another study reported the presence of phloroglucinol triacetate (**191**) and diphlorethol pentaacetate (**192**) along with a new phlorotannin derivative characterized as fucodiphlorethol G (**193**) from *E. cava* (Young et al. 2007). From *Eisenia bicycles*, phlorotannins were isolated and identified as phlorofucofuroeckol A (**188**) and dieckol (**190**). A metabolomic approach was conducted to detect the polyphenolics of four brown macroalgae namely *Laminaria japonica*, *Undaria pinnatifida*, *Sargassum fusiforme*, and *Ascophyllum nodosum*. The results revealed the presence of 12 compounds, identified as phenolic acids: 4-hydroxybenzoic acid (**194**), vanillic acid (**195**), gallic acid (**175**), caffeic acid (**176**), and ferulic acid (**174**); the flavonoid epicatechin (**196**); and phlorotannins: phloroglucinol (**186**) and its derivatives as phloroglucinol trimer and pentamer, also fuhalol derivatives as bifuhalol, trifuhalol, and tetrafuhalol (Shen et al. 2021). In another study, three brown seaweeds, namely *Himanthalia elongata*, *Laminaria saccharina* and *L. digitata*, were investigated for their antioxidant constituents using LC-MS analysis. The results revealed the presence of cyanidin-3-*O*-glucoside (**197**), fucoxanthin (**198**), violaxanthin (**199**), along with β-carotene, chlorophyll a derivatives, and chlorophyll b derivatives as major constituents in their lipophilic extract (Rajauria 2019). Using LC-MS analysis, the polyphenolic of two brown seaweeds *Ecklonia* sp. and *Sargassum* sp. were characterized and identified as 4-hydroxybenzoic acid 4-*O*-glucoside (**200**), ellagic acid glucoside (**201**) and *p*-hydroxybenzoic acid (**194**) (Zhong et al. 2020). In another study, HPLC analysis was performed to detect the phenolic constituents in extracts of *Dictyota dichotoma* and *Padina pavonica.* The results revealed that the significant phenolic acid in *D. dichotoma* and *P. pavonica* extracts was *trans*-ferulic acid (**174**) and protocatechuic acid (**202**). In general, hydroxycinnamic acid derivatives *p*-coumaric (**203**), *o*-coumaric (**204**), and *t*-ferulic acid (**174**) showed high concentrations in *D. dichotoma* extract. At the same time, high concentrations of protocatechuic (**202**) and *p*-hydroxybenzoic (**194**) acids were shown in the *P. pavonica* extract (Mekinić et al. 2021b).

13.3.5 Miscellaneous Compounds

Miscellaneous compounds isolated from different genera of brown algae such as benzofurans and glycolipids are illustrated in Figure 13.8. Rehman et al. (2019) reported a bioassay-guided isolation of the methanol extract of *Dictyopteris hoytii.* It resulted in the isolation of a new metabolite, ethyl methyl 2-bromobenzene 1,4-dioate (**205**), and a new natural metabolite, diethyl-2-bromobenzene 1,4-dioate (**206**), with notable α-glucosidase inhibitory activity. Another study reported the anti-inflammatory activity of disulphides isolated from *D. membranacea.* Three disulphides were isolated and identified as bis(5-methylthio-3-oxo-undecyl) disulphide (**207**), 5-methylthio-1-(3-oxo-undecyl) disulphanylundecan-3-one (**208**), and 3-hexyl-4,5-dithiocycloheptanone (**209**) (Daskalaki et al. 2020). Three compounds were isolated from the methanolic extract of *D. hoytii* for the first time from a natural source. They were elucidated as dimethyl 2-bromoterepthalate (**210**), dimethyl 2,6-dibromoterepthalate (**211**), and (*E*)-3-(4-(dimethoxymethyl) phenyl) acrylic acid (**212**). This was in addition to known compounds identified as (*E*)-3-(2-formyl phenyl) acrylic

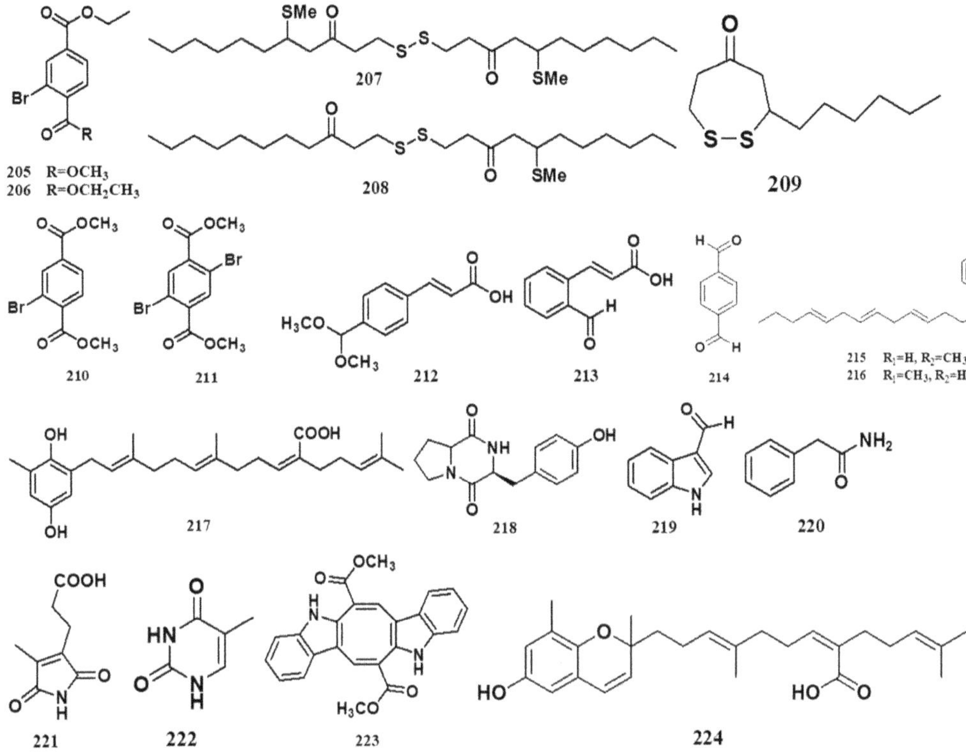

FIGURE 13.8 The structures of miscellaneous compounds (**205–224**) isolated from brown algae.

acid (**213**) and 1,4-benzenedicarboxaldehyde (**214**) (Rafiq et al. 2021). Phytochemical investigation of the crude alcoholic extract of *Sargassum cinereum* revealed isolation of new aryl cresol, namely, 4-(1-(4,7,11-pentadecenyl)-*o*-cresol (**215**) and 4-(1-(4,7,11-pentadecenyl)-*m*-cresol (**216**) (Alzarea et al. 2021). From the ethanolic extract of *S. macrocarpum* a plastoquinone was isolated and elucidated as sargahydroquinoic acid (**217**) (Joung et al. 2021). A recent study revealed the isolation and identification of five nitrogenous compounds from *S. fusiforme*, namely, cyclo (L-Pro-L-Tyr) (**218**), 1*H*-indole-3-carbaldehyde (**219**), 2-phenylacetamide (**220**), isohematinic acid (**221**), and thymine (**222**) (Wang et al. 2020). Caulerpin (**223**) was a bis-indole alkaloid isolated from the dichloromethane fraction of *S. platycarpum* (Abdelrheem et al. 2021). The phytochemical investigation of 70% ethanol extract of *S. horneri* afforded identification of sargachromenol (**224**) (Nagahawatta et al. 2022).

13.4 CONCLUSION AND FUTURE PERSPECTIVE

Marine ecosystems provide unique sources of natural products due to the harsh marine environmental conditions. Thus marine ecosystems provoke the production of potential metabolites with interesting applications in different fields, with particular emphasis on drug discovery. Brown algae are considered one of the oldest marine sources utilized by mankind; traditionally they were used as a major food source in coastal areas due to the high energy content provided by the polysaccharides as well as their iodine and vitamin D content. Recently, phytochemical, and biological investigations have provided significant evidence for brown algae as a potential resource in the drug discovery process. Phenolics represent major constituents widely distributed in different species of brown algae; phlorotannins and flavonoids constitute the main phenolic content of brown algae, supporting their antioxidant potential, especially their applications in skin care products. Sulphated polysaccharides, another major constituent produced by brown algae, found their applications in anticoagulant and anti-inflammatory drug research. Diterpenes and sesquiterpenes represent additional unique constituents of brown algae that may play a role in brown algae's anti-inflammatory and antimicrobial activities. Increasing research interest towards the chemical biology of brown algae is a growing trend in terms of isolating new metabolites and developing new drug products targeting common health problems such as diabetes and resistant microbial infections. Finally, investment in research on brown algae regarding their medicinal applications represents a promising area of research potentially resulting in effective and safe drug products targeting human welfare.

REFERENCES

Abdallah, S.H., Mostafa, N.M., Mohamed, M.A., Nada, A.S. and Singab, A.N.B. 2022. UPLC-ESI-MS/MS profiling and hepatoprotective activities of Stevia leaves extract, butanol fraction and stevioside against radiation-induced toxicity in rats. *Natural Product Research*, 36(21):5619–5625.

Abdelwahab, R. 2017. Therapeutic and pharmaceutical application of seaweeds. In: Nabti, E. (ed) *Biotechnological Applications of Seaweeds*. Hauppauge, NY: Nova Science Publishers, 85–116.

Abdelrheem, D.A., Abd El-Mageed, H.R., Mohamed, H.S., Rahman, A.A., Elsayed, K.N. and Ahmed, S.A. 2021. Bis-indole alkaloid caulerpin from a new source *Sargassum platycarpum*: isolation, characterization, in vitro anticancer activity, binding with nucleobases by DFT calculations and MD simulation. *Journal of Biomolecular Structure and Dynamics*, 39(14):5137–5147.

Abou-El-Wafa, G.S., Shaaban, M., Shaaban, K.A., El-Naggar, M.E., Maier, A., Fiebig, H.H. and Laatsch, H. (2013). Pachydictyols B and C: new diterpenes from *Dictyota dichotoma* Hudson. *Marine Drugs*, 11(9):3109–3123.

Ads, E.N., Hassan, S.I., Rajendrasozhan, S., Hetta, M.H., Aly, S.H., and Ali, M.A. 2022. Isolation, structure elucidation and antimicrobial evaluation of natural pentacyclic triterpenoids and phytochemical investigation of different fractions of *Ziziphus Spina-Christi* (L.) stem bark using LCHRMS analysis. *Molecules*, 27(6):1805.

Ahmed, S.A., Rahman, A.A., Elsayed, K.N., Abd El-Mageed, H.R., Mohamed, H.S. and Ahmed, S.A. 2021. Cytotoxic activity, molecular docking, pharmacokinetic properties and quantum mechanics calculations of the brown macroalga *Cystoseira trinodis* compounds. *Journal of Biomolecular Structure and Dynamics*, 39(11):3855–3873.

Albratty, M., Alhazmi, H.A., Meraya, A.M., Najmi, A., Alam, M.S., Rehman, Z. and Moni, S.S. 2021. Spectral analysis and antibacterial activity of the bioactive principles of *Sargassum tenerrimum* J. Agardh collected from the Red Sea, Jazan, Kingdom of Saudi Arabia. *Brazilian Journal of Biology*, 83:1–10.

Alghazeer, R., Enaeli, M. and Howell, N.K. 2018. Anticancer and antioxidant activities of some algae from the western Libyan coast. *Natural Sciences*, 10(7):232–246.

Al-Madhagy, S.A., Mostafa, N.M., Youssef, F.S., Awad, G.E.A., Eldahshan, O.A. and Singab, A.N.B. 2019. Metabolic profiling of a polyphenolic-rich fraction of *Coccinia grandis* leaves using LC-ESI-MS/MS and in vivo validation of its antimicrobial and wound healing activities. *Food Function*, 10:6267–6275.

Aly, S.H., Elissawy, A.M., Allam, A.E., Farag, S.M., Eldahshan, O.A., Elshanawany, M.A. and Singab, A.N.B. 2022. New quinolizidine alkaloid and insecticidal activity of *Sophora secundiflora* and *Sophora tomentosa* against *Culex pipiens* (Diptera: Culicidae). *Natural Product Research*, 36:2722–2734.

Aly, S.H., Elissawy, A.M., Eldahshan, O.A., Elshanawany, M.A., Efferth, T. and Singab, A.N.B. 2019. The pharmacology of the genus *Sophora* (Fabaceae): an updated review. *Phytomedicine*, 64:153070.

Aly, S.H., Elissawy, A.M., Fayez, A.M., Eldahshan, O.A., Elshanawany, M.A. and Singab, A.N.B. 2021. Neuroprotective effects of *Sophora secundiflora*, *Sophora tomentosa* leaves and formononetin on scopolamine-induced dementia. *Natural Product Research*, 35(24):5848–5852.

Alzarea, S.I., Elmaidomy, A.H., Saber, H., Musa, A., Al-Sanea, M.M., Mostafa, E.M., Hendawy, O.M., Youssif, K.A., Alanazi, A.S., Alharbi, M. and Sayed, A.M. 2021. Potential anticancer lipoxygenase inhibitors from the red sea-derived brown algae *sargassum cinereum*: an in-silico-supported in-vitro study. *Antibiotics*, 10(4):416.

Amico, V., Oriente, G., Piattelli, M., Ruberto, G. and Tringali, C. 1981. Novel acyclic diterpenes from the brown alga *Cystoseira crinita*. *Phytochemistry*, 20(5):1085–1088.

Antony, T., Chakraborty, K. and Joy, M. 2021. Antioxidative dolabellanes and dolastanes from brown seaweed *Padina tetrastromatica* as dual inhibitors of starch digestive enzymes. *Natural Product Research*, 35(4):614–626.

Apak, R., Özyürek, M., Güçlü, K. and Çapanoğlu, E. 2016. Antioxidant activity/capacity measurement. 1. Classification, physicochemical principles, mechanisms, and electron transfer (ET)-based assays. *Journal of Agricultural and Food Chemistry*, 64(5):997–1027.

Araújo, N., Viegas, C.S., Zubía, E., Magalhães, J., Ramos, A., Carvalho, M.M., Cruz, H., Sousa, J.P., Blanco, F.J., Vermeer, C. and Simes, D.C. 2020. Amentadione from the alga *Cystoseira usneoides* as a novel osteoarthritis protective agent in an ex vivo co-culture OA model. *Marine Drugs*, 18(12):624.

Ashmawy, A., Mostafa, N. and Eldahshan, O. 2019. GC/MS analysis and molecular profiling of lemon volatile oil against breast cancer. *Journal of Essential Oil-Bearing Plants*, 22:903–916.

Awad, N.E., Selim, M.A., Metawe, H.M. and Matloub, A.A. 2008. Cytotoxic xenicane diterpenes from the brown alga *Padina pavonia* (L.) Gaill. *Phytotherapy Research*, 22(12):1610–1613.

Ayrapetyan, O.N., Obluchinskaya, E.D., Zhurishkina, E.V., Skorik, Y.A., Lebedev, D.V., Kulminskaya, A.A. and Lapina, I.M. 2021. Antibacterial properties of fucoidans from the brown algae *Fucus vesiculosus* L. of the Barents Sea. *Biology*, 10(1):67.

Ayyad, S.E.N., Abdel-Halim, O.B., Shier, W.T. and Hoye, T.R. 2003a. Cytotoxic hydroazulene diterpenes from the brown alga *Cystoseira myrica. Zeitschrift für Naturforschung C*, 58(1–2):33–38.

Ayyad, S.E.N., Sowellim, S.Z., El-Hosini, M.S. and Abo-Atia, A. 2003b. The structural determination of a new steroidal metabolite from the brown alga *Sargassum asperifolium. Zeitschrift für Naturforschung C*, 58(5–6):333–336.

Ayyad, S.E.N., Makki, M.S., Al-Kayal, N.S., Basaif, S.A., El-Foty, K.O., Asiri, A.M., Alarif, W.M. and Badria, F.A. 2011. Cytotoxic and protective DNA damage of three new diterpenoids from the brown alga *Dictoyota dichotoma. European Journal of Medicinal Chemistry*, 46(1):175–182.

Azam, M.S., Choi, J., Lee, M.S. and Kim, H.R. 2017. Hypopigmenting effects of brown algae-derived phytochemicals: a review on molecular mechanisms. *Marine Drugs*, 15(10):297.

Balboa, E.M., Conde, E., Moure, A., Falqué, E. and Domínguez, H. 2013. In vitro antioxidant properties of crude extracts and compounds from brown algae. *Food Chemistry*, 138(2–3):1764–1785.

Begum, R., Howlader, S., Mamun-Or-Rashid, A.N.M., Rafiquzzaman, S.M., Ashraf, G.M., Albadrani, G.M., Sayed, A.A., Peluso, I., Abdel-Daim, M.M. and Uddin, M. 2021. Antioxidant and signal-modulating effects of brown seaweed-derived compounds against oxidative stress-associated pathology. *Oxidative Medicine and Cellular Longevity*, 2021:1–22.

Brownlee, I., Fairclough, A., Hall, A. and Paxman, J. 2012. The potential health benefits of seaweed and seaweed extract. In: Pomin, V.H. (ed) *Seaweed: Ecology, Nutrient Composition and Medicinal Uses. Marine Biology: Earth Sciences in the 21st Century*. Hauppauge, NY: Nova Science Publishers, 119–136.

Budhiyanti, S.A., Raharjo, S., Marseno, D.W. and Lelana, I.Y. 2012. Antioxidant activity of brown algae *Sargassum* species extracts from the coastline of java island. *American Journal of Agricultural and Biological Science*, 7(3):337–346.

Caamal-Fuentes, E., Moo-Puc, R., Freile-Pelegrín, Y. and Robledo, D. 2014. Cytotoxic and antiproliferative constituents from *Dictyota ciliolata*, *Padina sanctae-crucis* and *Turbinaria tricostata. Pharmaceutical Biology*, 52(10):1244–1248.

Castro-Silva, E.S., Bello, M., Rosales-Hernández, M.C., Correa-Basurto, J., Hernández-Rodríguez, M., Villalobos-Acosta, D., Méndez-Méndez, J.V., Estrada-Pérez, A., Murillo-Álvarez, J. and Muñoz-Ochoa, M. 2021. Fucosterol from *Sargassum horridum* as an amyloid-beta (Aβ1-42) aggregation inhibitor: in vitro and in silico studies. *Journal of Biomolecular Structure and Dynamics*, 39(4):1271–1283.

Chiboub, O., Sifaoui, I., Lorenzo-Morales, J., Abderrabba, M., Mejri, M., Fernández, J.J., Piñero, J.E. and Díaz-Marrero, A.R. 2019. Spiralyde A, an antikinetoplastid dolabellane from the brown alga *Dictyota spiralis. Marine drugs*, 17(3):192.

Cuevas, B., Arroba, A.I., de Los Reyes, C., Gómez-Jaramillo, L., González-Montelongo, M.C. and Zubía, E. 2021. Diterpenoids from the brown alga *Rugulopteryx okamurae* and their anti-inflammatory activity. *Marine Drugs*, 19(12):677.

Dang, T.T., Bowyer, M.C., Van Altena, I.A. and Scarlett, C.J. 2018. Comparison of chemical profile and antioxidant properties of the brown algae. *International Journal of Food Science & Technology*, 53(1):174–181.

Daskalaki, M.G., Bafiti, P., Kikionis, S., Laskou, M., Roussis, V., Ioannou, E., Kampranis, S.C. and Tsatsanis, C. 2020. Disulfides from the brown alga *Dictyopteris membranacea* suppress M1 macrophage activation by inducing AKT and suppressing MAPK/ERK signaling pathways. *Marine Drugs*, 18(11):527.

de Los Reyes, C., Zbakh, H., Motilva, V. and Zubía, E. 2013. Antioxidant and anti-inflammatory meroterpenoids from the brown alga *Cystoseira usneoides*. *Journal of Natural Products*, 76(4):621–629.

de Sousa, C.B., Gangadhar, K.N., Morais, T.R., Conserva, G.A., Vizetto-Duarte, C., Pereira, H., Laurenti, M.D., Campino, L., Levy, D., Uemi, M. and Barreira, L. 2017. Antileishmanial activity of meroditerpenoids from the macroalgae *Cystoseira baccata*. *Experimental Parasitology*, 174:1–9.

Deutschländer, M.S., Van de Venter, M., Roux, S., Louw, J. and Lall, N. 2009. Hypoglycaemic activity of four plant extracts traditionally used in South Africa for diabetes. *Journal of Ethnopharmacology*, 124(3):619–624.

Dhara, S. and Chakraborty, K. 2021. Anti-inflammatory xenicane-type diterpenoid from the intertidal brown seaweed *Sargassum ilicifolium*. *Natural Product Research*, 35(24):5699–5709.

Edmond, M.P., Mostafa, N.M., El-Shazly, M. and Singab, A.N.B. 2021. Two clerodane diterpenes isolated from *Polyalthia longifolia* leaves: comparative structural features, anti-histaminic and anti-*Helicobacter pylori* activities. *Natural Product Research*, 35:5282–5286.

El Hattab, M., Genta-Jouve, G., Bouzidi, N., Ortalo-Magne, A., Hellio, C., Maréchal, J.P., Piovetti, L., Thomas, O.P. and Culioli, G. 2015. Cystophloroketals A–E, unusual phloroglucinol–meroterpenoid hybrids from the brown alga *Cystoseira tamariscifolia*. *Journal of Natural Products*, 78(7):1663–1670.

Elkhawas, Y.A., Elissawy, A.M., Elnaggar, M.S. Mostafa, N.M., Kamal, E.M., Bishr, M.M. Singab, A.N.B. and Salama, O.M. 2020. Chemical diversity in species belonging to soft coral genus *Sacrophyton* and its impact on biological activity: a review. *Marine Drugs*, 18(1):41.

El-Nashar, H.A., Eldehna, W.M., Al-Rashood, S.T., Alharbi, A., Eskandrani, R.O. and Aly, S.H. 2021a. GC/MS analysis of essential oil and enzyme inhibitory activities of *Syzygium cumini* (Pamposia) grown in Egypt: chemical characterization and molecular docking studies. *Molecules*, 26(22):6984.

El-Nashar, H.A., Mostafa, N.M., El-Badry, M.A., Eldahshan, O.A. and Singab A.N.B. 2020b. Chemical composition, antimicrobial and cytotoxic activities of essential oils from *Schinus polygamous* (Cav.) cabrera leaf and bark grown in Egypt. *Natural Product Research*, 9:1–4.

El-Nashar, H.A., Mostafa, N.M., Eldahshan, O.A. and Singab, A.N.B. 2020a. A new antidiabetic and anti-inflammatory biflavonoid from *Schinus polygama* (Cav.) Cabrera leaves. *Natural Product Research* 1–9.

El-Nashar, H.A.S., Mostafa, N.M., El-Shazly, M. and Eldahshan, O.A. 2021b. The role of plant-derived compounds in managing diabetes mellitus: a review of literature from 2014 to 2019. *Current Medicinal Chemistry*, 28:4694–4730.

El-Sheekh, M.M., Mousa, A.S.H. and Farghl, A.A. 2020. Antibacterial efficacy and phytochemical characterization of some marine brown algal extracts from the Red Sea, Egypt. *Romanian Biotechnological Letters*, 25(1):1160–1169.

El Shoubaky, G.A. and Salem, E.A. 2014. Terpenes and sterols composition of marine brown algae *Padina Pavonica* (Dictyotales) and *Hormophysa Triquetra* (Fucales). *International Journal of Pharmacognosy and Phytochemical Research*, 6(4):894–900.

El-Zahar, H., Menze, E.T., Handoussa, H., Osman, A.K., El-Shazly, M., Mostafa, N.M. and Swilam N. 2022. UPLC-PDA-MS/MS profiling and healing activity of polyphenol-rich fraction of *Alhagi maurorum* against oral ulcer in rats. *Plants*, 11:455.

Faggio, C., Morabito, M., Minicante, S.A., Piano, G.L., Pagano, M. and Genovese, G. 2015. Potential use of polysaccharides from the brown alga *Undaria pinnatifida* as anticoagulants. *Brazilian Archives of Biology and Technology*, 58(5):798–804.

Fritsch, F.E. 1935. *Structure and Reproduction of the Algae*. Vol. I. London: Cambridge University Press.

Germoush, M.O., Elgebaly, H.A., Hassan, S., Kamel, E.M., Bin-Jumah, M. and Mahmoud, A.M. 2019. Consumption of terpenoids-rich *Padina pavonia* extract attenuates hyperglycemia, insulin resistance and oxidative stress, and upregulates PPARγ in a rat model of type 2 diabetes. *Antioxidants*, 9(1):22.

Ghazali, F.C., Aroyehun, A.Q.B. and Razak, S.A. 2017. Acknowledging sulfated polysaccharides from marine macroalgae multi-functional properties. *Haya: The Saudi Journal of Life Sciences*, 2(8):269–283.

Gheda, S., Naby, M.A., Mohamed, T., Pereira, L. and Khamis, A. 2021. Antidiabetic and antioxidant activity of phlorotannins extracted from the brown seaweed *Cystoseira compressa* in streptozotocin-induced diabetic rats. *Environmental Science and Pollution Research*, 28(18):22886–22901.

Gouveia, V.L., Seca, A.M., Barreto, M.C., Neto, A.I., Kijjoa, A. and Silva, A.M. 2013. Cytotoxic meroterpenoids from the macroalga *Cystoseira abies-marina. Phytochemistry Letters*, 6(4):593–597.

Hakim, M.M. and Patel, I.C. 2020. A review on phytoconstituents of marine brown algae. *Future Journal of Pharmaceutical Sciences*, 6(1):1–11.

Hame, S.L. and Alexander, R.A. 2013. Knee osteoarthritis in women. *Current Reviews in Musculoskeletal Medicine*, 6(2):182–187.

Hamrun, N., Oktawati, S., Haryo, H.M., Syafar, I.F. and Almaidah, A.N. 2020. Effectiveness of fucoidan extract from brown algae to inhibit bacteria causes of oral cavity damage. *Systematic Reviews in Pharmacy*, 11(10):686–693.

Hodhodi, A., Babakhani, A. and Rostamzad, H. 2022. Effect of different extraction conditions on phlorotannin content and antioxidant activity of extract from brown algae (*Sargassum angustifolium*). *Journal of Food Processing and Preservation* 46(3):e16307.

Hoppe, H.A. 1979. Marine algae and their products and constituents in pharmacy. In Hoppe, H.A., Levring, T. and Tanaka, Y. (eds) *Marine Algae in Pharmaceutical Science*. Berlin: Walter de Gruyter & Co., 25–120.

Ioannou, E., Vagias, C. and Roussis, V. 2013. Isolation and structure elucidation of three new dolastanes from the brown alga *Dilophus spiralis. Marine Drugs*, 11(4):1104–1112.

Ishibashi, F., Sato, S., Sakai, K., Hirao, S. and Kuwano, K. 2013. Algicidal sesquiterpene hydroquinones from the brown alga *Dictyopteris undulata. Bioscience, Biotechnology, and Biochemistry*, 77(5):1120–1122.

Ismail, G.A. 2017. Biochemical composition of some Egyptian seaweeds with potent nutritive and antioxidant properties. *Food Science and Technology*, 37:294–302.

Joung, E.J., Cao, L., Lee, B., Gwon, W.G., Park, S.H. and Kim, H.R. 2021. Sargahydroquinoic acid, a cyclooxygenase-2 inhibitor, attenuates inflammatory responses by regulating NF-κB inactivation and Nrf2 activation in lipopolysaccharide-stimulated cells. *Inflammation*, 44(5):2120–2131.

Karthick, V., Akhila, C., Kumar, V.G., Subashini, D., Dhas, T.S., Govindaraju, K. and Vasanth, K. 2019. In vitro anticancer activity of *Sargassum* sp. polysaccharides against MCF-7 cell lines. *Indian Journal of Geo-Marine Sciences* 48(8):1267–1273.

Khallil, A.M. and Daghman, I.M. and Fady, A.A. 2015. Antifungal potential in crude extracts of five selected brown seaweeds collected from the Western Libya Coast. *Journal of Microbiology and Modern Techniques*, 1(1):103.

Khan, M.I., Shin, J.H. and Kim, J.D. 2018. The promising future of microalgae: current status, challenges, and optimization of a sustainable and renewable industry for biofuels, feed, and other products. *Microbial Cell Factories*, 17(1):1–21.

Lee, J.H., Ko, J.Y., Kim, C.Y., Kim, Y.T., Kim, J. and Jeon, Y.J. 2019. Saringosterol acetate from a popular edible seaweed, *Hizikia fusiformis* attenuated proliferation of human lung adenocarcinoma epithelial cells (A549 cells) via apoptosis signaling pathway. *Preprints* 2019:2019110239.

Lee, J.I., Kwak, M.K., Park, H.Y. and Seo, Y. 2013. Cytotoxicity of meroterpenoids from *Sargassum siliquastrum* against human cancer cells. *Natural Product Communications*, 8(4):431–432.

Li, X., Zhou, D., Li, F., Jiang, Y., Dai, Y. and Jeon, Y. 2022. Saringosterol acetate isolated from *Sargassum fusiformis* induces mitochondrial-mediated apoptosis in MCF-7 breast cancer cells. *Chemistry & Biodiversity*, 19(3):e202100848.

Lomartire, S. and Gonçalves, A.M. 2022. An overview of potential seaweed-derived bioactive compounds for pharmaceutical applications. *Marine Drugs*, 20(2):141.

Lourenço-Lopes, C., Fraga-Corral, M., Jimenez-Lopez, C., Carpena, M., Pereira, A.G., Garcia-Oliveira, P., Prieto, M.A. and Simal-Gandara, J. 2021. Biological action mechanisms of fucoxanthin extracted from algae for application in food and cosmetic industries. *Trends in Food Science & Technology*, 117:163–181.

Lu, X., Dissanayake, A.A., Xiao, C., Gao, J., Zhao, M. and Nair, M.G. 2022. The edible seaweed *Laminaria japonica* contains cholesterol analogues that inhibit lipid peroxidation and cyclooxygenase enzymes. *PLoS One*, 17(1):e0258980.

Martens, N., Schepers, M., Zhan, N., Leijten, F., Voortman, G., Tiane, A., Rombaut, B., Poisquet, J., Sande, N.V.D., Kerksiek, A., et al. 2021. 24(*S*)-saringosterol prevents cognitive decline in a mouse model for Alzheimer's disease. *Marine Drugs*, 19(4):1–17.

Mekinić, I.G., Čagalj, M., Tabanelli, G., Montanari, C., Barbieri, F., Skroza, D., and Šimat, V. 2021a. Seasonal changes in essential oil constituents of *Cystoseira Compressa*: first report. *Molecules*, 26(21):6649.

Mekinić, I.G., Šimat,V., Botić, V., Crnjac, A., Smoljo, M., Soldo, B., Ljubenkov, I., Čagalj, M., and Skroza, D. 2021b. Bioactive phenolic metabolites from adriatic brown algae *Dictyota dichotoma* and *Padina pavonica* (Dictyotaceae). *Foods*, 10(6):1187.

Michalak, I. 2020.Seaweed resources of Poland. *Botanica Marina*, 63(1):73–84.

Moghadam, M.H., Firouzi, J., Saeidnia, S., Hajimehdipoor, H., Jamili, S., Rustaiyan, A. and Gohari, A.R. 2013. A cytotoxic hydroperoxy sterol from the brown alga, *Nizamuddinia Zanardinii. DARU, Journal of Pharmaceutical Sciences*, 21(1):2–5.

Mokrini, R., Mesaoud, M.B., Daoudi, M., Hellio, C., Maréchal, J.P., El Hattab, M., Ortalo-Magné, A., Piovetti, L., and Culioli, G. 2008. Meroditerpenoids and derivatives from the brown alga *Cystoseira baccata* and their antifouling properties. *Journal of Natural Products*, 71(11):1806–1811.

Moni, S.S., Alam, M.F., Makeen, H.A., Alhazmi, H.A., Sultan, M., Siddiqui, R., Jabeen, A., et al. 2021. Solvent extraction, spectral analysis and antibacterial activity of the bioactive crystals of *Sargassum aquifolium* (Turner) C. Agardh from Red Sea. *Natural Product Research*, 35(8):1379–1383.

Mostafa, N.M., Edmond, M.P., El-Shazly, M., Fahmy, H.A., Sherif, N.H. and Singab, A.N.B. 2022. Phytoconstituents and renoprotective effect of *Polyalthia longifolia* leaves extract on radiation-induced nephritis in rats via TGF-β/smad pathway. *Natural Product Research*, 36(16):4187–4192.

Mostafa, N.M., Mostafa, A.M., Ashour, M.L. and Elhady, S.S. 2021. Neuroprotective effects of black pepper cold-pressed oil on scopolamine-induced oxidative stress and memory impairment in rats. *Antioxidants*, 10:1993.

Moussa, H., Quezada, E., Viña, D., Riadi, H., and Gil-Longo, J. 2020. Redox-active phenolic compounds mediate the cytotoxic and antioxidant effects of *Carpodesmia Tamariscifolia* (=*Cystoseira Tamariscifolia*). *Chemistry and Biodiversity*, 17(7):e2000121.

Mwangi, H.M., Njue, W.M., Onani, M.O., Thovhoghi, N. and Mabusela, W.T. 2017. Phlorotannins and a sterol isolated from a brown alga *Ecklonia maxima*, and their cytotoxic activity against selected cancer cell lines HeLa, H157 and MCF7. *Interdisciplinary Journal of Chemistry*, 2(2):1–6.

Nagahawatta, D.P., Kim, H.S., Jee, Y.H., Thilina, U.J., Ahn, G., Namgung, J., Yeo, I.K., Sanjeewa, K.K.A. and Jeon. Y.J. 2022. Sargachromenol isolated from *Sargassum horneri* inhibits particulate matter-induced inflammation in macrophages through Toll-like receptor-mediated cell signaling pathways. *Marine Drugs*, 20(1):28.

Navarro, G., Fernández, J.J. and Norte, M. 2004. Novel meroditerpenes from the brown alga *Cystoseira* sp. *Journal of Natural Products*, 67(3):495–499.

Nigam, S., Singh, R., Bhardwaj, S.K., Sami, R., Nikolova, M.P., Chavali, M. and Sinha, S. 2022. Perspective on the therapeutic applications of algal polysaccharides. *Journal of Polymers and the Environment*, 30(3):785–809.

Okeke, E.S., Nweze, E.J., Chibuogwu, C.C., Anaduaka, E.G., Chukwudozie K.I. and Ezeorba. T.P.C. 2021. Aquatic phlorotannins and human health: bioavailability, toxicity, and future prospects. *Natural Product Communications*, 16(12):1934578X2110561.

Osman, N.A., Siam, A.A., El-Manawy I. and Jeon, Y.J. 2019. Antimicrobial and anti-diabetic activity of six seaweeds collected from the Red Sea, Egypt. *Catrina: The International Journal of Environmental Sciences*, 19(1):55–60.

Pacheco, D., García-Poza, S., Cotas, J., Gonçalves A., and Pereira L. 2020. Fucoidan—a valuable source from the ocean to pharmaceutical. *Frontiers in Drug, Chemistry and Clinical Research*, 3:1–4.

Palta, S., Saroa, R. and Palta, A. 2014. Overview of the coagulation system. *Indian Journal of Anaesthesia*, 58(5):515.

Pati, M.P., Sharma, S., Nayak L., and Panda C.R. 2016. Uses of seaweed and its application to human welfare: a review. *International Journal of Pharmacy and Pharmaceutical Sciences*, 8(10):12–20.

Peng, Y., Huang, R.M. Lin, X.P. and Liu, Y.H. 2018. Norisoprenoids from the brown alga *Sargassum Naozhouense* Tseng et Lu. *Molecules*, 23(2):1–9.

Pooja, S. 2014. Algae used as medicine and food – a short review. *Journal of Applied Pharmaceutical Science and Research*, 6(162):33–35.

Rafiq, K., Khan, A., Ur Rehman, N., Halim, S.A., Khan, M., Ali, L., Al-Balushi A.H., Al-Busaidi H.K. and Al-Harrasi A. 2021. New carbonic anhydrase-II inhibitors from marine macro brown alga *Dictyopteris hoytii* supported by in silico studies. *Molecules* 26 (23):7074.

Rajamani, K., and Thirugnanasambandan, S.S. 2018. Polyphenols from brown alga, *Padina boergesenii* (Allendar & Kraft) decelerates renal cancer growth involving cell cycle arrest and induction of apoptosis in renal carcinoma cells. *Environmental Toxicology*, 33(11):1135–1142.

Rajauria, G. 2019. In-vitro antioxidant properties of lipophilic antioxidant compounds from 3 brown seaweed. *Antioxidants*, 8(12):596.

Rashad, S. and El-Chaghaby, G.A. 2020. Marine algae in Egypt: distribution, phytochemical composition and biological uses as bioactive resources (a review). *Egyptian Journal of Aquatic Biology and Fisheries*, 24(5):147–160.

Rashed, Z.E., Grasselli, E., Khalifeh, H., Canesi L. and Demori I. 2020. Brown-algae polysaccharides as active constituents against nonalcoholic fatty liver disease. *Planta Medica*, 88(1):9–19.

Rehman, N.U., Rafiq, K., Khan, A., Halim, S.A., Ali, L., Al-Saady, N., Al-Balushi, A.H., Al-Busaidi, H.K. and Al-Harrasi, A. 2019. α-Glucosidase inhibition and molecular docking studies of natural brominated metabolites from marine macro brown alga *Dictyopteris hoytii*. *Marine Drugs*, 17(12):666.

Rushdi, M.I., Abdelraheem, I., Saber, H., Attia, E.Z. and Abdelmohsen, U.R. 2022. The natural products and pharmacological biodiversity of brown algae from the genus *Dictyopteris*. *Journal of the Mexican Chemical Society*, 66(1):154–80.

Rushdi, M.I., Abdel-Rahman, I.A., Saber, H., Attia, E.Z., Abdelraheem, W.M., Madkour, H.A., Hassan, H.M., Elmaidomy A.H. and Abdelmohsen U.R. 2020. Pharmacological and natural products diversity of the brown algae genus *Sargassum*. *RSC Advances*, 10(42):24951–24972.

Sánchez-Machado, D.I., López-Hernández, J., Paseiro-Losada, P., and López-Cervantes, J. 2004. An HPLC method for the quantification of sterols in edible seaweeds. *Biomedical Chromatography: BMC*, 18(3):183–190.

Sanger, G., Rarung, L.K., Wonggo, D., Dotulong, V., Damongilala, L.J. and Tallei, T.E. 2021. Cytotoxic activity of seaweeds from North Sulawesi marine waters against cervical cancer. *Journal of Applied Pharmaceutical Science*, 11(09):066–073.

Sangha, J.S., Kelloway, S., Critchley, A.T. and Prithiviraj, B. 2014. Seaweeds (Macroalgae) and their extracts as contributors of plant productivity and quality. the current status of our understanding. *Advances in Botanical Research*, 71:189–219.

Sanjeewa, K.K.A., Fernando, I.P.S., Kim, E.A., Ahn, G., Jee, Y. and Jeon, Y.J. 2017. Anti-inflammatory activity of a sulfated polysaccharide isolated from an enzymatic digest of brown seaweed *Sargassum horneriin* RAW 264.7 cells. *Nutrition Research and Practice*, 11(1):3.

Shanthi, N., Arumugam, P., Murugan, M., Sudhakar M.P. and Arunkumar K. 2021. Extraction of fucoidan from *Turbinaria decurrens* and the synthesis of fucoidan-coated AgNPs for anticoagulant application. *ACS Omega*, 6(46):30998–31008.

Shen, P., Gu, Y., Zhang, C., Sun, C., Qin, L., Yu, C., and Qi, H. 2021. Metabolomic approach for characterization of polyphenolic compounds in *Laminaria japonica*, *Undaria pinnatifida*, *Sargassum fusiforme* and *Ascophyllum nodosum*. *Foods*, 10(1):192.

Silva, A., Rodrigues, C., Garcia-Oliveira, P., Lourenço-Lopes, C., Silva, S.A., Garcia-Perez, P., Carvalho, A.P., Domingues, V.F., Barroso, M.F., Delerue-Matos, C., Simal-Gandara, J. and Prieto, M.A. 2021. Screening of bioactive properties in brown algae from the Northwest Iberian Peninsula. *Foods*, 10(8):1915.

Singkoh, M.F., Katili, D.Y. and Rumondor, M.J. 2021. Phytochemical screening and antibacterial activity of brown algae (*Padina australis*) from Atep Oki Coast, East Lembean of Minahasa Regency. *Aquaculture, Aquarium, Conservation & Legislation*, 14(1):455–461.

Ślusarczyk, J., Adamska E. and Czerwik-Marcinkowska J. 202. Fungi and algae as sources of medicinal and other biologically active compounds: a review. *Nutrients* 13(9):3178.

Sudha, G. and Balasundaram, A. 2018. Analysis of bioactive compounds in *Padina pavonica* using HPLC, UV-VIS and FTIR techniques. *Journal of Pharmacognosy and Phytochemistry*, 7(3):3192–3195.

Sugiura, Y., Matsuda, K., Yamada, Y., Nishikawa, M., Shioya, K., Katsuzaki, H., Imai K., and Amano, H. 2007. Anti-allergic phlorotannins from the edible brown alga, *Eisenia arborea*. *Food Science and Technology Research*, 13(1):54–60.

Thambi, A. and Chakraborty, K. 2022. Brown and red marine macroalgae as novel bioresources of promising medicinal properties. *Journal of Aquatic Food Product Technology*, 31(3):227–241.

Thomas, N.V. and Kim, S.K. 2013. Beneficial effects of marine algal compounds in cosmeceuticals. *Marine Drugs*, 11(1):146–164.

Wang, S.Y., Xiang, J., Huang, X.J., Wei, X., Zhang, Q., Du, X.H., Hu, J.S., Wang, Y.M. and Zhang, C.X. 2020. Chemical constituents from *Sargassum fusiforme* (Harv.) Setch. *Chemistry and Biodiversity*, 17(6):1–8.

Widyastuti, S., Geraldine, B.A., Sunarwidhi, A.L., Ariyana, M.D., Prasedya E.S. and Sunarpi H. 2019. The use of brown algae extract to extend shelf life and improve post-harvest quality of tomato fruit. *AIP Conference Proceedings*, 2199(1):070008.

Wu, J., Xi, Y., Li, G., Zheng, Y., Wang, Z., Wang, J., Fang, C., Sun, Z., Hu, L., Jiang, W., et al. 2021. Hydroazulene diterpenes from a *Dictyota* brown alga and their antioxidant and neuroprotective effects against cerebral ischemia-reperfusion injury. *Journal of Natural Products*, 84(4):1306–1315.

Yan, Y., Niu, Z., Wang, B., Zhao, S., Sun, C., Wu, Y., Li, Y., Ying, H., and Liu, H. 2021. Saringosterol from *Sargassum fusiforme* modulates cholesterol metabolism and alleviates atherosclerosis in ApoE-deficient mice. *Marine Drugs*, 19 (9):485.

Yang, Y.I., Ahn, J.H., Choi, Y.S. and Choi, J.H. 2015. Brown algae phlorotannins enhance the tumoricidal effect of cisplatin and ameliorate cisplatin nephrotoxicity. *Gynecologic Oncology*, 136(2):355–364.

Yoon, N.Y., Chung, H.Y., Kim, H.R. and Choi, J.S. 2008. Acetyl- and butyrylcholinesterase inhibitory activities of sterols and phlorotannins from *Ecklonia Stolonifera. Fisheries Science*, 74 (1):200–207.

Young, M.H., Jong, S.B., Jin, W.H. and Nam, H.L. 2007. Isolation of a new phlorotannin, fucodiphlorethol G, from a brown alga *Ecklonia cava. Bulletin of the Korean Chemical Society*, 28(9):1595–1597.

Zbakh, H., Zubía, E., de Los Reyes, C., Calderón-Montaño, J.M. and Motilva, V. 2020. Anticancer activities of meroterpenoids isolated from the brown alga *Cystoseira usneoides* against the human colon cancer cells HT-29. *Foods*, 9(3):300.

Zhong, B., Robinson, N.A., Warner, R.D., Barrow, C.J., Dunshea, F.R. and Suleria, H.A.R. 2020. LC-ESI-QTOF-MS/MS characterization of seaweed phenolics and their antioxidant potential. *Marine Drugs*, 18(6):1–21.

14 Phytochemical and Bioactive Compounds of Green Algae (Chlorophyta) and Their Applications

Taruni Bajaj,[1] *Hina Alim,*[2] *Ahmad Ali,*[2] *and Nimisha Patel*[1,*]
[1] Department of Life Sciences, J.C. Bose University of Science and Technology, YMCA, Faridabad, Haryana, India
[2] Department of Life Sciences, University of Mumbai, Mumbai, Maharashtra, India
* Corresponding author (Nimisha Patel): nimishapatel@jcboseust.ac.in

14.1 INTRODUCTION

India, a tropical country rich in biodiversity and cultural legacy with a land size of over 2 million square kilometres, has a 7500-kilometre exclusive economic zone that displays enormous biological marine resources (Jayalakshmi et al. 2021). Seaweeds account for over 90% of the marine flora in this zone, contributing about 50% of global energy generation through photosynthesis (Dring 1982; John 1994). The oceanic environment, which has over 10,000 species, demonstrates the incredible richness of seaweeds (Phillips 2001). Algae are heterogeneous organisms with a high level of complexity in nature, consisting primarily of photoautotrophs living in various watery habitats (Round 1981; Wetzel 1983). Unicellular organisms classified as microalgae are usually 1–400 micrometres in length and invisible to the naked eye (Nethravathy et al. 2019; Rashad et al. 2019), and multicellular organisms classified as macroalgae, also known as seaweeds, are highly aqua-prone organisms (Khan et al. 2018), which are distinct from each other making them polyphyletic (Aditya et al. 2016). Rhodophyta, Phaeophyta, and Chlorophyta, namely red, brown, and green seaweeds, respectively, are a vast community of microbenthic and photosynthetic flora found in saline environments (www.biologyonline.com/dictionary/seaweedd3). Both microalgae and macroalgae have high growth and biomass production rates that do not require fresh water for culture, despite their capacity to thrive in a variety of water types, including fresh water, seawater, brackish water, and even metropolitan and other wastewater (Ding et al. 2018; Morales et al. 2019).

Chlorodendrophyceae, *Pedinophyceae*, and the UTC clade, which includes *Trebouxiophyceae*, *Chlorophyceae*, and *Ulvophyceae*, are ancient Chlorophyta species with a tremendous morphological and cytological diversity of photosynthetic eukaryotes (De Clerck et al. 2012; Leliaert et al. 2012). Marine macroalgae are capable of surviving drastic environmental changes using various adaptation strategies and processes, which are influenced by a variety of environmental parameters such as changes in climatic conditions, pH, salinity, exposure to sunlight, other physiological conditions, and carbon dioxide supply (Trivedi et al. 2015; Ji et al. 2019). The structural core of secondary bioactive compounds and metabolites is made up of various substances such as cyclic peptides, alkaloids, polyketides, glycerol, lipids, phlorotannin, polysaccharides, diterpenoids,

DOI: 10.1201/9781003256830-14

sterols, and quinones (Dahms and Dobretsov 2017; Jesus et al. 2019; Lever et al. 2020). Biogenic substances are chemically active metabolites or biochemical molecules that help defend algae from other settling creatures by releasing polysaccharides, alkaloids, lipoproteins, aldehydes, terpenoids, and alcohols (Bhadury and Wright 2004; Smith 2004; Athbi et al. 2009).

Chlorophyta is the most common phylum in the global ecosystem, and it performs a vital role in freshwater, marine, and subaquatic habitats (O'Kelly 2007; Leliaert et al. 2012; Dittami et al. 2017; Leliaert 2019). Chlorophyta has been in demand as a wellspring of foodstuffs for human consumption with high nutritional value due to their widespread distribution and a vast amount of biomass, such as the alga *Cladophora crispate*, which is highly rich in protein concentration of around 40.3% (Athbi et al. 2009), whereas the alga *Ulothrix cylindricum* is a good source of fatty acids such as oleic acids and palmitic acids (Athbi et al. 2011). Another algae species, *Chlorella vulgaris*, has a high concentration of amino acids, including aspartic acid, leucine, isoleucine, alanine, and tryptophan, as well as 50.5% concentration of protein (Athbi et al. 2009). Green algae carbohydrates are abundant in physiochemically active chemicals that can be employed for many biological effects including anticoagulation, immunomodulation, antioxidant, anti-cancerous, and anti-inflammation activities (Paulert et al. 2009, 2010; Prasad et al. 2009; Alves et al. 2010; Ciancia et al. 2012; Araujo et al. 2013; Samara et al. 2013). Compounds originating from the genus *Caulerpa* belonging to the Chlorophyta are interesting in terms of researching its possible anticancer actions (Mehra et al. 2019).

Antibacterial agents are made from many algae species, and their effectiveness as natural antibacterial agents are still being studied. *Ulva lactuca*, *Enteromorpha compressa*, *Padina pavonica*, *Ceramium rubrum*, *Hydroclathrus clathrus*, *Ulva pertus*, *Ulva prolifera*, and other seaweeds are effective against a variety of bacteria (Elnabris et al. 2013; Vimala and Poonghuzhali 2017).

Seaweeds are commonly employed for their antioxidant properties because of the presence of numerous compounds such as carbohydrates, carotenoids, vitamins and their precursors, and polyphenols, which block the oxidation process (Y. Kumar et al. 2019). Antioxidant activities are found in ethanol extracts from *Capsosiphon fulvescens*, *Enteromorpha compressa*, *Ulva pertusa*, *Chaetomorpha moniligera*, and other bioactive substances such as flavonoids and phenolics (Cho et al. 2010). *Hijikia fusiformis* and phlorotannins in *Sargassum kjellamanianum* are two more macroalgae that contain metabolically active chemicals with antioxidant properties, both of which are in high demand in the drug industry (Yan et al. 1999).

The presence of macromolecules ranging from polyunsaturated fatty acids (PUFAs), vitamins, dietary fibres, essential amino acids, and other natural biologically active molecules in *Caulerpa racemose* contributes to its antimutagenic, antioxidant, antibacterial, anticancer, and anticoagulant characteristics (M. Kumar et al. 2011; Nagappan et al. 2014; Tanna et al. 2018; Yap et al. 2019). The fermentation of seaweeds to produce wine results in the synthesis of several secondary metabolically active chemicals, such as polyphenols, which are released into the aqueous ethanolic solution of wine during the process and have a variety of therapeutic characteristics that can be used by humans (Rathi 2018).

The production of high-molecular-weight hydrophilic compounds has been of significant importance in commercial industries as a result of the farming of edible species of macroalgae leading to the formation of various seaweed products from carrageenan, alginate, and agar. The acquired compounds have been used for various medical, pharmacological, and biochemical research having previously been limited to their conventional uses until the 1950s (Lincoln et al. 1991). Physical properties such as semi-solid jelly-like formation, water retention, and use as an emulsifying agent are primarily exploited commercially (Renn 1997).

This chapter emphasizes the variety of metabolites obtained from green algae (Chlorophyta). It summarizes the potential biological functions that varying algal strains perform, which are beneficial to humans in various ways and whose bioavailability can be increased at an industrial level and for commercial applications to establish a significant role in the ecosystem.

14.2 BIOSYNTHESIS OF SECONDARY METABOLITES IN GREEN ALGAE

Green seaweeds have been discovered to be a source of a wide range of biologically active metabolic substances with various properties in many areas. Around 72,500 marine species have been found, forming a vast population of macroalgae (Guiry 2012); approximately 3200 natural products have been reported to make up around 13% of bioactive compounds gathered from marine algae (Leal et al. 2013). Because of the high concentration of terpenoid molecules in marine green algae, polyphenols, sterols, diterpenoids, caulerpin, and other substances have been found (Blunt et al. 2006). Ulvan is a sulphated polysaccharide derived from Chlorophyta species such as *Ulva rigida* and *Ulva lactuca*, with the principal ingredients being rhamnose (12.73–45%), uronic acids (6.50–25.96%), sulphate (12.80–23%), and xylose (2–12%) (Yaich et al. 2017). The ulvanobiouronic acid 3-sulphate structure is made up of major and minor recurrent units made up of either glucuronic or iduronic acid, as well as units of sulphated xylose that substitute the glucuronic or uronic acid as a branch on O-2 of the rhamnose 3 sulphate, respectively (Lahaye and Ray 1996; Lahaye et al. 1997). Ulvans are widely employed for their many biochemical properties, such as antiviral, anticancer, anticoagulant, and antihyperlipidaemic actions, which depend on the quantity of sulphated content present in their structure (Yaich et al. 2017).

14.2.1 Alkaloids

Alkaloids have a variety of biological actions due to the presence of nitrogen in their chemical formation. The following is a list of alkaloids generated from Chlorophyta species:

14.2.1.1 Binsidole Alkaloids

Chlorophyta has been recognized as a key source of secondary metabolites, including the alkaloid binsidole. *Caulerpa racemose*, for example, is formed of two unique binsidole alkaloids, racemosin A and racemosin B, and the pigment caulerpin (Figure 14.1) (A.H. Liu et al. 2013). *Caulerpa racemosa*, *C. lamourouxii*, *C. serrulate*, *C. sertularioides*, and other green algae species generate caulerpin. Caulerpin is an abisindole alkaloid identified in *Caulerpa serrulate* (Güven et al. 2010). It is made up of two antiparallel indole nuclei. Caulerchlorin, an alkaloid isolated from *Caulerpa racemose* (Figure 14.1), has been found to have antifungal activity against the 32609 strain of *Cryptococcus neoformans* (D.Q. Liu et al. 2012); meanwhile, caulerpin, an algal lipoid extract of the same species *C. racemose*, has been found to have anti-inflammatory behaviour in mice, and the bisindolic pharmacophoric nuclei (Souto et al. 2011).

14.2.1.2 Other Alkaloids

The first study on marine seaweeds producing compound prenylated para-xylene in the green algae species *Caulerpa racemose* resulted in the introduction of two prenylated para-xylenes: caulerprenylol A (Figure 14.1), which shows the antifungal activity of weak intensity against *Candida glabrata* (537), *Trichophyton rubrum*, *Cryptococcus neoformans* (32609), and cauler (A.H. Liu et al. 2013).

A pyrrolopiperazine-2,5-dione alkaloid, a type of alkaloid derived from the alga *Ulva prolifera*, is recognized for its antialgal efficacy against Rhodophyta species' severely hazardous chemicals (Jiang et al. 2013). Various alkaloids and prenylated compounds derived from green algae with their varied bioactivities have been listed in Table 14.1. Prenylated bromohydroquinones such as 3′-methoxy-7-hydroxycymopolone, 7-hydroxycymopochromanone (PBQI), 7-hydroxycymopolone (PBQ2), 3-hydroxycymopolone (Figure 14.1), and many others with chemotherapeutic and antimutagenic activity against *Salmonella typhimurium* are found in *Cymopolia* (Dorta et al. 2002; Gallimore et al. 2009).

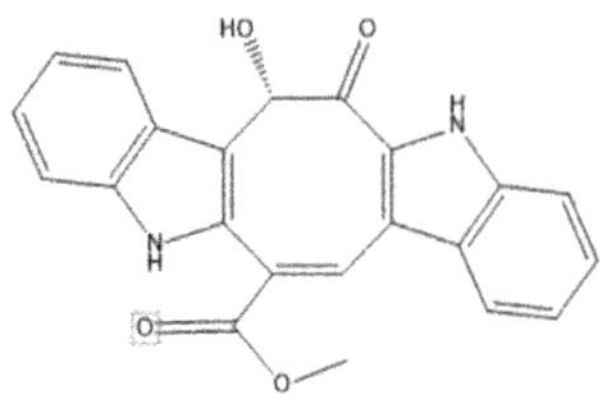

1

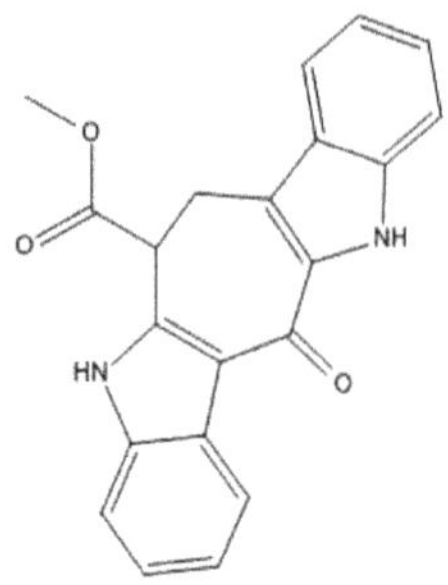

2

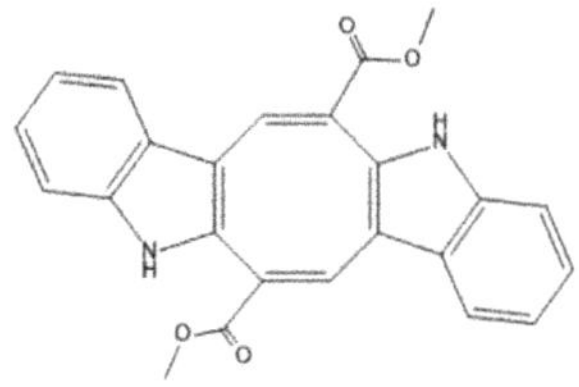

3

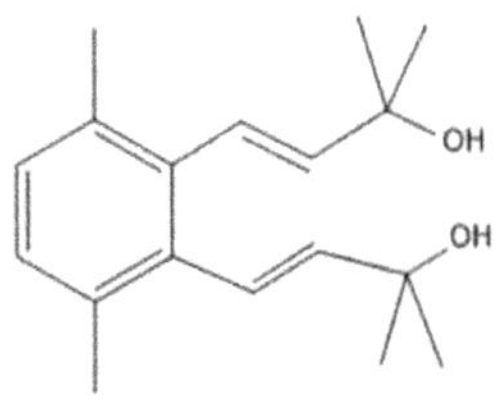

4

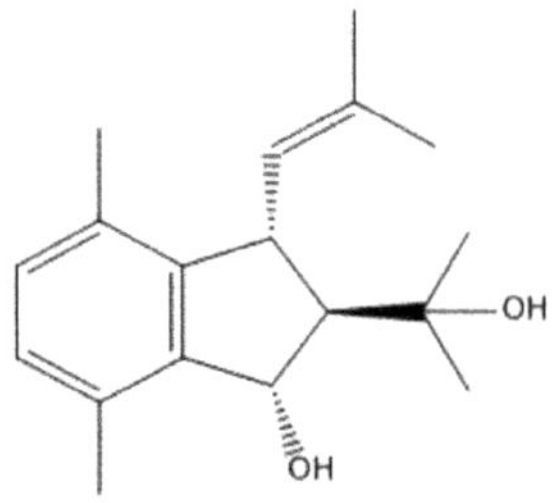

5

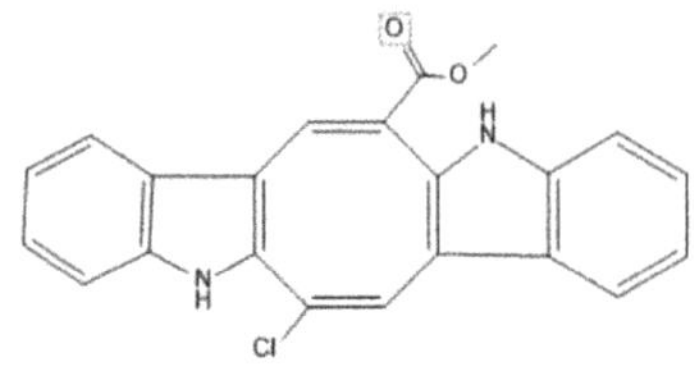

6

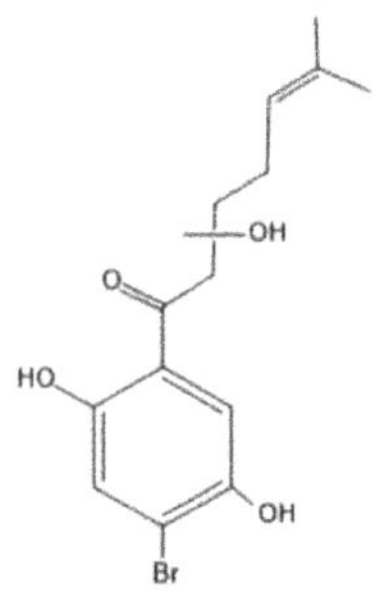

7

FIGURE 14.1 Structures of alkaloid compounds: (1) racemosin C, (2) caulersin, (3) caulerpin, (4) caulerprenylol A, (5) caulerprenylol B, (6) caulerchlorin, and (7) 3-hydroxycymopolone.

14.3 TERPENES

Terpenoids are a class of secondary metabolites composed primarily of isoprene units that exhibit a wide range of structural and functional variations depending on the number of isoprene units present (Dorta et al. 2002). Carotenoids are terpenoids with eight isoprene units in their structure (Balboa et al. 2013).

14.3.1 Monoterpenoids

Hydro-distillation of terrestrial plants and marine algae yields two isoprene units, which make up monoterpenes. Monoterpenes can be classified as cyclic or linear compounds because carbon

TABLE 14.1
Bioactivity of Various Alkaloids and Prenylated Substances Obtained from Distinct Chlorophyta Species

Species	Compound	Bioactivity	Reference
Caulerpa racemose	Racemosin A, B	Neuroprotective	A.H. Liu et al. 2013
Caulerpa racemose	Racemosin C (**1**)	Significant PTP1B inhibitor	H. Yang et al. 2014
Caulerpa serrulate	Caulersin (**2**)	PTP1B inhibitor	Güven et al. 2010
Caulerpa racemosa and *Caulerpa genus*	Caulerpin (**3**) and caulerpic acid	Strong PTP1B inhibitor	Ornaro et al. 2014
Caulerpa racemose	Caulerprenylols A, B (**4, 5**)	Antifungal	A.H. Liu et al. 2013
Caulerpa racemose	Caulerchlorin (**6**)	Weak antifungal	D.Q. Liu et al. 2012
Ulva prolifera	Pyrrolopipera-zine-2,5-dione	Antialgal	Jiang et al. 2013
Cymopolia barbata	7-hydrocymo-pochromanone (PBQ1), 7-hydroxycymo-polone (PBQ2)	Chemotherapeutic	Dorta et al. 2002
Cymopolia barbata	7-dyhydroxycymo-pochromenol, 3,7-hydroxycymopolone	Antimutagenic against *S. typhimurium*	Dorta et al. 2002
Cymopolia barbata	3-hydroxycymopolone (**7**)	Antimutagenic against *S. typhimurium*	Dorta et al. 2002

For structure of compounds 1–7 see Figure 14.1.

skeletons are organized into peculiar rings or straight chains. The characteristic volatile chemicals in essential oils are a broad set of monoterpenes. For instance, the green alga *Caulerpa taxifolia* produces the aldehyde monoterpene taxifoloal D (Guerriero et al. 1992). Numerous varieties of monoterpenes are well known for their biochemical properties, which also include the ability to reject and depress the immune system as well as impede germination. Additionally, it displays the distinct biosynthetic route for synthesizing organohalogens (Kladi et al. 2004).

14.3.2 Carotenoids

Eight isoprene units comprise carotenoid terpenoids (Balboa et al. 2013). Numerous microalgae produce a variety of antioxidants that are known for their ability to combat free radicals in human bodies, which are drawn from the photosynthetic pigment carotenoid (Zaho et al. 2004; Hajimahmoodi et al. 2009; Plaza et al. 2009; Goh et al. 2010; Goiris et al. 2012). The primary source of the carotenoid pigments, which are frequently used as food colourings, is a number of freshwater green algae species, including *Chlorella vulgaris*, *Chlorella fusca*, *Selenestrum capricornutum*, *Botryococcus sudeticus*, *Chlorococcum* spp., and *Pandorina morum*. Among these, *C. vulgaris* is recognized for producing significant amounts of carotenoids (Othman et al. 2018).

14.3.3 Sesquiterpenes

The structural portion of sesquiterpenes comprises three isoprene units that are connected by a C15 carbon backbone. The Chinese green alga *Caulerpa taxifolia* was discovered to have a unique aromatic carbon skeleton of valernane-type compounds caulerpal A and caulerpal B, as well as one recognized biomolecule, caulerpin (Mao et al. 2006). Both caulerpal A and B have a weaker mode of action against hPTP1B (human protein tyrosine phosphatase 1B), an enzyme found in a wide variety of human genome expression during biochemical processes that aid in controlling cell response to external stimuli and which performs an unusual role in the pathogenesis of numerous inherited and acquired human diseases (Tonks 2006).

TABLE 14.2
Different Diterpenoid Compounds and Their Bioactivity Obtained from Chlorophyta

Species	Compound	Bioactivity	Reference
Ulva fasciata	Labda-14-ene-8-ol, Ent-labda-13(16),14-diene-3α-ol	Antibacterial	Cengiz et al. 2011
Caulerpa racemosa	Racemobutenolids A, B	–	P. Yang et al. 2015
Ulva fasciata	Labda-14-ene-8α,9α-diol	Antibacterial	Cengiz et al. 2011
Caulerpa racemosa	An α-tocopheroid, α-tocoxylenoxy, 4,5-dehydrodiodictyonema A	PTP1B inhibitor	P. Yang et al. 2015

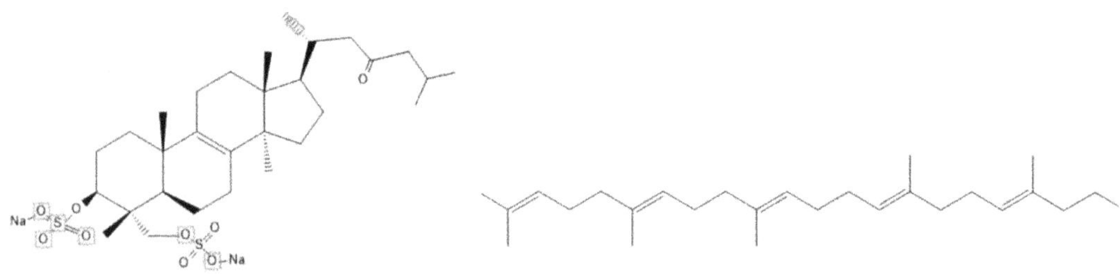

8 9

FIGURE 14.2 Structure of (8) lanostane-type triterpenoid disulphate, (9) squalene.

14.3.4 Diterpenoids

Diterpenoids consist of four isoprene units held together by a C20 carbon backbone. Chakraborty et al. had shown seven types of diterpenoids, namely labda-14-ene-3α,8α-diol; labda-14-ene-8-ol; labda-14-ene8α,9α-diol; ent-labda-13(16),14-diene-2-one; ent-labda-13(16),14-diene-3α-ol; labda-14-ene-8α-hydroxy-3-one; ent-labda-13(16) and 14-diene-3α-ol (Chakraborty et al. 2010) which were classified as the chief components present in green alga *Ulva fasciata* (Cengiz et al. 2011). Many forms of diterpenoid metabolites with diverse bioactivities have been listed in Table 14.2. The sesquiterpene compounds are mostly isolated from *Caulerpa* species, including feeding preference, ichthyotoxicity, antibacterial, and feeding deterrents (Máximo et al. 2018).

14.3.5 Triterpenoids

Certain triterpenoids have been found in green macroalgae; e.g. *Tydemania expeditionis* has four sulphate-conjugated triterpenoids, including one unique lanostane-type triterpenoid disulphate and three well-known cycloartane-type triterpenoid disulphates (Jiang et al. 2008). The structure of some forms of triterpenoids is shown in Figure 14.2.

The anticancer effects of disulphated natural products like lanosta 8-en 3,29-diol-23-oxo-3,29 disodium sulphate are lesser; nonetheless, the same natural product has been proven to exhibit moderate cytotoxicity for malignant cells in invertebrates.

Furthermore, only caulerpin has been shown to have antifungal properties against the marine pathogen *Lindra thalassiae* at the usual proportion of whole tissue among these natural compounds. Various other forms of terpenoids derived from algal species and their bioactivities have been depicted in Table 14.3.

Compounds such as capisterone A and B, which are triterpene sulphate esters derived from the tropical algal species *Penicillus capitatus* in the tropical Atlantic Ocean, have the potential

TABLE 14.3
Terpenoids Derived from Chlorophyta

Species	Compound	Bioactivity	Reference
Penicillius capitatus	Capisterones B, capisterones A	Potent antifungal	Puglisi et al. 2004
Tydemania expeditionis	Lanosta-8-en-3,29-diol-23-oxo-3,29-disodium sulphate (lanostane-type triterpenoid disulphate) (**8**)	Cytotoxic cancerous cell	Jiang et al. 2008
Codium dwarkense	Dwarrkenoic acid	Alpha-glucosidase inhibitor	Ali et al. 2015
C. racemose	Squalene (**9**)	–	Ragasa et al. 2015

For structure of compounds **8** and ` see Figure 14.2.

to have antifungal properties that protect the alga from the indiscriminate marine microorganism *L. thallasiae* (Puglisi et al. 2004).

14.3.6 Steroids and Fatty Acids

The largest source of sterols derived from the marine environment is marine Chlorophyta (Mouritsen et al. 2017). Green algae contain a variety of steroids, which are made up of six isoprene units, including triterpenoid chemicals (Salehi et al. 2019). These steroidal compounds are known to have a distinct method of action, in which they inhibit phospholipase A2 with lipocortin-1 synthesis by inhibiting eicosanoid development (Souto et al. 2011). Compared to non-steroidal anti-inflammatory medications, steroidal compounds are thought to have stronger anti-inflammatory activities among the identified category of molecules.

Ulva fasciata, a green alga, produces three types of PUFAs, according to researchers:

- Hexadeca-4,7,10,13-tetraenoic acid (HDTA).
- Octadeca-6,9,12,15-tetraenoic acid (ODTA).
- α-Linolenic acid. which was first isolated from *Caulerpa racemosa* and had a strong algicidal effect on *Heterosigma akashiwo* (Ragasa et al. 2015).

Several other types of steroids and fatty acid molecules generated by green algae and their bioactivities have been listed in Table 14.4. Furthermore, compounds derived from the marine green seaweed *Codium iyengarii*, such as iyengaroside A and B (Figure 14.3), have been reported to be effective against *Klebsiella pneumoniae* (Ali et al. 2002).

Various fatty acid compounds extracted from many species of green seaweeds have been recognized to have inhibitory properties; e.g. Yang and his team isolated a compound known as 28-oxostigmastic steroid, (23*E*)-3*n*-hydroxy-stigmasta-5,23 dien-28-one from the algal species *Caulerpa racemose* in 2014, which was found to have the most active PTP1B inhibitory properties (P. Yang et al. 2015).

14.3.7 Glycerol and Lipids

Green alga *U. prolifera* produces monoglycerides such as 9-hexadecenoic acid 2,3-dyhdroxypropyl ester (glyceryl palmitate) (Figure 14.4) and glycerol monopalmitate. Several dinoflagellates (red tide microalgae) are sensitive to *U. prolifera*'s antifungal activities (Sun et al. 2016). Other

TABLE 14.4
Sterols and Fatty Acids Derived from Chlorophyta

Species	Compound	Bioactivity	Reference
Ulva prolifera	Cholest-5-en-3-ol (**10**)	Antialgal	Mao et al. 2011
Caulerpa racemose	(8*E*,12*Z*,15*Z*)-10-hydroxy-8,12,15-octadecatrien-4,6-diynoic acid	–	Alamsjah et al. 2005
Ulva fasciata	α-Linolenic acid (**11**)	Algicidal	Ragasa et al. 2015
Codium iyengarii	Iyengaroside A and B (**12, 13**)	Antibacterial	Ali et al. 2002
Tydermania expeditionis	3-hydroxy-octadeca-4(*E*),6(*Z*),15(*Z*)-trienoic acid	Antitumour	Jiang et al. 2008
Ulva australis	(24*R*)-5,28-stigmastadiene-3β,24-diol-7-one (**14**), 24*R* and 24*S*-vinylcholesta-3β,5α,6β,24-tetraol	Inhibitor of aldose reductase	Li et al. 2017
Caulerpa racemose	(23*E*)-3b-hydroxy-stigmasta-5,23 dien-28-one	PTP1B inhibitor	P. Yang et al. 2015
Codium iyengarii	Clerosterol galactoside (**15**)	-	Ali et al. 2002
Caulerpa racemose	β-Sitosterol (**16**)	-	Ragasa et al. 2015
Ulva fasciata	Hexadeca 4,7,10,13 tetraenoic acid (HDTA)	Algicidal	Mao et al. 2011
Ulva lactuca	Amide derivative, keto-type fatty acid	ARE activators	R. Wang et al. 2013

For structure of compounds **10–16** see Figure 14.3.

Caulerpa racemosa species have been identified as sources of compounds such as monogalactosyl diacylglycerol and 1-eicosapentaenoyl-2-linolenoyl 3-galactosylglycerol, as well as *β*-sitosterol, chlorophyll a, and unsaturated hydrocarbons (Ragasa et al. 2015). Other species that produce glycerol and lipids and their bioactivity have been summarized in Table 14.5.

Compounds such as capsofulvesins A, B, and C, which were derived from the green algae *Capsosiphon fulvescens*, have been reported to have acetylcholinesterase (AChE) inhibitory activity with half-maximal inhibitory concentration (IC_{50}) values of 53.13 ± 2.83, 51.38 ± 0.90, and 82.54 ± 0.88 M, respectively (Fang et al. 2012).

14.4 SIGNIFICANCE OF GREEN ALGAE (CHLOROPHYTA)

14.4.1 Biochemical Compounds in Green Algae for Formation of Sex Steroid Hormones

Recent studies have revealed that several substances derived from Chlorophyta can synthesize steroid molecules, such as cholesterol synthesis, which are linked to hormonal actions, particularly in women. After cholesterol is synthesized in granulosa cells, pregnenolone, progesterone, androstenedione, and testosterone are formed, and hormones are produced in theca cells of the ovarian organs in females. Oestrone and 17-oestradiol are formed once these hormones are aromatized (Janczyk et al. 2006; Kozlova et al. 2020). According to certain research, *Ulva lactuca* extracts contain steroid chemicals that aid in the growth, differentiation, and creation of gonads, as well as proper reproduction in mice (Yulistiyanto et al. 2020). Furthermore, certain substances acquired from the same species, such as testosterone, aid in the proliferation and maturation of the sex glands. Similarly, another well-known species, *Chlorella vulgaris*, aids in improving rabbit fertility in both males and females (Okab et al. 2013).

For the proper maturation of the oocytes, the transportation of fatty acids is helpful in the process of vitellogenesis during the menstrual cycle. Bioactive compounds like fucosterol and isofucosterol

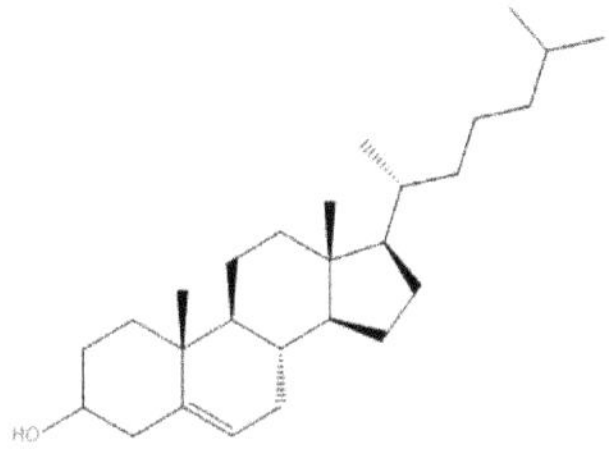

10

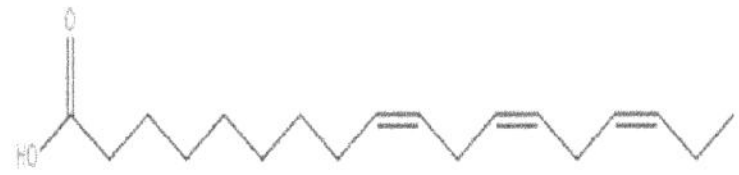

11

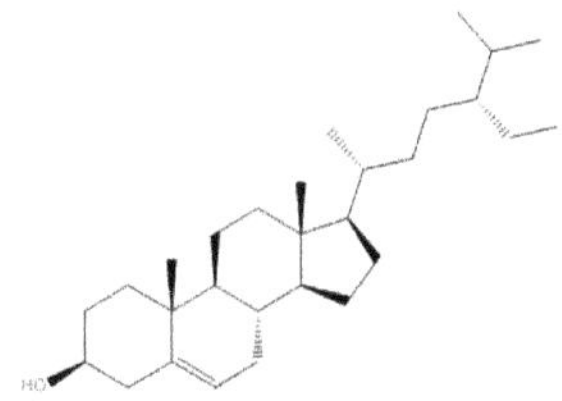

12

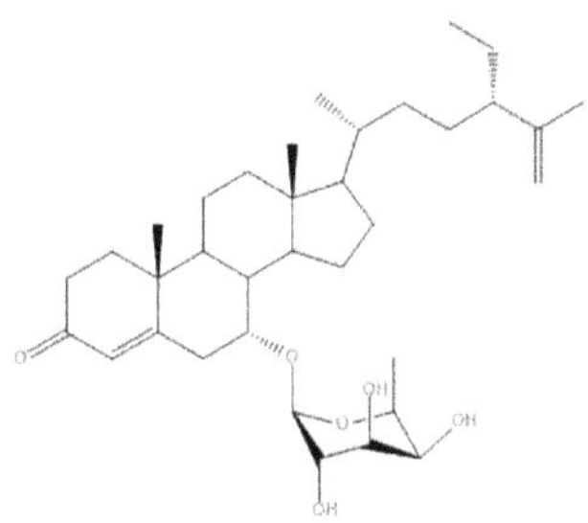

13

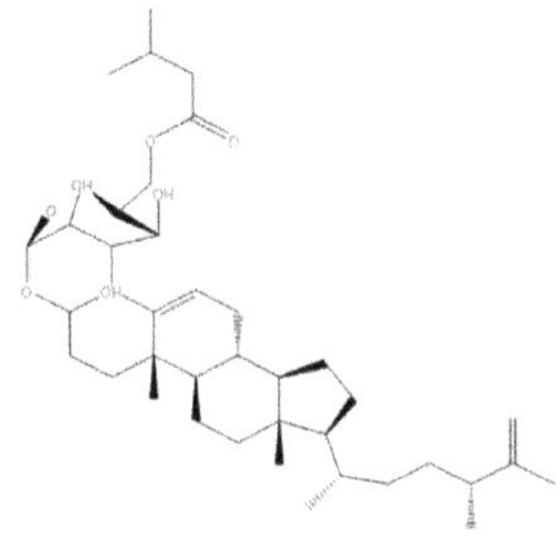

14

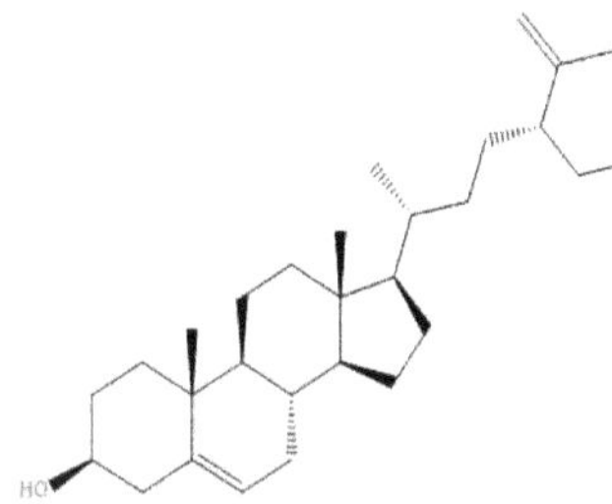

15

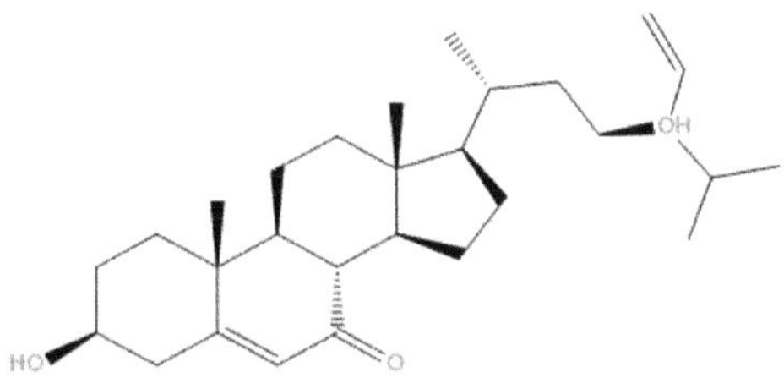

16

FIGURE 14.3 Structure of compounds: (10) cholest-5-en-3-ol, (11) α-linolenic acid, (12) iyengaroside A, (14) iyengaroside B, (14) (24*R*)-5,28-stigmastadiene-3β,24-diol-7-one, (15) clerosterol galactoside, and (16) β-sitosterol.

are synthesized by green algae (Corral-Rosales et al. 2019). In contrast, other compounds, such as phytosterols, play a neuro-transmission role in steroid substances required for producing steroidal hormones that are also obtained from green algae (Lorensi et al. 2019). It is clear now how the synthesis of essential steroid metabolites in algal species is known to play a critical function in the continuation of female reproductivity.

14.4.2 Significance of Steroidal Compounds in Reproductive Health

Research has reported that various algal species producing steroids and sterol compounds greatly impact reproductivity and female fertility. Generally, steroid compounds and their biosynthesis have been observed to be beneficial in production, the proliferation of hormones and the reproductive

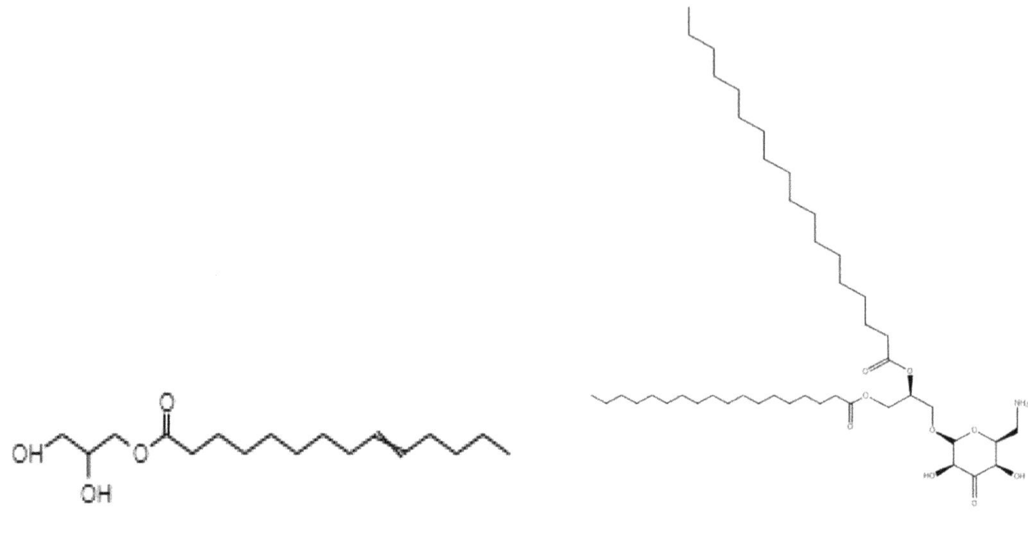

FIGURE 14.4 Structures of (A) glyceryl palmitate, and (B) avrainvilloside.

TABLE 14.5
Lipids and Glycerol Derived from Green Seaweeds

Species	Compound	Bioactivity	Reference
Ulva prolifera	1-*O*-Palmitoyl-2-ooleoyl-3-*O*-β-d-galactopyranosyl glycerol, 1-*O*-octadecanoic acid-3-*O*-β-d-galactopyranosyl glycerol, 9-hexadecenoic acid, 2,3-dihydroxy propyl ester (Figure 14.4)	Antialgal	Sun et al. 2016
Caulerpa racemose	1-Eicosapentaenoyl-2-linolenoyl-3-galacto-sylglycerol	Anti-inflammatory	Ragasa et al. 2015
Ulva pertusa	Galactosylglycerolipid (GGL)	–	Williams et al. 2007; Ishibashi et al. 2014
Capsosiphon fulvescens	Capsofulvesins	Acetylcholinesterase (ache) inhibitor	Fang et al. 2012
Avrainvillea nigricans	Avrainvilloside (Figure 14.4)	Inactive cytotoxic	Andersen and Taglialatela-Scafati 2005
Caulerpa racemose	Sulphoquinovosyl diacylglycerol (SQDG)	Antiviral against herpes simplex virus 2 (HSV-2)	H. Wang et al. 2007

organs, cell damage repair, antioxidant activity, inhibitory activity, development of gonads, neurotransmission, and as a source of supplements as well. For instance, sterol compounds like phytosterols, obtained from *Nauphoeta cinera* cockroaches, of the species *Prasiolacrispa* are beneficial in the neurotransmission of inner steroidal molecules (Lorensi et al. 2019). Likewise, steroids like the hormones oestrogen and testosterone (Figure 14.5) synthesized in *Chlorella vulgaris*, hormones like oestrogen and gonadotropin (luteinizing hormone and follicle-stimulating hormone) (Figure 14.5) from *Chlorella vulgaris* and *Spirulina platensis*, and compounds like fungisterol,

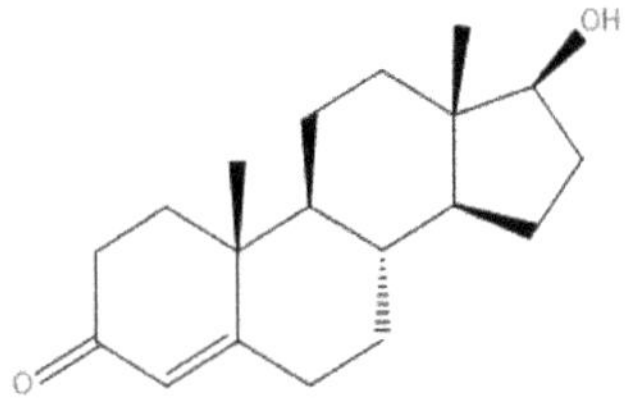

17

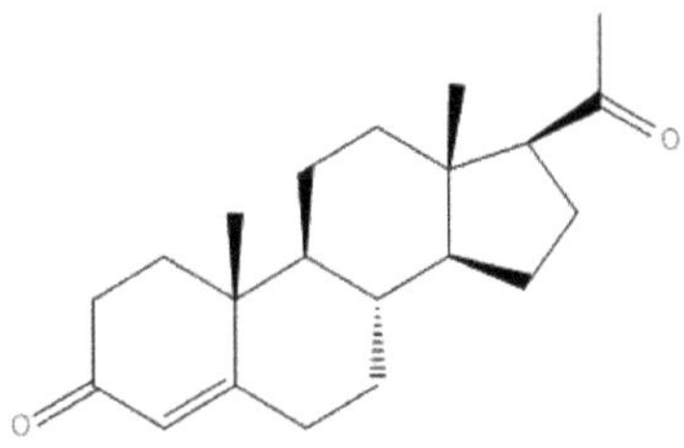

18

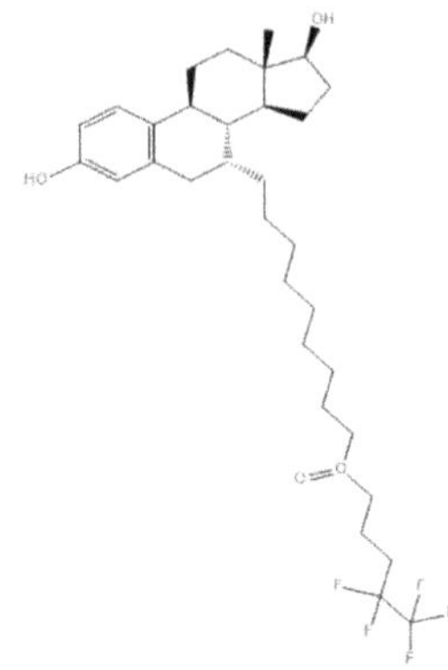

19

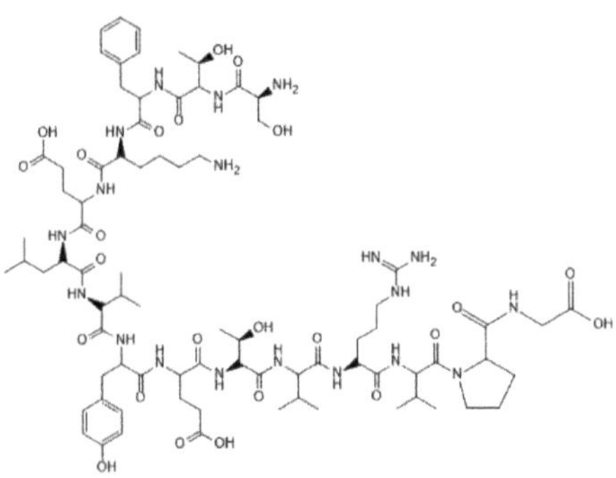

20

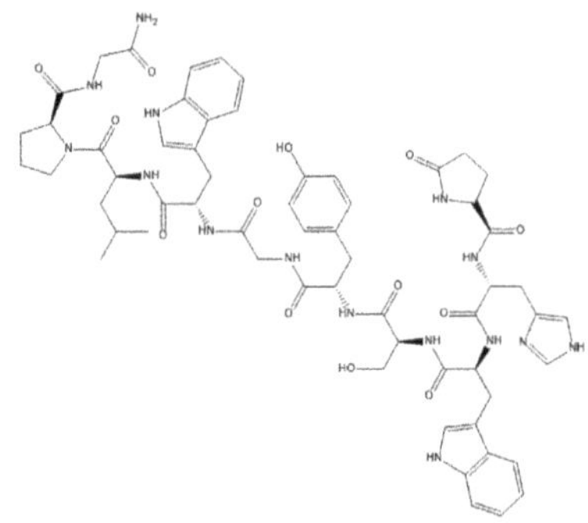

21

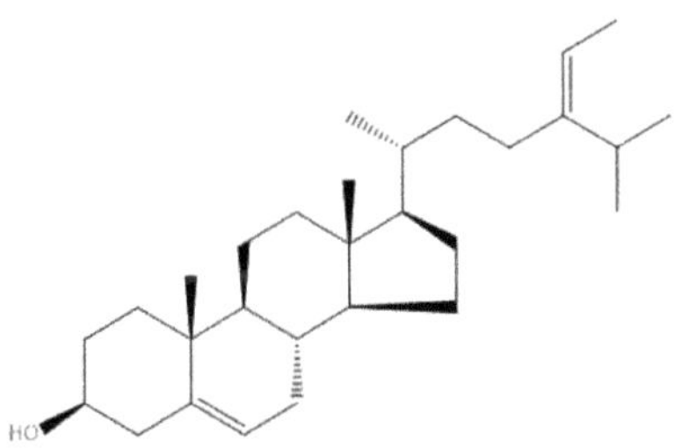

22

FIGURE 14.5 Structures of steroidal hormonal compounds: (17) testosterone, (18) oestrogen, (19) progesterone, (20) follicle-stimulating hormone, (21) luteinizing hormone, and (22) fucosterol.

chodrillasterol, and 22-ddihydro chondrillasterol produced in *Scenedesmus quadricauda* and *Chlamydomonas globose* have a great advantage in increasing female fertility and male fertility (through increased seminal fluid volume) (Okab et al. 2013), repairing damaged ovarian cells (Abdel-Aziem et al. 2018), and proliferation and differentiation of reproductive organs (Piepho et al. 2010), respectively.

TABLE 14.6
Bioactive Steroid Compounds Advantageous for Reproductive Health and Fertility

Algae species	Sample	Sterol compounds	Advantage	Reference
Spirulina Arthrospira plantensis sp.	Female albino Wister rats	–	Algae are functional in defending the hazardous chemical constituents that are exposed to multiple cells or reproductive organ, like ovarian sacs (follicle) in females, also act as an antioxidant for ovary	Yener et al. 2013
Ulva lactuca L.	12-week-old *Rattus norvegicus*	Steroid-like testosterone hormone (**17**)	Growth and maturation of gonads and its fertility	Yulistiyanto et al. 2020
Dunaliella salina	Goat	Steroid hormones such as oestrogen, progesterone, follicle-stimulating hormone, and luteinizing hormone (**18,19,20,21**)	Oestrus cycle and the reproductive cycle	Senosy et al. 2017
Spirogyra	Algal culture and rats	Sterol-enriched fraction	Cell death, reduction of reactive oxygen species generation, and nitric oxide production	L. Wang et al. 2020
Ulva clathrate	Shrimp (*Litopenaeus vannamei*)	Fucosterol and its isomer (isofucosterol) (**22**)	Helps in the transportation of fatty acid during vitellogenesis, influences the reproductive performance of shrimp and maturation of oocyte	Corral-Rosales et al. 2019

For structure of compounds **17–22** see Figure 14.5.

Other bioactive compounds synthesized from the pure culture of the same species, *Chlorella vulgaris*, is also used as a source of supplements and natural medicines (Novianti et al. 2019). Ergosterol and 7-dehydroporiferasterol obtained from the algal species *Chlamydomonas reinhardti* (Miller and Sheridan 2014) and fucosterol (Figure 14.5), stigmasterol, β-sitosterol, 28-isofucosterol, and cholesterol from the species *Ulva prolifera* (Geng et al. 2019) play a major role in the biosynthesis of hormones whereas compounds like ethylcholesterol and isofucoterol (Figure 14.5) from the species *U. australis* has an inhibitory action on the enzyme activity which has been decreased that is responsible for holding metabolic functions in the body (Li et al. 2017). Multiple species of green algae (Chlorophyta) are capable of producing numerous sterol compounds, and the synthesis of steroid compounds from them has been observed to have various benefits for reproductive health and fertility. Some of them have been described in Table 14.6.

14.5 THERAPEUTIC USE OF CHLOROPHYTA

14.5.1 Anti-cancerous Properties of Chlorophyta

Green algae, which are mostly found in the ocean, are known to produce bioactive compounds with anticancer characteristics. Astaxanthin, a good oxidant obtained from Chlorophyta species such as *Chlorococcum* spp. and *Haematococcus pluvialis* (Palozza et al. 2009), is also helpful in apoptotic activity; nigricanosides A and B compound, a type of glycolipid produced from the species *Avrainvillea nigricans*, is effective against breast cancer cells (Williams et al. 2007). In contrast, a polysaccharide (unidentified) produced from the species *Capsosiphon fluvescens* (Kwon and Nam 2007) was discovered to trigger stomach cancer cell killing. Another species, *Bryopsis* sp. (Blunt et al. 2003; Haefiner 2003; Cragg and Newman 2009; Molinski et al. 2009), and *Avrainvillea rawsonii* produce isorawsonol (Nagle et al. 2004) that shows good effect on the movement of cancerous cells.

Other bioactive chemicals such as udoteal and halimedatrial (diterpenoids), and rhypocephanal (sesquiterpenoid) are derived from the species *Udotea flabellum*, *Halimeda* spp., and *Rhipocephalus phoenix*, respectively, and show good anticancer action (Fenical and Paul 1984).

An unsaturated fatty acid compound, 3(ζ)-hydroxy-octadeca-4(*E*),6(*Z*),15(*Z*)-trienoic acid, and 3(ζ)-hydroxy-hexadeca-4(*E*),6(*Z*)-dienoic acid produced from the species *Tydemania expeditionis* demonstrate good activity against proliferating cells with the potency of IC_{50} 15 μM. (Jiang et al. 2008).

Furthermore, methanol extracts from various Chlorophyta species were effective against melanoma cells. For example, methanol extracts from *Enteromorpha intestinalis* and *Rhizoclonium ripariums* have antiproliferative activity against cervical cancer cells, *Udotea flabellum* and *U. conglutinate* are effective against human melanoma cell lines, and *Caulerpa* spp. have been observed to act against tumour cell migration, which is dependent on the suppression of the concentration of metastatic MDA-MB-231 cells (Moo-Puc et al. 2009; Paul et al. 2013). Furthermore, *Cladophoropsis vaucheriaeformis* has been shown to have tumorigenic properties against murine lymphoid leukaemia L1210 cells, and the species *Chaetomorpha compressa* has been shown to have better anti-cancerous properties against breast carcinoma cells in humans as well as human colon cancer cells HCT-116 (Cinar et al. 2019; Acharya et al. 2020). In conclusion, green algae are an important element of nature because they produce a variety of physiologically active chemicals with anticancer properties.

14.5.2 Antioxidant Properties

According to studies, green algae have a strong antioxidant defence system, allowing them to resist oxidative damage even when exposed to reactive oxygen species. Several bioactive metabolites produced by green algae are known to protect cells against a variety of illnesses and oxidative damage that occurs during the ageing process (Kelman et al. 2012). Antioxidants derived from algae are divided into two categories: those that dissolve in water and those that dissolve in fat (Zouaoui et al. 2017).

Chaetomorf aerea, *C. brachygona*, *C. linum*, and *C. crassa* are widely known for their antioxidant activity at low inhibitory concentrations of IC_{50} (1.484 ± 0.168 mg/ml) (Farasat et al. 2013). Astaxanthin is a bioactive chemical found in a few species of *Haematococcus* seaweeds as well as several Chlorella species. *Chlamydomonas nivalis* is well known for its antioxidant qualities (Chen 2019) and *Ulva fasciata* and *U. lactuca* exhibit antioxidant capabilities for sesquiterpenoids and flavonoids, respectively (Meenakshi et al. 2009; Chakraborty et al. 2010; Roy 2020).

14.5.3 Antiviral Properties

Carrageenans, alginates, agarans, laminarans, fucans, galactans, and other sulphated polysaccharides derived from seaweeds have shown resistance to virus growth by inhibiting the replication of enveloped viruses such as the human immunodeficiency virus, dengue virus, influenza A and B, and others (Damonte et al. 1994, 2004; Ponce et al. 2003).

Sulpholipids, metabolically active chemicals produced in the alga *U. fasciata*, can suppress the growth and replication of viruses such as herpes simplex virus type 1 (HSV-1) (El Baz et al. 2013). Many additional green seaweeds are known to generate antiviral compounds that operate differently in different types of viruses. Table 14.7 lists some of these along with their functions.

TABLE 14.7
List of Antiviral Compounds from Green Seaweeds and Their Functions

Species	Antiviral compound	Function	Reference
Ulva lactuca	*N*-Palmitoyl-2-amino 1,3,4,5-tetyrahydroxyoactadecane, sphingosine	Protects from Semliki Forest virus (SFV)	Garg et al. 1992
Halimeda tuna	Halitunal – type of diterpene	Protects from murine coronavirus A59	Koehn et al. 1991
Caulerpa racemose	Sulquinovosyl diacylglycerol	Protects from herpes simplex virus 2 (HSV-2)	H. Wang et al. 2007
Monostroma latissinum	Rhamnan sulphate	HCMV, HIV-1, HSV-1	Lee et al. 1999
Caulerpa	Caulerpin	Protects against SARS CoV-2 by destabilizing spikes protein	Ahmed et al. 2020

14.5.4 Antibacterial Properties

Antimicrobial chemicals work by destroying the phospholipid bilayer of the cell membrane, degrading enzyme activity, and disrupting the DNA of microorganisms (Abu-Ghannam and Rajauria 2013). Secondary metabolically active compounds found in many seaweeds, such as quinones, tannins, flavonols, flavones, flavonoids, and phlorotannins, hinder membrane function by affecting cell membrane permeability and integrity, resulting in cell death (Bajpai et al. 2008; Rosa et al. 2019).

Most of the algal abstracts isolated and targeted for their antibacterial properties are composed of oral microbes. Bromophenols isolated from the algae species *Avrainvillea* have been shown to inhibit *Bacillus subtilis* and *Staphylococcus aureus* as well as *Pseudomonas aeruginosa*, *Serratia marcesens*, *Candida albicans*, and *Escherichia coli* (Jesus et al. 2019). For example, metabolically bioactive compounds like GLA (gamma-linolenic acid) and SA (stearidonic acid) synthesized in *Ulva linza* have a strong antibacterial potential against *Porphyromonas gingivalis* and *Porphyromonas intermedia*, with minimum inhibitory concentration (MIC) values of 9.76 g/ml and 39.06 g/ml, respectively (Park et al. 2013). Other species, such as *Ulva lactuca*, have been shown to kill *E. coli* bacteria, while *Ascophyllum nodosum* has a strong antibacterial activity against *Micrococcus luteus* and *Brochothrix thermosphacta*. Fatty acids produced by the *Utricularia rigida* species are also beneficial in the development of antibacterial compounds that are effective against human diseases caused by marine microbes (Ismail et al. 2018).

14.5.5 Anti-inflammatory, Antilipidaemic, and Hypocholesterolaemic Activity

In general, macrophages are the cells that release inflammatory factors when the endotoxic component of the cell wall, lipopolysaccharides, triggers them. These lipopolysaccharides activate the intracellular signalling cascade, which produces proinflammatory cytokines (Wu et al. 2016).

Green seaweeds are the most outstanding examples for synthesizing a wide range of secondary metabolic compounds with human-beneficial characteristics. Compounds like 2-(20, 40-dibromophenoxy)-4,6-dibromoanisole, discovered from *Cladophora vagabunda* (formerly *Cladophora fascicularis*), demonstrate exceptional anti-inflammatory potential by hindering the productivity of microorganisms like *S. aureus*, *B. subtilis*, and *E. coli* (Kobayashi and Ishabashi 1993). Other algae species, their bioactive chemicals, and bioactivities are included in Table 14.8.

Furthermore, bioactive substances such as ethanol extract from the edible seaweed *Caulerpa racemosa* exhibit an effective hypocholesterolaemic and hypolipidaemic response by absorbing

TABLE 14.8
List of Bioactive Compounds from Green Seaweeds, Their Bioactivities and Modes of Action

Species	Compound name	Bioactivity	Mode of action	Reference
Ulva lactuca	Sulphated polysaccharides	Anti-inflammatory, antinociceptive activity	Decreases mouse oedema after 4 days when experimented in an animal	Margret et al. 2009
Caulerpa racemosa	Sulphated polysaccharides	Anti-inflammatory activity	Interacts with secretory phospholipase A2 obtained from *Crolatus durissus* venum causing an increase in bactericidal and enzymatic activity of sPLA2 in Sprague Dawley (SD) rats	Pires et al. 2013
Enteromorpha prolifera	Sulphated polysaccharides	Antihyperlipidaemic activity	EPPs reduces body weight gain, liver weights, liver TG, cholesterol level and liver TC	Teng et al. 2013
Ulva pertusa	Ulvans	Antihyperlipidaemic activity	Reduces low-density lipoprotein (LDL)-cholesterol and serum total cholesterol in rats	Yu et al. 2003
Ulva pertusa	High sulphate content, ulvans, and acetylated ulvans	Strong antihyperlipidaemic activity	Decrease the concentration of LDL and triglyceride cholesterol in mice	Qi et al. 2012a, 2012b

molecules such as cholesterol and faecal material by eliminating it from the digestive system (Kothawala et al. 2018).

14.5.6 Anticoagulant Properties

Marine Chlorophyta has been noted as a significant source of sulphate polysaccharides in abundance with more effective anticoagulant properties than those obtained from Phaeophytes and Rhodophytes (commonly termed as sulphated fucoidans and carrageenan) (Maeda et al. 1991; Shanmugam et al. 2001).

Anticoagulants are mostly made from algae belonging to the *Codium* species, which block thrombin generation. Sulphated polysaccharides isolated from members of the same species, such as *C. indicum*, *C. dwarkense*, *C. geppi*, and *C. tomentosum*, are more effective in inhibiting thrombin than other compounds, such as heparin or dermatan sulphate; *C. dwarkense* is known to have the most blood anticoagulant property (Hayakawa et al. 2000; Shanmugam et al. 2002).

When purified by chromatography, the sulphated polysaccharides extract from *M. nitidum* showed very strong anticoagulant action, with a potency six times higher than ordinary heparin (Maeda et al. 1991). These sulphated polysaccharides also serve as a potent thrombin inhibitor via heparin cofactor II and a weak coagulating factor Xa inhibitor via antithrombin III potentiation (Mao et al. 2009). Other studies have found that *Monostroma latissimum* (Zhang et al. 2008), *Enteromorpha linza*, and *Enteromorpha clathrata* (Qi et al. 2012c; X. Wang et al. 2013) exhibit effective anticoagulant properties in thrombin time and activated partial thromboplastin time (APTT). Four fractions of sulphated hetergalactan with different sulphate/sugar ratios derived from *Caulerpa cupressoides* var. *flabellate* and three fractions of crude sulphated polysaccharides derived from *Caulerpa cupressoides* var. *lycopodium* were also found to be effective in APTT activities (Rodrigues et al. 2011; Costa et al. 2012). Marine green seaweeds are sources of sulphated polysaccharides with anticoagulant characteristics that are used in medicine. Researchers are constantly looking for more bioactive chemicals like these.

14.5.7 Use in Cosmetics

Green seaweeds have been targeted as a significant source for various cosmetic colourants, protecting agents, anti-stretch mark cream, body ointments, skin moisturizers, masks, creams, sunscreens, balms, gels, immune stimulants, whitening agents, oral care, anti-allergic, anti-septic, anti-acne, and other therapeutic purposes due to the presence of numerous metabolically active substances (Pereira 2018; Jesumani et al. 2019).

14.5.8 Natural Colouring Agent in Mojosari Ducks

Recent studies on freshwater green algae have revealed that freshwater Chlorophyta can be used as a natural pigmenting agent in laying ducks. When around 8% of green algae was added to ducks' diet, the yolk colour value was determined to be the best at 15.0, with no negative impact on duck productivity (Indarsih et al. 2015).

14.6 CONCLUSION AND FUTURE PERSPECTIVES

Green algae have been discovered to be a rich source of secondary compounds such as proteins, carotenoids, fatty acids, sterols and steroid compounds, hormones, polysaccharides, primarily sulphated polysaccharides, and polyphenols, among others, paving the way for further research into pharmaceutical, nutraceutical, and other therapeutic applications. More research is being done to look into other biologically active chemicals responsible for various medicinal and other functions. Secondary metabolites generated in green seaweeds, such as sulphated polysaccharides, ulvans, and polyphenolic chemicals, have proved crucial in various sectors. The various categories of secondary metabolites related to chlorophytes of various species and genera (primarily *Ulva*, *Caulerpa*, *Tydermania*, *Penicillus*, *Avrainvillea*, *Codium*, and others) and their applications as antioxidants, antimicrobials, anti-inflammatory, anticancer, and cosmetic substances have been discussed in this chapter. Furthermore, steroidal substances have been described as a source of biosynthesis for various hormones relevant to female fertility and reproductive health. Marine algae, in particular, have been discovered to be a rich source of anticoagulant chemicals, which could be very useful in medicine. They are also a good food source with medicinal uses and several additional health advantages. Various new ways to modify large-scale manufacturing and isolate bioactive compounds from green seaweeds are being developed to address the hurdles encountered during their biosynthesis. The protein tyrosine phosphatase 1B (PTP1B) mode of action is currently being studied to develop a new curative class for treating cancerous cells. Although, rather than isolating active molecules responsible for their capabilities, researchers are currently concentrating on observing the potential of biological compounds from crude extracts of macroalgae. Once the discovery of new bioactive substances reaches a peak, it is clear that green algae would be in great demand in the market for their medicinal and therapeutic applications. However, there are still a lot more potential applications for these chlorophytes to be discovered.

REFERENCES

Abdel-Aziem, S.H., H.A.M. Abd El-Kader, F.M. Ibrahim, H.A. Sharaf, and A.I. El makawy. 2018. Evaluation of the alleviative role of *Chlorella vulgaris* and *Spirulina platensis* extract against ovarian dysfunctions induced by monosodium glutamate in mice. *Journal of Genetic Engineering and Biotechnology* 16:653–660.

Abu-Ghannam, N., and G. Rajauria. 2013. Antimicrobial activity of compounds isolated from algae. In: H. Domínguez (ed) *Functional Ingredients from Algae for Foods and Nutraceuticals*. Sawston, UK: Woodhead Publishing, 287–306.

Acharya, D., S. Satapathy, P. Somu, U.K. Parida, and G. Mishra. 2020. Apoptotic effect and anticancer activity of biosynthesized silver nanoparticles from marine algae *Chaetomorpha linum* extract against human colon cancer cell HCT-116. *Biological Trace Element Research* 199:1812–1822.

Aditya, T., G. Bitu, and E.E. Mercy. 2016. The role of algae in pharmaceutical development. *Research & Reviews: Journal of Pharmaceutics and Nanotechnology* 4:82–89.

Ahmed, S.A., D.A. Abdelrheem, H.R.A. El-Mageed, H.S. Mohamed, A.A. Rahman, K.N.M. Elsayed, and S.A. Ahmed. 2020. Destabilizing the structural integrity of COVID-19 by caulerpin and its derivatives along with some antiviral drugs: an in-silico approaches for a combination therapy. *Structural Chemistry* 31:2391–2412.

Alamsjah, M.A., S. Hirao, F. Ishibashi, and Y. Fujita. 2005. Isolation and structure determination of algicidal compounds from *Ulva fasciata*. *Bioscience, Biotechnology and Biochemistry* 69:2186–2192.

Ali, L., A.L. Khan, L. Al-Kharusi, J. Hussain, and A. Al-Harrasi. 2015. New α-glucosidase inhibitory triterpenic acid from marine macro green alga *Codium dwarkense* Boergs. *Marine Drugs* 13:4344–4356.

Ali, M.S., M. Saleem, R. Yamdagni, and M.A. Ali. 2002. Steroid and antibacterial steroidal glycosides from marine green alga *Codium iyengarii* Borgesen. *Natural Product Reports* 16:407–413.

Ali, M.S., M. Saleem, R. Yamdagni, and M.A. Ali. 2002. Steroid and antibacterial steroidal glycosides from marine green alga *Codium iyengarii* Borgesen. *Natural Product Reports* 16:407–413.

Alves, A., S.G. Caridade, J.F. Mano, R.A. Sousa, and R.L. Reis. 2010. Extraction and physicochemical characterization of a versatile biodegradable polysaccharide obtained from green algae. *Carbohydrate Research* 345:2194–2200.

Andersen, R.J. and O. Taglialatela-Scafati. 2005. Avrainvilloside, a 6-deoxy-6-aminoglucoglycerolipid from the green alga *Avrainvillea nigricans*. *Journal of Natural Products* 68:1428–1430.

Antony, T. and Chakraborty, K., 2019. Pharmacological properties of seaweeds against progressive lifestyle diseases. *Journal of Aquatic Food Product Technology*, 28(10):1092-1104.

Araujo, L. and M.J. Stadnik. 2013. Cultivar-specific and ulvan-induced resistance of apple plants to glomerella leaf spot are associated with enhanced activity of peroxidases. *Acta Scientific Agriculture* 35:287–293.

Athbi, A., A. Muhsin, D. Salman, and A. Al-Lafta Noory. 2009. Extraction and identification of total proteins and some of fatty acids from alga *Cladophora crispata*. *Mesopotamian Journal of Marine Science* 2:140–147.

Athbi, A., A. Muhsin, D. Salman, and A. Al-Lafta Noory. 2011. Determination of total proteins and some amino acids in green alga *Chlorella vulgaris*. *Journal of Thi-Qar Science* 5:224–234.

Bajpai, V.K., A. Rahman, N.T. Dung, M.K. Huh and S.C. Kang. 2008. In vitro inhibition of food spoilage and foodborne pathogenic bacteria by essential oil and leaf extracts of *Magnolia liliflora* Desr. *Journal of Food Science* 3:M314–M320.

Balboa, E.M., E. Conde, A. Moure, E. Falqué, and H. Domínguez. 2013. In vitro antioxidant properties of crude extracts and compounds from brown algae. *Food Chemistry* 138:1764–1785.

Bhadury, P. and P.C. Wright. 2004. Exploitation of marine algae: biogenic compounds for potential anti fouling application. *Planta* 219:561–578.

Blunt, J.W., B.R. Copp, M.H.G. Munro, P.T. Northocte, and M.R. Prinsep. 2003. Marine natural products. *Natural Product Reports* 20:1–48.

Blunt, J.W., B.R. Copp, M.H.G. Munro, P.T. Northcote, and M.R. Prinsep. 2006. Marine natural products. *Natural Product Reports* 23:26–78.

Cengiz, S., L. Cavas, K. Yurdakoc, and G. Pohnert. 2011. The sesquiterpene caulerpenyne from *Caulerpa* spp. is a lipoxygenase inhibitor. *Marine Biotechnology* 13:321–326.

Chakraborty, K., A.P. Lipton, R. Paulraj, and R.D. Chakraborty. 2010. Guaiane sesquiterpenes from seaweed *Ulva fasciata* Delile and their antibacterial properties. *European Journal of Medicinal Chemistry* 45:2237–2244.

Chen, H., T. Li and Q. Wang. 2019. Ten years of algal biofuel and bioproducts: gains and pains. *Planta* 249, 195–219. Doi: 10.1007/s00425-018-3066-8

Cho, M.L., I.J. Kang, M.H. Won, H.S. Lee, and S.G. You. 2010. Antioxidant activities of ethanol extracts and their solvent partitioned fractions from various green macroalgae. *Journal of Medicinal Food* 13:232–1239.

Cho, M., G.M. Park, S.N. Kim, T. Amna, S. Lee and W.S. Shin. 2014. Glioblastoma-specific anticancer activity of pheophorbide a from the edible red seaweed Grateloupia elliptica. *Journal of Microbiology and Biotechnology*, 24(3), 346–353

Ciancia, M., J. Alberghina, A.P. Ximena, H. Benavides, F. Leliaert, H. Verbruggen, and J.E. Manuel Estevez. 2012. Characterization of cell wall polysaccharides of the coenocytic green seaweed *Byropsis plumose* (Bropsidaceae, Chlorophyta) from the Argentine cost. *Journal of Phycology* 48:326–335.

Cinar, E., E. Taskin, D. Tasdemir, E. Ozkale, U. Grienke, and D. Firsova. 2019. Anti-acetylcholinesterase, antiprotozoal and cytotoxic activities of some Turkish marine algae. *Fresenius Environmental Bulletin* 28:3991–4000.

Corral-Rosales C., D. Ricque-Marie, L.E. Cruz-Suárez, O. Arjona, and E. Palacios. 2019. Fatty acids, sterols, phenolic compounds, and carotenoid changes in response to dietary inclusion of *Ulva clathrata* in shrimp. *Litopenaeus Vannamei* Broodstock. *Journal of Applied Phycology* 31:4009–4020.

Costa, M.S.S.P., L.S. Costa, S.L. Cordeiro, J. Almeida-Lima, N. Dantas-Santos, K.D. Magalhaes, D.A. Sabry, I.R. Lopes Albuquerque, M.R. Pereira, E.L. Leite, and H.A.O Rocha. 2012. Evaluating the possible anticoagulant and antioxidant effects of sulfated polysaccharides from the tropical green alga *Caulerpa cupressoides* var. *flabellata*. *Journal of Applied Phycology* 24:1159–1167.

Cragg, G.M. and D.J. Newman. 2009. Nature: a vital source of leads for anticancer drug development. *Phytochemistry Reviews* 8:313–331.

Dahms, H.U. and S. Dobretsov. 2017. Antifouling compounds from marine macroalgae. *Marine Drugs* 15:265.

Damonte, E.B., M.C. Matulewicz, and A.S. Cerezo. 2004. Sulfated seaweed polysaccharides as antiviral agents. *Current Medicinal Chemistry* 11:2399–2419.

Damonte, E., J. Neyts Pujol, C.A. Snoeck, R. Andrei, G. Ikeda, S. Witvrouw, M. Reymen, D. Haines, H. Matulewicz, M.C. Cerezo, C.E. Coto, and E. De Clerco. 1994. Antiviral activity of a sulphated polysaccharide from the red seaweed *Nothogenia fastigiate*. *Biochemical Pharmacology* 47: 2187–2192.

De Clerck, O., K.A. Bogaert, and F. Leliaert. 2012. Diversity and evolution of algae. 778 Primary endosymbiosis. *Advances in Botanical Research* 64:55–86.

Ding, L., G.E. Chan, J. Cheng, A. Xia, R. O'Shea, A.J. Guneratnam, and J.D. Murphy. 2018. Assessment of continuous fermentative hydrogen and methane co-production using macro- and micro-algae with increasing organic loading rate. *Energy* 151:760–770.

Dittami, S.M., S. Heesch, J.L. Olsen, and J. Collen. 2017. Transitions between marine and freshwater environments provide new clues about the origins of multicellular plants and algae. *Journal of Phycology* 53:731–745.

Dorta, E., J. Darias, A. San Martín, and M. Cueto. 2002. New prenylated bromoquinols from the green alga *Cymopolia barbata*. *Journal of natural Products* 65:329–333.

Dring, M.J. 1982. *The Biology of Marine Plants*. London: Edward Arnold publishers Limited.

El Baz, F.K., G.S. El Baroty, H.H.A. Abd El Baky, O.I. Abd El-Salam, and E.A. Ibrahim. 2013. View of structural characterization and biological activity of sulfolipids from selected marine algae. *Grasas y aceites* 64:561–571.

Elnabris, K.J., A.A. Elmanama, and W.N. Chihadeh. 2013. Antibacterial activity of four marine seaweeds collected from the coast of Gaza Strip, Palestine. *Mesopotamian Journal of Marine Science* 28:81–92.

Fang, Z., S.Y. Jeong, H.A. Jung, J.S. Choi, B.S. Min, and M.H. Woo. 2012. Capsofulvesins A–C, cholinesterase inhibitors from *Capsosiphon fulvescens*. *Chemical and Pharmaceutical Bulletin* 60:1351–1358.

Farasat, M., R.A. Khavari-Nejad, S.M.B. Nabavi, and F. Namjooyan. 2013. Antioxidant properties of some filamentous green algae (*Chaetomorpha* genus). *Brazilian Archives of Biology and Technology* 56:921–927.

Fenical, W. and V.J. Paul. 1984. Antimicrobial and cytotoxic terpenoids from tropical green algae of the family Udoteaceae. *Hydrobiologia* 116:135–114.

Gallimore, W.A., T. Sambo, and T. Campbell. 2009. Debromocymopolone from the green alga, *Cymopolia barbata*. *Journal of Chemical Research* 3:160–161.

Garg, H.S., M. Sharma, D.S. Bhakuni, B.N. Pramanik, and A.K. Bose. 1992. An antiviral sphingosine derivative from the green alga *Ulva fasciata*. *Tetrahedron Letters* 33:1641–1644.

Geng, H., R. Yu, Q. Zhang, T. Yan, F. Kong, and M. Zhou, 2019. Tracing the settlement region of massive floating green algae in the Yellow Sea. *Journal of Oceanology and Limnology* 37:1555–1565.

Goh, S.H., F.M. Yusoff, and S.P.A. Loh. 2010. A comparison of the antioxidant properties and total phenolic content in a diatom *Chaetoceros* sp. and a green microalga, *Nannochloropsis* sp. *Journal of Agricultural Science* 2:123–130.

Goiris, K., K. Muylaert, I. Fraeye, I. Foubert, J. De Brabanter, and L. De Cooman. 2012. Antioxidant potential of microalgae in relation to their phenolic and carotenoid content. *Journal of Applied Phycology* 24:1477–1486.

Guerriero, A., A. Meinesz, M. D'Ambrosio, and F. Pietra. 1992. Isolation of toxic and potentially toxic sesqui- and monoterpenes from the tropical green seaweed *Caulerpa taxifolia* which has invaded the region of Cap Martin and Monaco. *Helvetica Chimica Acta* 75:689–695.

Guiry, M.D. 2012. How many species of algae are there? *Journal of Phycology* 48:1057–1063.

Güven, K.C., A. Percot, and E. Sezik. 2010. Alkaloids in marine algae. *Marine Drugs* 8:269–284.

Haefner, B. 2003. Drugs from the deep: marine natural products as drug candidates. *Drug Discovery Today* 8:536–544.

Hajimahmoodi, M., M.A. Faramarzi, N. Mohammadi, N. Soltani, M.R. Oveisi, and N. Nafissi-Varcheh. 2009. Evaluation of antioxidant properties and total phenolic contents of some strains of microalgae. *Journal of Applied Phycology* 22:43–50.

Hayakawa, Y., T. Hayashi, J.B. Lee, P. Srisomporn, M. Maeda, T. Ozawa, and N. Sakuragawa. 2000. Inhibition of thrombin by sulfated polysaccharides isolated from green algae. *Biochimica et Biophysica Acta (BBA) – Protein Structure and Molecular Enzymology* 1543:86–94.

Indarsih B., A. Asnawi and D.K. Purnamasari. 2015. Freshwater green algae (Chlorophyta) as a natural pigment for Mojosari ducks. *Journal of the Indonesian Tropical Animal Agriculture* 40:45–51.

Ishibashi, Y., Y. Nagamatsu, T. Miyamoto, N. Matsunaga, N. Okino, K. Yamaguchi, and M. Ito. 2014. A novel ether-linked phytol-containing digalactosyl glycerolipid in the marine green alga, *Ulva pertusa. Biochemical and Biophysical Research Communications* 452:873–880.

Ismail, A., L. Ktari, Y.B.R. Romdhane, B. Aoun, S. Sadok, A. Boudabous, and M. El Bour. 2018. Antimicrobial fatty acids from green alga *Ulva rigida* (Chlorophyta). *BioMed Research International* 2018:3069595.

Janczyk, P., M. Langhammer, U. Renne, V. Guiard, and W.B. Souffrant. 2006 Effect of feed supplementation with *Chlorella vulgaris* powder on mice reproduction. *Archiva Zootechnica* 9:122–344.

Jayalakshmi, L., J. Gomathy, J. Jayanthi, and M.G. Raghunathan. 2021. Phytochemical analysis, in vitro antioxidant and antibacterial efficacy of aqueous extracts of seaweeds *Enteromorpha intestinalis* (L.) and *Gracilaria edulis* (Gmelin) (Silva) collected from Pulicat lake, Tamil Nadu. *Uttar Pradesh Journal of Zoology* 42:66–73.

Jesumani, V., H. Du, M. Aslam, P. Pei, and N. Huang. 2019. Potential use of seaweed bioactive compounds in skincare—a review. *Marine Drugs* 17:688.

Jesus, A., M. Correia-da-Silva, C. Afonso, M. Pinto, and H. Cidade. 2019. Isolation and potential biological applications of haloaryl secondary metabolites from macroalgae. *Marine Drugs* 17:73.

Ji, X., H. Li, J. Zhang, H. Saiyin, and Z. Zheng. 2019. The collaborative effect of *Chlorella vulgaris-Bacillus licheniformis* consortia on the treatment of municipal water. *Journal of Hazardous Materials* 365:483–493.

Jiang, H.P., B.B. Gao, W.H. Li, M. Zhu, C.F. Zheng, Q.S. Zheng, and C.H. Wang. 2013. Physiological and biochemical responses of *Ulva prolifera* and *Ulva linza* to cadmium stress. *Science World Journal* 2013;289537.

Jiang, R.W., M.E. Hay, C.R. Fairchild, J. Prudhomme, K. Le Roch, W. Aalbersberg, and J. Kubanek. 2008. Antineoplastic unsaturated fatty acids from Fijian macroalgae. *Phytochemistry* 69:2495–2500.

John, D.M. 1994. Biodiversity and conservation: an algal perspective. *The Phycologist* 38:3–15.

Kelman, D., E.K. Posner, K.J. McDermid, K.T. Nicole, R.W. Patrick, and D.W. Anthony. 2012. Antioxidant activity of Hawaiian marine algae. *Marine Drugs* 10:403–416.

Khan, M.I., J.H. Shin, and J.D. Kim. 2018. The promising future of microalgae: current status, challenges, and optimization of a sustainable and renewable industry for biofuels, feed, and other products. *Microbial Cell Factories* 17:36.

Kladi, M., C. Vagias, and V. Roussis. 2004. Volatile halogenated metabolites from marine red algae Rhodophyceae. *Phytochemistry Reviews* 3:337–366.

Kobayashi, I. and M. Ishibashi. 1993. Bioactive metabolites of symbiotic marine microorganisms. *Chemical Reviews* 93:1753–1769.

Koehn, F.E., S. P. Gunasekera, D.N. Niel and S.S. Cross. 1991. Halitunal, an unusual diterpene aldehyde from the marine alga *Halimeda tuna*. *Tetrahedron Letter* 32:169–172.

Kothalawala, S.G., D.U. Chathurangi, and L.N. Yatiwella. 2018. Brief overview of bioactive compounds in seaweeds, their properties and practical applications in functional foods. *International Journal of Science and Research* 8:8.

Kozlova, T.A., B.P. Hardy, and D.B. Levin. 2020. Effect of fish steroids 17β-estradiol and 17, 20β-dihydroxy-4-pregnen-3-one on growth, accumulation of pigments, and fatty acid profiles in the microalgae *Scenedesmus quadricauda* (CPCC-158). *Renewable Energy* 148:798–806.

Kumar, M., P. Kumari, N. Trivedi, M.K. Shukla, V. Gupta, C. Reddy, and B. Jha. 2011. Minerals, PUFAs and antioxidant properties of some tropical seaweeds from Saurashtra coast of India. *Journal of Applied Phycology* 23:797–810.

Kumar, Y., A. Tarafdar, D. Kumar, and P.C. Badgujar. 2019. Effect of Indian brown seaweed *Sargassum wightii* as a functional ingredient on the phytochemical content and antioxidant activity of coffee beverage. *Journal of Food Science and Technology* 56:4516–4525.

Kwon, M.J. and T.J. Nam. 2007. A polysaccharide of the marine alga *Capsosiphon fulvescens* induces apoptosis in AGS gastric cancer cells via an IGF-IR-mediated PI3 K/Akt pathway. *Cell Biology International* 31:768–775.

Lahaye, M. and B. Ray. 1996. Cell-wall polysaccharides from the marine green alga *Ulva 'rigida'* (Ulvales, Chlorophyta)—NMR analysis of ulvan oligosaccharides. *Carbohydrate Research* 283:161–173.

Lahaye, M., M. Brunel, and E. Bonnin. 1997. Fine chemical structure analysis of oligosaccharides produced by an ulvanlyase degradation of the water-soluble cell-wall polysaccharides from *Ulva* sp. (Ulvales, Chlorophyta). *Carbohydrate Research* 304:325–333.

Leal, M.C., M.H.G. Munro, J.W. Blunt, J. Puga, B. Jesus, R. Calado, R. Rosa, and C. Madeira. 2013. Biogeography and biodiscovery hotspots of macroalgal marine natural products. *Natural Product Reports* 30:1380–1390.

Lee, J.B., K. Hayashi, T. Hayashi, U. Sankawa, and M. Maeda. 1999. Antiviral activities against HSV-1, HCMV, and HIV-1 of rhamnan sulfate from *Monostroma latissimum*. *Planta Medica* 65:439–441.

Leliaert, F. 2019. Green algae: Chlorophyta and Streptophyta. In: T.M. Schmidt (ed) *Encyclopedia of Microbiology*. Oxford: Academic Press, 457–468.

Leliaert, F., D.R. Smith, H. Moreau, M.D. Herron, H. Verbruggen, C.F. Delwiche, and O. De Clerck. 2012. Phylogeny and molecular evolution of the green algae. *Critical Reviews in Plant Sciences* 31:1–46.

Lever, J., R. Brkljaca, G. Kraft, and S Urban. 2020. Natural products of marine macroalgae from South Eastern Australia, with emphasis on the Port Phillip Bay and heads regions of Victoria. *Marine Drugs* 18:142.

Li, G.L., W.J. Guo, G.B. Wang, R.R. Wang, Y.X. Hou, K. Liu, Y. Liu, and W. Wang. 2017. Sterols from the green alga *Ulva Australis*. *Marine Drugs* 15:299.

Lincoln, R.A., K. Strupinski, and J.M. Walker. 1991. Bioactive compounds from algae. *Life Chemistry Reports* 8:97–183.

Liu, A.H., D.Q. Liu, T.J. Liang, X.Q. Yu, M.T. Feng, L.G. Yao, Y. Fang, B. Wang, L.H. Feng, M.X. Zhang, et al. 2013. Caulerprenylols A and B, two rare antifungal prenylated para-xylenes from the green alga *Caulerpa racemosa*. *Bioorganic & Medicinal Chemistry Letters* 23:2491–2494.

Liu, D.Q., S.C. Mao, X.Q. Yu, L.H. Feng, and X.P. Lai. 2012. Caulerchlorin, a novel chlorinated bisindole alkaloid with antifungal activity from the Chinese green alga *Caulerpa racemosa*. *Heterocycles* 85:661–666.

Lorensi, G.H., R.S. Oliveira, A.P. Leal, C.G.M. Zanatta, Y.C. de Almeida Barreto, M.E. Rosa, V.P. de Brum, R.C.J. Brito, V.F. de Carvalho, A.B. Pereira, et al. 2019. Entomotoxic activity of *Prasiolacrispa* (Antarctic algae) in *Nauphoeta cinerea* cockroaches: identification of main steroidal compounds. *Marine Drugs* 17:5.

Maeda, M., T. Uehara, N. Harada, M. Sekiguchi, and A. Hiraoka. 1991. Heparinoid-active sulfated polysaccharide from *Monostroma-nitidum* and their distribution in the Chlorophyta. *Phytochemistry* 30:3611–3614.

Mao, S.C., D.Q. Liu, X.Q. Yu, and X.P. Lai. 2011. A new polyacetylene fatty acid and other secondary metabolites from the Chinese green alga *Caulerpa racemosa* (Caulerpaceae) and their chemotaxonomic significance. *Biochemical Systematics and Ecology* 39:253–257.

Mao, S.C., Y.W. Guo, and X. Shen. 2006. Two novel aromatic valerenane-type sesquiterpenes from the Chinese green alga *Caulerpa taxifolia*. *Bioorganic & Medicinal Chemistry Letters* 16:2947–2950.

Mao, W., H. Li, Y. Li, H. Zhang, X. Qi, H. Sun, Y. Chen, and S. Guo. 2009. Chemical characteristic and anti-coagulant activity of the sulfated polysaccharide isolated from *Monostroma latissimum* (Chlorophyta). *International Journal of Biological Macromolecules* 44:70–74.

Margret, R.J., S. Kumaresan, and S. Ravikumar. 2009. A preliminary study on the anti-inflammatory activity of methanol extract of *Ulva lactuca* in rat. *Journal of Environmental Biology* 30:899–902.

Máximo, P., L.M. Ferreira, P. Branco, P. Lima, and A. Lourenço. 2018. Secondary metabolites and biological activity of invasive macroalgae of southern Europe. *Marine Drugs* 16:265.

Meenakshi, S., D.M. Gnanambigai, S.T. Mozhi, M. Arumugam, and T. Balasubramanian. 2009. Total flavonoid and in vitro antioxidant activity of two seaweeds of Rameshwaram Coast. *Global Journal of Pharmacology* 3:59–62.

Mehra, R., S. Bhushan, F. Bast, and S. Singh. 2019. Marine macroalga Caulerpa: role of its metabolites in modulating cancer signaling. *Molecular Biology Reports* 46:3545–3555.

Molinski, T., D.S. Dalisay, L.L. Sarah, and J.P. Saludes. 2009. Drug development from natural products. *Nature Reviews Drug Discovery* 8:69–85.

Moo-Puc, R., D. Robledo, and Y. Freile-Pelegrín. 2009. In vitro cytotoxic and antiproliferative activities of marine macroalgae from Yucatán, Mexico. Actividad citotóxica y antiproliferativa in vitro de macroalgas marinas de Yucatán, México. *Ciencias Marinas* 35:345–358.

Morales, M., P. Collet, L. Lardon, L. Hélias, J.P. Steyer, O. Bernard. 2019. Lifecycle assessment of microalgal-based biofuel. In: Pandey, A., J.-S. Chang, C.R. Soccol, D.-J. Lee, Y. Chisti (eds) *Biofuels from Algae*, 2nd edn. Amsterdam: Elsevier, 507–550.

Mouritsen, O.G., L.A. Bagatolli, L. Duelund, O. Garvik, J.H. Ipsen, and A.C. Simonsen. 2017. Effects of seaweed sterols fucosterol and desmosterol on lipid membranes. *Chemistry and Physics of Lipids* 205:1–10.

Nagappan, T. and C.S. Vairappan. 2014. Nutritional and bioactive properties of three edible species of green algae, genus *Caulerpa* (Caulerpaceae). *Journal of Applied Phycology* 26:1019–1027.

Nagle, D.G., Y.D. Zhou, F.D. Mora, K.A. Mohammed, and Y.P. Kim. 2004. Mechanism targeted discovery of antitumor marine natural products. *Current Medicinal Chemistry* 11:1725–1756.

Nethravathy, M.U., J.G. Mehar, S.N. Mudliar and A.Y. Shekh. 2019. Recent advances in microalgal bioactives for food, feed, and healthcare products: commercial potential, market space, and sustainability. *Comprehensive Reviews in Food Science and Food Safety* 18:1882.

Novianti, T., M. Zainuri, and I. Widowati. 2019. Aktivitas Antioksidan Dan Identifikasi Golongan Senyawa Aktif Ekstrak Kasar Mikroalga Chlorella Vulgaris Yang Dikultivasi Berdasarkan Sumber Cahaya Yang Berbeda *Barakuda 45* 1:72–87.

Okab, A.B., E.M. Samara, K.A. Abdoun, J. Rafay, L. Ondruska, V. Parkanyi, J. Pivko, M.A. Ayoub, A.A. Al-Haidary, R.S. Aljumaah, M. Peter, and N. Lukac. 2013. Effects of dietary seaweed (*Ulva lactuca*) supplementation on the reproductive performance of buck and doe rabbits. *Journal of Applied Animal Research* 41:347–355.

O'Kelly, C.J. 2007. The origin and early evolution of green plants. In: P.G. Falkowski and A.H. Knoll (eds) *Evolution of Primary Producers in the Sea*. Vol. 13. Burlington, MA: Elsevier Academic Press, 287–309.

Ornano, L., Y. Donno, C. Sanna, M. Ballero, M. Serafini, and A. Bianco. 2014. Phytochemical study of *Caulerpa racemosa* (Forsk.) J. Agarth, an invading alga in the habitat of la Maddalena archipelago. *Natural Product Research* 28:1795–1799.

Othman R., N. Hidayah, A.M. Farah, F.A.M. Hatta and M.A. Jamaludin. 2018. Natural carotenoid pigments of 6 chlorophyta freshwater green algae species. *Journal of Pharmacy and Nutrition Sciences* 8:1–5.

Palozza, P., C. Torelli, A. Boninsegna, R. Simone, A. Catalano, M.C. Mele, and N. Picci. 2009. Growth-inhibitory effects of the astaxanthin-rich alga *Haematococcus pluvialis* in human colon cancer cells. *Cancer Letters* 283:108–117.

Park, N.H., J.S. Choi, S.Y. Hwang, Y.C. Kim, Y.K. Hong, K.K. Cho, and I.S. Choi. 2013. Antimicrobial activities of stearidonic and gamma-linolenic acids from the green seaweed *Enteromorpha linza* against several oral pathogenic bacteria. *Botanical Studies* 54:39.

Paul, S. and R. Kundu. 2013. Antiproliferative activity of methanolic extracts from two green algae, *Enteromorpha intestinalis* and *Rizoclonium riparium* on HeLa cells. *DARU Journal of Pharmaceutical Sciences* 21:72.

Paulert, R., D. Ebbinghaus, C. Urlass, and B.M. Moerschbacher. 2010. Priming of the oxidative burst in rice and wheat cell cultures by ulvan, a polysaccharide from green macroalgae, and enhanced resistance against powdery mildew in wheat and barley plants. *Plant Pathology* 59:634–642.

Paulert, R., V. Talamini, J.E.F. Cassolato, M.E.R. Duarte, M.D. Noseda, A. Jr. Smania, and M.J. Stadnik. 2009. Effects of sulfated polysaccharide and alcoholic extracts from green seaweed *Ulva fasciata* on anthracnose severity and growth of common bean (*Phaseolus vulgaris* L.). *Journal of Plant Diseases and Protection* 116:263–270.

Pereira, L. 2018. Seaweeds as source of bioactive substances and skin care therapy – cosmeceuticals, algotherapy, and thalassotherapy. *Cosmetics* 5:68.

Phillips, J. 2001. Marine macroalgal biodiversity hotspots: why is there high species richness and endemism in southern Australian marine benthic flora? *Biodiversity and Conservation.* 10:1555–1577.

Piepho, M., D. Martin-Creuzburg, and A. Wacker. 2010. Simultaneous effects of light intensity and phosphorus supply on the sterol content of phytoplankton. *PLoS One* 5:e15828.

Pires, C.L., S.D. Rodrigues, D. Bristot, H.H. Gaeta, D.D.O. Toyama, W.R. L Farias, and M.H. Toyama. 2013. Sulfated polysaccharide extracted from the green algae *Caulerpa racemosa* increase the enzymatic activity and paw edema induced by sPLA2 from *Crotalus durissus terrificus* venom. *Revista Brasileira de Farmacognosia* 23:635–643.

Plaza, M., M. Herrero, A. Cifuentes, and E. Ibáñez. 2009. Innovative natural functional ingredients from microalgae. *Journal of Agricultural and Food Chemistry* 5:7159–7170.

Ponce, N.M., C.A. Pujol, E.B. Damonte, M.L. Flores, and C.A. Stortz. 2003. Fucoidans from the brown seaweed *Adenocystis utricularis*: extraction methods, antiviral activity and structural studies. *Carbohydrate Research* 338:153–165.

Prasad, G., K. Prasad, R. Meena, and A.K. Siddhanta. 2009. Facile preparation of *Chaetomorpha antenna* based porous polysaccharide-PMMA hybrid material by radical polymerization under microwave irradiation. *Journal of Materials Science* 44:4062–4068.

Puglisi, M.P., L.T. Tan, P.R. Jensen, and W. Fenical. 2004. Capisterones A and B from the tropical green alga *Penicillus capitatus*: unexpected anti-fungal defenses targeting the marine pathogen *Lindra thallasiae. Tetrahedron* 60:7035–7039.

Qi, H., L. Huang, X. Liu, D. Liu, Q. Zhang, and S. Liu. 2012a. Antihyperlipidemic activity of high sulfate content derivative of polysaccharide extracted from Ulva pertusa (Chlorophyta). *Carbohydrate Polymers* 87:1637–1640.

Qi, H., X. Liu, J. Zhang, Y. Duan, X. Wang, and O. Zhang. 2012b. Synthesis and antihyperlipidemic activity of acetylated derivative of ulvan from Ulva pertusa. *International Journal of Biological Macromolecules* 50:270–272.

Qi, X., W. Mao, Y. Gao, Y. Chen, C. Zhao, N. Li, C. Wang, M. Yan, C. Lin, and J. Shan. 2012c. Chemical characteristic of an anticoagulant-active sulfated polysaccharide from *Enteromorpha clathrata. Carbohydrate Polymers* 90:1804–1810.

Ragasa, C.Y., V.D. Ebajo, N. Lazaro-Llanos, R. Brkljaca, and S. Urban. 2015. Secondary metabolites from *Caulerpa racemosa. Der Pharmacia Lettre* 7:122–125.

Rashad, S., G.A. El-Chaghaby, and M.A. Elchaghaby. 2019. Antibacterial activity of silver nanoparticles biosynthesized using spirulina platensis microalgae extract against oral pathogens. *Egyptian Journal of Aquatic Biology and Fisheries* 23 :261–266.

Rathi, V. 2018. Herbal wine: a review. *Journal of Nutrition and Weight Loss* 3:113.

Renn, D.W. 1997. Biotechnology and the red seaweed polysaccharide industry: status, needs and prospects. *Tibtech* 15:9–14.

Rodrigues, J.A.G., I.N.L.D. Queiroz, A.L.G. Quinderé, B.C. Vairo, P.A. de Souza Mourão, and N. Benevides. 2011. An antithrombin-dependent sulphated polysaccharide isolated from the green alga *Caulerpa cupressoides* has in vivo anti- and prothrombotic effects. *Ciência Rural* 41:634–639.

Rosa, G.P., W.R. Tavares, P.M. Sousa, A.K. Pagès, A.M. Seca & D.C. Pinto. 2019. Seaweed secondary metabolites with beneficial health effects: An overview of successes in in vivo studies and clinical trials. *Marine Drugs* 18(1):9–14.

Round, F.E. 1981. *The Ecology of Algae*. Cambridge: Cambridge University Press.

Roy, S. 2020. Screening and partial characterization of natural antioxidants from seaweeds collected from Rameshwaram Southeast Coast of India. *Journal of Marine Science Research and Oceanography* 3:1–12.

Salehi, B., J. Sharifi-rad, A.M.L. Seca, D.C.G.A. Pinto, M. Izabela, T. Antonio, A.P. Mishra, M. Nigam, W. Zam, and N. Martins. 2019. Current trends on seaweeds: looking at chemical composition, phytopharmacology, and cosmetic applications. *Molecules* 24:4182.

Samara, E.M., A.B. Okab, K.A. Abdoun, A.M. El-Waziry, and A.A. Al-Haidary. 2013. Subsequent influences of feeding intact green seaweed *Ulva lactuca* to growing lambs on the seminal and testicular characteristics in rams. *Journal of Animal Science* 91:5654–5667.

Senosy, W., A.Y. Kassab, and A.A Mohammed 2017. Effects of feeding green microalgae on ovarian activity, reproductive hormones and metabolic parameters of Boer goats in arid subtropics. *Theriogenology* 96:16–22.

Shanmugam, M., B.K. Ramavat, K.H. Mody, R.M. Oza, and A. Tewari. 2001. Distribution of heparinoid-active sulphated polysaccharides in some Indian marine green algae. *Indian Journal of Marine Sciences* 30:222–227.

Shanmugam, M., K.H. Mody, B.K. Ramavat, A.S.K. Murthy, and A.K Siddhanta. 2002. Screening of *Codiacean* algae (Chlorophyta) of the Indian coasts for blood anticoagulant activity. *Indian Journal of Marine Sciences* 31:33–38.

Smith, A.J. 2004. Medicinal and pharmaceutical uses of seaweed natural products: a review. *Journal of Applied Phycology* 16:245–262.

Souto, A.L., J.F. Tavares, M.S. Da Silva, M.F.F.M. De Diniz, P.F. De Athayde-Filho, and J.M. Barbosa Filho. 2011. Anti-inflammatory activity of alkaloids: an update from 2000 to 2010. *Molecules* 16: 8515–8534.

Sun, Y. Y, H. Wang, G. Guo, Y. lin, Y. Pu, Y. fang, B. lun, and C. Wang. 2016. Isolation, purification, and identification of antialgal substances in green alga *Ulva prolifera* for antialgal activity against the common harmful red tide microalgae. *Environmental Science and Pollution Research International* 23:1449–1459.

Tanna, B., B. Choudhary, and A. Mishra. 2018. Metabolite profiling, antioxidant, scavenging and anti-proliferative activities of selected tropical green seaweeds reveal the nutraceutical potential of *Caulerpa* spp. *Algal Research* 36:96–105.

Teng, Z., L. Qian, and Y. Zhou. 2013. Hypolipidemic activity of the polysaccharides from *Enteromorpha prolifera. International Journal of Biological Macromolecules* 62:254–256.

Tonks, N.K. 2006. Protein tyrosine phosphatases: from genes, to function, to disease. *Nature Reviews Molecular Cell Biology* 7:833–846.

Trivedi, J., M. Aila, D.P. Bangwal, S. Kaul, and M.O. Garg. 2015. Algae based biorefinery—how to make sense? *Renewable and Sustainable Energy Reviews* 47:295–307.

Vimala T. and T.V. Poonghuzali. 2017. In vitro antimicrobial activity of solvent extracts of marine brown alga, *Hydroclathrus clathratus* (C. Agardh) M. Howe from Gulf of Mannar. *Journal of Applied Pharmaceutical Science* 7 :157–162.

Wang, H., Y.L. Li, W.Z. Shen, W. Rui, X.J. Ma, and Y.Z. Cen. 2007. Antiviral activity of a sulfo quinovosyl diacylglycerol (SQDG) compound isolated from the green alga *Caulerpa racemose. Botanica Marina* 50:185–190.

Wang, L., Y.J. Jeon, and J.I. Kim. 2020. In vitro and in vivo anti-inflammatory activities of a sterol-enriched fraction from freshwater green alga, spirogyra sp. *Fisheries and Aquatic Sciences* 23:1–9.

Wang, R., V.J. Paul, and H. Luesch. 2013. Seaweed extracts and unsaturated fatty acid constituents from the green alga *Ulva lactuca* as activators of the cytoprotective Nrf2-ARE pathway. *Free Radical Biology and Medicine* 57:141–153.

Wang, X., Z. Zhang, Z. Yao, M. Zhao, and H. Qi. 2013. Sulfation, anticoagulant and antioxidant activities of polysaccharide from green algae *Enteromorpha linza. International Journal of Biological Macromolecules* 58:225–230.

Wetzel, R.G. 1983. *Limnology*. Philadelphia, PA: Saunders.

Williams, D.E., C.M. Sturgeon, M. Roberge, and R.J. Andersen. 2007. Nigricanosides A and B, antimitotic glycolipids isolated from the green alga *Avrainvillea nigricans* collected in Dominica. *Journal of the American Chemical Society* 129:5822–5823.

Wu, G.J., S.M. Shiu, M.C. Hsieh, and G.J. Tsai, 2016. Anti-inflammatory activity of a sulfated polysaccharide from the brown alga *Sargassum cristaefolium. Food Hydrocolloids* 53:16–23.

Yaich, H., A.M. Amira, F. Abbes, B. Mohamed, S. Besbes, A. Richel, C. Blecker, H. Attia, and H. Garna. 2017. Effect of extraction procedures on structural, thermal and antioxidant properties of ulvan from *Ulva lactuca* collected in Monastir coast. *International Journal of Biological Macromolecules* 105:1430–1439.

Yan, X.J., Y. Chuda, M. Suzuki, and T. Nagata. 1999. Fucoxanthin as the major antioxidant in *Hijikia fusiformis. Bioscience, Biotechnology and Biochemistry* 63:605–607.

Yang, H., D.Q. Liu, T.J. Liang, J. Li, A.H. Liu, P. Yang, K. Lin, X.Q. Yu, Y.W. Guo, S.C. Mao, and B. Wang. 2014. Racemosin C, a novel minor bisindole alkaloid with protein tyrosine phosphatase-1B inhibitory activity from the green alga *Caulerpa racemosa. Journal of Asian Natural Products Research* 16:1158–1165.

Yang, P., D.Q. Liu, T.J. Liang, J. Li, H.Y. Zhang, A.H. Liu, Y.W. Guo, and S.C. Mao 2015. Bioactive constituents from the green alga *Caulerpa racemosa. Bioorganic and Medicinal Chemistry* 23:38–45.

Yap, W.F., V. Tay, S.H. Tan, Y.Y. Yow, and J. Chew. 2019. Decoding antioxidant and antibacterial potentials of Malaysian green seaweeds: *Caulerpa racemosa* and *Caulerpa lentillifera. Antibiotics* 8:152.

Yener, N.A., O. Sinanoglu, E. Ilter, A. Celik, G. Sezgin, A. Midi, U. Devici, and F. Aksungar. 2013. Effect of spirulina on cyclophosphamide-induced ovarian toxicity in rats: biochemical and histomorphometric evaluation of the ovary. *Biochemistry Research International* 2013:764262.

Yu, P.Z., Q.B. Zhang, N. Li, Z.H. Xu, Y.M. Wang, and Z.E. Li. 2003. Polysaccharides from *Ulva pertusa* (Chlorophyta) and preliminary studies on their antihyperlipidemia activity. *Journal of Applied Phycology* 15:21–27.

Yulistiyanto, A.C., M. Hersasanti, R.Y. Hartantyo, L. Fitria, A.R. Chasani, and M. Mulyati. 2020. *Ulva lactuca* Linnaeus potentially promotes reproductive indices and depressive-like behavior of hypertriglyceridemia male Wistar rats (*Rattus norvegicus* Berkenhout, 1769). *Journal of Tropical Biodiversity and Biotechnology* 5:228–238.

Zaho, X., C.H. Xue, Z.J. Li, Y.P. Cai, H.Y Liu, and H.T. Qi. 2004. Antioxidant and hepatoprotective activities of low molecular weight sulfated polysaccharide from *Laminaria japonica. Journal of Applied Phycology* 16:111–115.

Zhang, H.-J., W.-J. Mao, F.F., Hong-Yan Li, H.-H. Sun, Y. Chen, and X.-H. Qi. 2008. Chemical characteristics and anticoagulant activities of a sulfated polysaccharide and its fragments from *Monostroma latissimum. Carbohydrate Polymers* 71:428–434.

Zouaoui, B., R. Bachir, F. Sahnouni, A. Hariri, G. Bouhadi, and M. Benchohra. 2017. Antioxidant activities of sulfated polysaccharide obtained from red algae *Corallina officinalis. International Journal of Pharmacognosy* 4:88–91.

15 Chemical Composition and Biological Activity of Red Algae (Rhodophyta)

Limenew Abate,[1,2,3] Archana Bachheti,[4] Mesfin Getachew Tadesse,[2,3] D.P. Pandey,[5] Azamal Husen,[6] and Rakesh Kumar Bachheti[1,2,3,]*

[1] Nanotechnology Center of Excellence, Addis Ababa Science and Technology University, Ethiopia

[2] Bio-process and Biotechnology Center of Excellence, Addis Ababa Science and Technology University, Ethiopia

[3] Department of Industrial Chemistry, Addis Ababa Science and Technology University, Ethiopia

[4] Department of Environment Science, Graphic Era University, Dehradun, Uttarakhand, India

[5] Department of Chemistry, Govt. P.G. College, Uttarkashi, India

[6] Wolaita Sodo University, Wolaita, Ethiopia

[*] Corresponding author (Rakesh Kumar Bachheti): rakesh.kumar@aastu.edu.et; rkbachheti@gmail.com

15.1 INTRODUCTION

Algae are a specific group of mostly photosynthetic, aquatic, and nucleus-bearing organisms that lack the true leaves, stems, roots, and specialized multicellular reproductive systems other plants have (Xu et al. 2022). They are multicellular or single-celled organisms frequently found in fresh and saltwater and are a source of biomolecules (Bachheti et al. 2021). Since the beginning of civilization, particularly in Asian countries like Korea, Japan, and China, they have been consumed as food (Sudhakar et al. 2018). Water, carbohydrates, proteins, and lipids comprise algae's principal metabolites (Babich et al. 2022). There are two types of algae in the world: micro- and macroalgae. Microalgae (microphytes) are represented by diatoms algae (Bacillaryophyta), golden algae (Ochrophyta и Chrysophyta), yellow-green algae (Ochrophyta и Xanthophyta), blue-green algae (Cyanobacteria), and green algae (Chlorophyta). However, Rhodophyta (red algae), Chlorophyta (green algae), and Ochrophyta (brown algae) are classified under the macroalgae type (Lee et al. 2020).

Rhodophyta, or red algae, are the biggest phylum of algae in terms of species variety and number (Cian et al. 2015). These species have large concentrations of carotenoids, including xanthophylls, astaxanthin, fucoxanthin, and β-carotene, as well as chlorophyll (a and d), and other pigments like allophycocyanin, phycocyanin, and phycoerythrin (Cian et al. 2015). When compared to brown and green algae, red algae often have more extensive contents of minerals, carbohydrates, and proteins (Belghit et al. 2017). Their inclusion in the diet is primarily a result of their supply of minerals (Fe, Ca, I, K, and Se), vitamins (B_{12}, A, and C), and long-chain fatty acids like ω-3 (FAO 2020).

DOI: 10.1201/9781003256830-15

Compared to other algal species, Rhodophyta is the most important producer of physiologically active metabolites (Kasanah et al. 2015). Many studies have shown that red algae contain significant bioactive chemicals (Leal et al. 2013). Bromophenols were detected in red algae families such as Rhodomelaceae, Corallinaceae, Ceramiaceae, Rhodophyllaceae, Bonnemaisoniaceae, and Delesseriaceae (Pedersén et al. 1974). The halogenated substances showed different biological properties like insecticidal, cytotoxic, anti-inflammatory, antifungal, antibacterial, and cytotoxic activities. The red algae also produce acetogenin, polyether, and terpenoid, in addition to certain derivatives of nucleic acids, shikimate, acetate, and amino acids (Maschek and Baker 2008; Ayyad et al. 2011).

Regarding their characteristics, both *in vitro* and *in vivo* research have characterized their immunomodulatory, anticancer, antihyperlipidaemic, antiviral, antitumour, anti-inflammatory, and antioxidant actions (Carpena et al. 2021). Significant substances derived from Rhodophyta, such as phenolic compounds, amino acids, lectins, pigments, and polysaccharides, are responsible for anticancer, antibiotic, and antioxidant properties (Yanshin et al. 2021). Additionally, they are effective at preventing and treating diet-related metabolic diseases like obesity and type 2 diabetes (Kim et al. 2019; Sørensen et al. 2019). Red algae can also be utilized for the manufacture of biofuel (Kumar et al. 2013) and natural colorants (Azeem et al. 2019) and bioremediation (Corey et al. 2012). The purpose of this chapter is to present a summary of the existing knowledge regarding the potential uses of Rhodophyta in the treatment of various diseases by reviewing the primary therapeutic targets of those diseases and then the availability of substances or extracts having bioactive activities against the derived extracts.

15.2 CHEMICAL COMPOSITION OF RED ALGAE

15.2.1 Primary Metabolites

Numerous free sugars are present in Rhodophyta, including glucose, galactose, mannose, xylose, and fucose (Gómez-Ordóñez et al. 2010). The two sulphated polysaccharides known as phycocollooids, carrageenan and agar, which together account for 40–50% of the dry weight of Rhodophyta, are the most significant sources of carbohydrates (Torres et al. 2019) (Figure 15.1); porphyrans, sulphated galactans, and xylans, are additional polysaccharides that are present in much lower amounts in red algae (Øverland et al. 2019).

Currently, seaweeds are an essential alternative raw material for proteins for the nutrition of both humans and animals (Mišurcová 2012). Among them, Rhodophyta has a higher content of protein than brown and green algae. Data from the literature show that the total protein content of several red algae ranges from 2.3% dry weight in *Calendula officinalis* to 47% dry weight in *Pyropia tenera* (Černá 2011; Pangestuti and Kim 2015; Gamero-Vega et al. 2020). The authors of one study suggest that a red algae species named *Porphyra umbilicalis* has high protein levels equivalent to those seen in soybeans (Palasí Mascarós 2015). The total protein content of some red algae species, including *Enantiocladia duperreyi*, *Amansia multifida*, and *Polysiphonia* spp., was reported to range from 19.5% to 31.3% dry weight (Guérin-Deremaux et al. 2011; Jang and Park 2020).The most prevalent proteins in red algae in terms of total protein content are phycobiliproteins, which can make up to 50% of the total protein composition and give these species their characteristic crimson hue (Niu et al. 2007).

Aspartic acid and glutamic acid are the most common residues in red algae proteins, accounting for up to 22–44% of the total amino acids present. This indicates that red algal proteins have many essential amino acids (Cian et al. 2015). One study indicated that the amounts of essential amino acids like leucine, methionine, and valine in red algae are comparable to those in egg albumin (Astorga-España et al. 2016). Levels of other amino acids, such as threonine and isoleucine, are present in red algae at similar levels to plants belonging to the legume family.

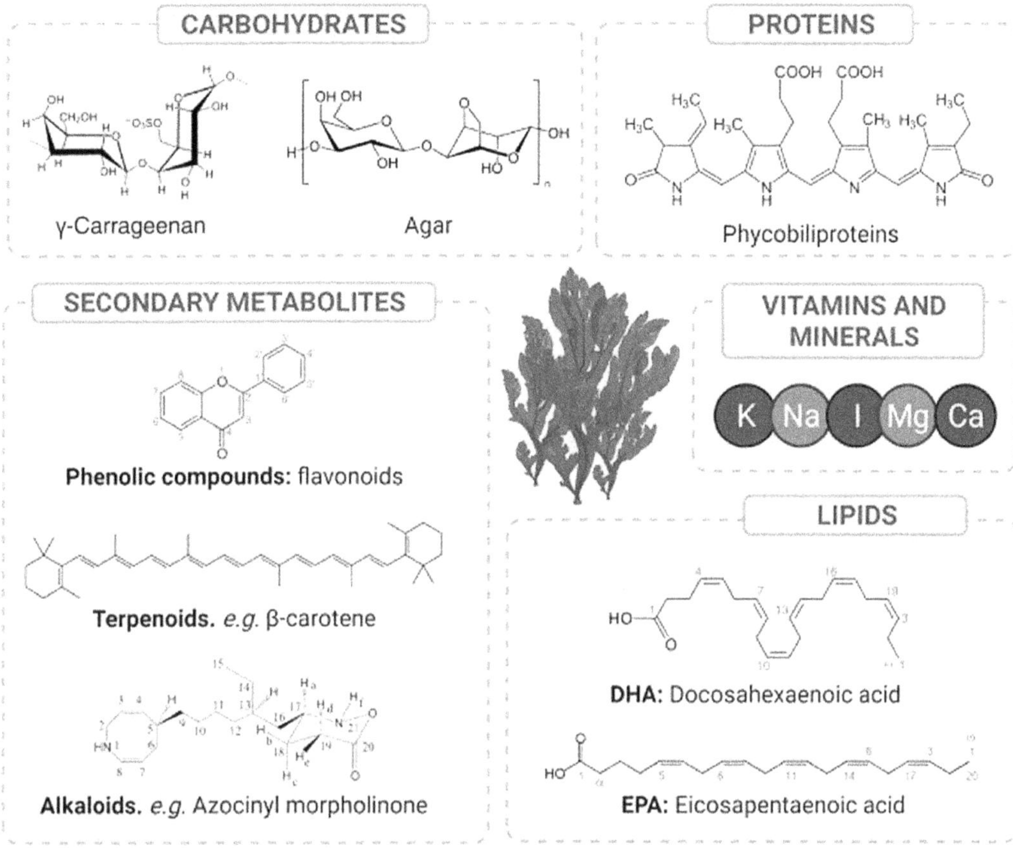

FIGURE 15.1 Main bioactive compounds of red algae. (From Carpena et al. (2022).)

From red algae species, a wide range of vitamins like pro-vitamin A (β-carotene), water-soluble vitamins like vitamin B_{12}, vitamin B_2, vitamin B_1, and vitamin C, and lipid-soluble vitamins like vitamin E have been extracted (Škrovánková 2011).

Rhodophyta can acquire high mineral amounts from salt water and organic micronutrients (Rosemary et al. 2019) and high concentrations of Mg, Ca, K, and Na have been found, ranging from 0.4 to 4 g/100 g (Rupérez Antón 2002). In this context, trace elements like Cu, Mn, Zn, and Fe have also been recorded as high as 10 mg/100 g in red seaweeds like *Nori* and *Chondrus* spp. Additionally, iodine has received particular attention because it has been discovered in *Gracilaria lemaeniformis* in substantial concentrations, supporting the enhancement of thyroid use (Wen et al. 2006).

15.2.2 Secondary Metabolites (Bioactive Compounds)

The phytochemical testing of methanolic extracts of *Botryocladia leptopoda* was studied by König et al. (1999). The findings revealed that various phytochemicals, including anthraquinones, tannins, flavonoids, phenols, glycosides, and triterpenoids, are present in the methanol extract (König et al. 1999). Phytochemicals including tannin, triterpenoids, alkaloids, fatty acids, saponins, steroids, flavonoids, phenols, anthraquinones, sterols, and glycosides are found in the red algae species *Champia parvula* (Kumar et al. 2008).

Numerous phytochemical substances, including steroids, saponins, alkaloids, flavonoids, and phenols, are present in red algae, such as *Acanthophora spicifera* (Shankhadarwar 2015).

Kazłowska et al. (2010) showed the presence of phenolic acids, including chlorogenic, hypogallic, salicylic, caffeic, and *p*-coumaric acids in red algae. Different flavonoids have been isolated in *Porphyra dentate*, such as catechin, rutin, and, mostly, flavan-3-ols and flavonols (Kazłowska et al. 2010). In another research work, phytochemicals like quercetin and catechin were found in *Euchema cottonii* (Namvar et al. 2012). Tetraterpenoids and sesquiterpenoids, among other terpenoids with varying degrees of isoprene monomer, have been discovered in Rhodophyta. In plant species, most terpenoids are produced by a process called biosynthesis in response to herbivores and pathogenic microbial attacks (Philippus et al. 2018).

Hexanoic extract of the *Gracilaria corticata* revealed five main substances: quinines, alkaloids, terpenoids, flavonoids, and polyphenols. Six main chemicals are found in the acetonic extract of *Gracilaria corticata*, including alkaloids, polyphenols, quinones, saponins, and tannins (Rafiquzzaman et al. 2016). In *Gracilaria fergusonii*, bioactive compounds such as saturated fatty acid, fatty acid, protein, fixed oil, xanthoprotein, amino acids, sugar and glycosides, tannins, steroids, saponin, quinones, phenol, flavonoid, coumarin, and alkaloids were present (Panneer and Balakrishnan 2017).

Bioactive substances including tannins, flavonoids, protein, coumarins, alkaloids, terpenoids, saponins, phenols, carbohydrates, and steroids, have been found in the methanol extract of *Gracillaria edulis*. In contrast, the aqueous extract contained tannins, glycosides, flavonoids, carbohydrates, protein, phenols, and saponins (Kiruba et al. 2015). Coumarins, anthraquinones, terpenes, saponins, tannins, and alkaloids were found in the red algae *Graciliaria latifolium*, on the western coast of Libya (Alghazeer et al. 2013).

Some bioactive compounds have been isolated from the red algae *Halymenia dilatata* by Dayuti (2018). According to their finding, the plant contains 17 compounds based on gas chromatography–mass spectrometry results. Some of the compounds found in plant species are oleic acid, ethanol, 11,13-dimethyl 12-tetradecen-1-ol acetate, lathosterol, hexacosane, 2-(9-octadecenyloxy)-, (*Z*)-, heptacosane, deoxyspergualin, 8,11 octadecadienoic acid, *n*-hexdecanoic, methyl ester, 9-hexadecenoic acid, methyl ester, hexadecanoic acid, 6,10,14-trimethylpentadecan-2-one, 1-hexadecanol, diethyl phthalate, and methyl ester.

15.3 BIOLOGICAL ACTIVITIES

Biological activity of some red algae species is summarized in Table 15.1.

15.3.1 Antioxidant Activities

According to numerous studies, extracts from various red algae species stimulate strong antioxidant activity via different mechanisms of action, such as metal chelation, prevention of lipid oxidation, and the scavenging of free radicals and reactive oxygen species (ROS) (Rodrigues et al. 2015). For instance, using hydroxyl radical (HO), 2,2'-azino-bis(3-ethylbenzothiazoline-6-sulfonic acid) (ABTS), Fe^{2+} ion chelating ability, 1,1-dipheny1–2-picrylhydrazyl (DPPH) free radical scavenging activities, and hydrogen peroxide (H_2O_2) scavenging potential, the antioxidant activities of three Rhodophyta, including *Hypnea musciformis*, *Hypnea valentiae*, and *Jania rubens*, taken from India's South Eastern Coast's Gulf of Mannar were assessed. The result showed they have excellent antioxidant activities (Chakraborty et al. 2015). Ethanol extracts of *Gracilaria tenuistipitata* (Yang et al. 2012) and *Callophyllis japonica* (Kang et al. 2005) Rhodophyta species are said to have antioxidant properties. For instance, ethanol extracts from *Callophyllis japonica* inhibited H_2O_2-induced cellular death and stimulated cellular antioxidant enzymes (Kang et al. 2005). According to research on the H1299 cell line, the administration of a *Gracilaria tenuistipitata* extracts obtained from aqueous solvent improved the ability of the cells to recover from DNA damage brought on by H_2O_2, suppresses cellular growth, and triggered G2/M arrest (Yang et al. 2012).

Some red algae, such as *Kappaphycus striatum* and *Kappaphycus alvarezii*, were assessed for their antioxidant properties using the DPPH and Trolox equivalent antioxidant capacity (TEAC) assays. The result showed that the two red algae showed high antioxidant properties (Diyana et al. 2015). Moreover, extracts from *Palmaria palmata* (Wang et al. 2010; Hardouin et al. 2014), *Gracilaria arcuata* (Agatonovic-Kustrin and Morton 2017), *Gracilaria verrucosa* (De Almeida et al. 2011), and *Porphyra tenera* (Onofrejová et al. 2010) showed excellent antioxidant activities. The extract from two red algae, *Pterocladia capillacea* (Fleita et al. 2015), and *Mastocarpus stellatus* (Gómez-Ordóñez et al. 2010) showed the antioxidant properties were due to the presence of polysaccharides. On the other hand, it was discovered that extracts from *Gracilaria birdiae*, *Mastocarpus stellatus*, and *Palmaria palmata* have strong antioxidant characteristics and show metal reduction and chelation power (Fidelis et al. 2014; Gómez-Ordóñez et al. 2014; Yuan et al. 2005).

Red algae *Eucheuma cottonii* was subjected to sequential extraction with petroleum ether and methanol, allowing the isolation of steroid compounds like cholesterol, β-sitosterol, campesterol, and stigmasterol (Saleh and Al Mariri 2017). The extracted bioactive compound showed antioxidant properties. Ethanol extract of *Grateloupia filicina*, aqueous extract of *Gracilaria textorii*, and methanol extract of *Polysiphonia japonica* and *Gracilaria verrucosa* exhibited the highest DPPH and hydroxide free radicals (Heo et al. 2006). Hmani et al. (2021) performed an antioxidant test in six species: *Sphaerococcus cornopifolius*, *Gracilaria gracilis*, *Pterocladiella capillacea*, *Laurencia obtusa*, *Asparagopsis armata*, and *Hypnea musciformis*. The result showed significant TAC and DPPH radical scavenging activities. In one study, based on DPPH methods, the antioxidant activities of extracts derived from ethanolic and aqueous extracts of *Laurencia* and *Polysiphonia* were evaluated. The result confirmed that the aqueous and ethanolic extract DPPH radical scavenging behaviour of *Polysiphonia* showed the lowest activities with half-maximal inhibitory concentration (IC_{50}) values of 9.57 ± 0.46 and 9.30 ± 0.18 mg/ml, compared to *Laurencia* with IC_{50} values of 6.34 ± 0.41 and 5.64 ± 0.68 and mg/ml, respectively. The above data showed that Rhodophyta species are high in antioxidant activities and support their use in the food industry, cosmetics, and dietary supplements (Al-Amro et al. 2019).

15.3.2 Antimicrobial Activity

Red algal extracts have been studied for their potential to fight pathogens in human, agricultural, and fish populations. Extracts of red algae such as *Turbinaria conoides*, *Sargassum wightii*, *Acanthophora spicifera*, *Gracilaria crassa*, *Caulerpa scalpelliformis*, and *Codium decorticatum* showed high antibacterial activities against different bacteria species such as *Pseudomonas aeruginosa*, *Klebsiella pneumoniae*, *Streptococcus pyogenes*, *Proteus mirabilis*, *Escherichia coli*, *Enterococcus faecalis*, *Staphylococcus aureus*, and *Vibrio parahaemolyticus* (Lavanya and Veerappan 2011). In one study, ethyl acetate, ethanol, and methanol extract of red algae *Grateloupia doryphora* demonstrated antimicrobial activity against *Enterococcus faecalis*, *Pseudomonas aeruginosa*, *Escherichia coli*, and *Bacillus subtilis Staphylococcus aureus*, using the well-cut diffusion technique (Abdel-Latif et al. 2018). The result demonstrated that all bacteria species respond to significant antibacterial activities. In another research work by Widowati et al. (2014), red algae *Gracilaria verrucosa* showed antibacterial activities against six bacterial strains: *Vibrio alginolyticus*, *Vibrio parahaemolyticus*, *Vibrio harveyi*, *Proteus mirabilis*, *Staphylococcus aureus*, and *Escherichia coli*. *Calorpha peltada* and *Gracilaria edulis* ethanolic extract was tested against some bacteria species isolated from fish such as *Staphylococcus aureus*, *Enterobacter aerogenes*, *Bacillus cereus*, *Escherichia coli*, *Pseudomonas aeruginosa*, and *Streptococcus faecalis*. The results showed that every test organism except *Enterobacter aerogenes* and *Bacillus cereus* was inhibited by the ethanolic extract of *Gracilaria edulis*. *Streptococcus faecalis*, *Staphylococcus aureus*, and *Escherichia coli* were all susceptible to ethanol extracts of *Calorpha peltada* (Kolanjinathan et al. 2009). The ethanol extract from red algae (*Gracilaria edulis*) had antibacterial properties against *Pseudomonas aeruginosa*, *Streptococcus faecalis*, *Staphylococcus aureus*, *Enterobacter aerogenes*, and *Escherichia coli* (Dayuti 2018).

The antibacterial properties of aqueous extracts of flour derived from *Gelidium* spp. are vulnerable to Gram-negative bacteria, including *Escherichia coli*, *Klebsiella pneumoniae*, *Vibrio alginolyticus*, and *Enterobacter aerogenes*, as well as Gram-positive bacteria like *Bacillus subtilis* and *Bacillus cereus* (Miranda et al. 2022). The antibacterial activity of numerous bacterial species, including *Pseudomonas fluorescence*, *Staphylococcus aureus*, *Vibrio cholera*, and *Proteus mirabilis*, against various *Kappaphycus* (red algae) species from the coast of Tamilnadu, India, was examined *in vitro* using the disc diffusion method. The results showed how effective the red algae's antibacterial properties are (Rajasulochana et al. 2009).

15.3.3 Anti-inflammatory Activity

Interleukin (IL)-8 inhibition experiments were essential to evaluate the anti-inflammatory potential of crude extracts from the Rhodophyta *Sphaerococcus coronopifolius*. In telomerase reverse transcriptase immortalized human umbilical vein endothelial cells (HUVECtert) stimulated by lipopolysaccharide (LPS) and tumour necrosis factor (TNF)-α, IL-8 release was dramatically reduced by *Sphaerococcus coronopifolius* extracts. According to this study, *Sphaerococcus coronopifolius* has unique anti-inflammatory capabilities (Salhi et al. 2018). Studies revealed that *Polytopes affinis* ethanol extract reduced asthmatic symptoms (Lee et al. 2011), a polysaccharide from *Porphyridium* spp. prevented retrovirus multiplication (Talyshinsky et al. 2002), and a *Gracilaria tenuistipitata* aqueous extract lessened virus-induced inflammation (Chen et al. 2013). A methanol extract of *Neorhodomela aculeata* had anti-inflammatory effects on neurological illnesses by preventing the production of cellular ROS, inducible nitric oxide (NO) synthase, and H_2O_2-induced lipid peroxidation (Lim et al. 2006). Fucoxanthin, a carotenoid derived from *Myagropsis myagroides*, has also shown anti-inflammatory characteristics (Heo et al. 2010). In another study, anti-inflammatory was performed by polyphenol phlorotannins derived from red algae species such as *Ecklonia kurome*, *Ecklonia cava*, and *Eisenia bicycles* (Kim and Himaya 2011), and sargachromanol G derived from *Sargassum siliquastrum* (Yoon et al. 2012). Brown algae contain large amounts of phloroglucinol that is said to have an antioxidative stress impact and to reduce the generation of inflammatory mediators in LPS-stimulated cells (Kim and Kim 2010). Using tandem mass spectrometry to analyse the methanolic extract of *Gracilaria change* confirmed it contained many unknown compounds, the known 10-hydroxypheophytin a, and chlorophyll proteins methyl 10-hydroxyphaeophorbide. *Gracilaria change* extract (10 g/ml) was administered to U937 cells during differentiation. This significantly decreased the amount of the TNF-α response and the synthesis of the TNF-α and IL-6 genes. The study's findings demonstrated that *Gracillaria change* has anti-inflammatory capabilities because bioactive chemicals are extracted using a methanolic solvent (Shu et al. 2013). He et al. (2019) showed the anti-inflammatory effect due to the major monosaccharide component (galactose) from red alga (*Chondrus verrucosus*) against RBL-2H3 cells. Through the reduction of vascular permeability, neutrophil migration, and histamine, a sulphated polysaccharide compounds isolated from *Cornea gracilaria* plant exhibits anti-inflammatory activities (Coura et al. 2015). In mouse macrophages of RAW264.7 cells, porphyrins produced from *Porphyra* spp. have anti-inflammatory effects on people, reduce NO generation, and prevent nuclear factor kappa-B activation (Jiang et al. 2012).

15.3.4 Anticancer/Antitumour Activities

Apoptosis assays, as well as cell viability, were used to examine the antitumoural effects of red alga *Sphaerococcus coronopifolius* extracts on pancreatic, breast, and human cervical cancer cells. *Sphaerococcus coronopifolius* extracts inhibited HeLa, MIA PaCa-2, and SKBR-3 cell growth in a dose- and time-dependent manner (Salhi et al. 2018). According to Yeh et al. (2012a) and (2012b), extracts derived from *Gracilaria tenuistipitata* by using both ethanol and methanol solvent prevented the growth of Ca9-22 oral cancer cells. They caused oxidative stress, DNA damage, and then cellular death. Studies have shown that the aqueous extracts of *Sargassum oligocystum* and *Gracilaria corticata* suppress the proliferation of human leukaemic cell lines (Zandi et al. 2010a, 2010b). In another work, using methanol extract of *Plocamium telfairiae* reduces the HT-29 colon cancer cell growth (Kim et al. 2007). Researchers have found that sulphated polysaccharides, like carrageenans, are effective anticancer treatments in a variety of red algae species, including *Jania rubens* (Gheda et al. 2018), *Hypnea mascifformis* (Souza et al. 2018), *Gracilaria caudata* (Costa et al. 2010), *Gelidium amansii* (Chen et al. 2004; Shao et al. 2013), and *Champia feldmannii* (Lins et al. 2009). Based on their anticancer activity, polyphenols from the extracts of *Sagittaria filiformis* (Chaves et al. 2018), *Euchema serra* (Sugahara et al. 2001), and *Eucheuma cottonii* (Namvar et al. 2012) have also been reported.

TABLE 15.1
Biological Activities of Some Red Algae Species

Family	Species	Activity	References
Bangiaceae	*P. tenera*	Antioxidant	Onofrejová et al. 2010
Bangiaceae	*P. yezoensis*	Antioxidant	Zhou et al. 2012
Bangiaceae	*P. columbina*	Inflammatory	Cian et al. 2015
Bonnemaisoniaceae	*D. pulchra*	Antimicrobial	Manefield et al. 2001
Ceramiaceae	*C. rubrum*	Antimicrobial	Rhimou et al. 2010
Ceramiaceae	*C. virgatum*	Antimicrobial	Horincar et al. 2014
Champiaceae	*C. feldmannii*	Antitumour	Lins et al. 2009
Corallinaceae	*G. furcate*	Antitumour	Shao et al. 2013
Corallinaceae	*C. mediterranea*	Antimicrobial	El-Din and El-Ahwany 2016
Corallinaceae	*J. rubens*	Antitumour	Gheda et al. 2018
Cystocloniaceae	*H. musciformis*	Antitumour	Souza et al. 2018
Galaxauraceae	*G. marginata*	Antimicrobial	Liao et al. 2003
Gelidiaceae	*G. pacificum*	Inflammatory	Cui et al. 2019
Gelidiaceae	*G. amansii*	Antitumour	Chen et al. 2004
Gelidiaceae	*G. attenatum*	Antimicrobial	Rhimou et al. 2010
Gelidiaceae	*G. micropterum*	Antimicrobial	Manilal et al. 2015
Gelidiaceae	*G. pulchellum*	Antimicrobial	Rhimou et al. 2010
Gelidiaceae	*G. pusillum*	Antimicrobial	Rhimou et al. 2010
Gelidiaceae	*G. spinulosum*	Antimicrobial	Rhimou et al. 2010
Gigartinaceae	*C. verrucosus*	Inflammatory	He et al. 2019
Gigartinaceae	*C. acicularis*	Antimicrobial	Rhimou et al. 2010
Gigartinaceae	*C. crispus*	Antimicrobial	Chambers et al. 2011
Gracilariaceae	*G. birdiae*	Antioxidant	Fidelis et al. 2014
Gracilariaceae	*G. arcuata*	Antioxidant	Agatonovic-Kustrin and Morton 2017
Gracilariaceae	*G. caudata*	Inflammatory	Chaves et al. 2018
Gracilariaceae	*G. salicornia*	Inflammatory	Antony and Chakraborty 2019
Gracilariaceae	*G. cornea*	Inflammatory	Coura et al. 2015
Gracilariaceae	*G. opuntia*	Inflammatory	Makkar and Chakraborty 2017
Gracilariaceae	*G. caudata*	Antitumour	Costa et al. 2010
Gracilariaceae.	*G. corticata*	Antimicrobial	Arulkumar et al. 2018
Halymeniaceae	*G. elliptica*	Antitumour	Cho et al. 2014
Palmariaceae	*P. palmata*	Antioxidant	Wang et al. 2010
Palmariaceae	*P. palmata*	Antioxidant	Suwal et al. 2019
Palmariaceae	*P. palmata*	Inflammatory	Banskota et al. 2014

(*continued*)

TABLE 15.1 (Continued)
Biological Activities of Some Red Algae Species

Family	Species	Activity	References
Phyllophoraceae	*M. stellatus*	Antioxidant	Gómez-Ordóñez et al. 2010
Pterocladiaceae	*P. capillacea*	Antioxidant	Fleita et al. 2015
Rhodomelaceae	*L. glandulifera*	Inflammatory	Makkar and Chakraborty 2018
Rhodomelaceae	*L. snackeyi*	Inflammatory	Wijesinghe et al. 2014
Rhodomelaceae	*L. microcladia*	Antitumour	Campos et al. 2012
Rhodomelaceae	*L. obtusa*	Antitumour	Iliopoulou et al. 2003
Rhodomelaceae	*L. popillose*	Antitumour	El Baz et al. 2013
Rhodomelaceae	*P. lanosa*	Antitumour	Shoeib et al. 2004
Rhodomelaceae	*A. corallinum*	Antimicrobial	Rhimou et al. 2010
Rhodomelaceae	*C. armata*	Antimicrobial	Al-Fadhli et al. 2006
Rhodophyta	*E. cottonii*	Antitumour	Namvar et al. 2012
Rutaceae	*G. cylindriea*	Antitumour	El Baz et al. 2013
Solieriaceae	*S. filiformis*	Antitumour	Chaves et al. 2018
Solieriaceae	*C. serratus*	Antimicrobial	Lane et al. 2009
Solieriaceae	*S. filiformis*	Inflammatory	De Arau´jo et al. 2011
Solieriaceae	*K. alvarezii*	Inflammatory	Makkar and Chakraborty 2018
Solieriaceae	*E. serra*	Antitumour	Sugahara et al. 2001
Solieriaceae	*E. serra*	Antimicrobial	Liao et al. 2003)

15.4 CONCLUSION

Rhodophyta, a significant class of macroalgae with about 7000 species, includes red algae. They contain various bioactive ingredients with various structural variations, including phenolic compounds, minerals, vitamins, polyunsaturated fatty acids, pigments, sulphated polysaccharides, and protein, all of which are important in terms of nutrition, medicine, and industry. Red algae are low in calories due to high quantities of dietary fibre and protein. They have up to 18.8 g/100 g more protein on average than brown and green algae. Red algae have large amounts of the bioactive components terpenoids, alkaloids, sterols, and phenolic compounds, which give them, among other things, antioxidant, antibacterial, and anti-inflammatory qualities. This chapter covers the critical aspects of red algae use in medicinal applications and secondary metabolites to give a general review of the available knowledge regarding the possible uses of red algae's chemical makeup and biological activities in the treatment of various diseases.

REFERENCES

Abdel-Latif, H., N. Shams El-Din and H. Ibrahim. 2018. Antimicrobial activity of the newly recorded red alga *Grateloupia doryphora* collected from the Eastern Harbor, Alexandria, Egypt. *Journal of Applied Microbiology* 125(5): 1321–1332.

Agatonovic-Kustrin, S. and D.W. Morton. 2017. High-performance thin-layer chromatography-direct bioautography as a method of choice for alpha-amylase and antioxidant activity evaluation in marine algae. *Journal of Chromatography A* 1530: 197–203.

Al-Amro, A., M. Al-Mutlaq, S. Al-Mother and N. Al-Tukhaifi. 2019. Antioxidant activity of rhodophyta algae *Polysiphonia* and *Laurencia* collected from the Arabian Gulf. *Asian Journal of Applied Sciences* 12: 71–75.

Al-Fadhli, A., S. Wahidulla and L. D'Souza. 2006. Glycolipids from the red alga *Chondria armata* (Kütz.) Okamura. *Glycobiology* 16(10):902–915.

Alghazeer, R., F. Whida, E. Abduelrhman, F. Gammoudi and S. Azwai. 2013. Screening of antibacterial activity in marine green, red and brown macroalgae from the western coast of Libya. *Natural Science* 5: 7–14.

Arulkumar, A., T. Rosemary, S. Paramasivam and R.B. Rajendran. 2018. Phytochemical composition, in vitro antioxidant, antibacterial potential and GC-MS analysis of red seaweeds (*Gracilaria corticata* and *Gracilaria edulis*) from Palk Bay, India. *Biocatalysis Agricultural Biotechnology* 15: 63–71.

Astorga-España, M.S., B. Rodríguez-Galdón, E.M. Rodríguez-Rodríguez and C. Díaz-Romero. 2016. Amino acid content in seaweeds from the Magellan Straits (Chile). *Journal of Food Composition Analysis* 53: 77–84.

Ayyad, S.-E.N., K.O. Al-Footy, W.M. Alarif, T.R. Sobahi, S.A. Bassaif, M.S. Makki, A.M. Asiri, A.Y. Al Halawani, A.F. Badria and F.A. Badria. 2011. Bioactive C15 acetogenins from the red alga *Laurencia obtusa*. *Chemical Pharmaceutical Bulletin* 59(10): 1294–1298.

Azeem, M., N. Iqbal, R.A. Mir, S. Adeel, F. Batool, A.A. Khan and S. Gul. 2019. Harnessing natural colorants from algal species for fabric dyeing: a sustainable eco-friendly approach for textile processing. *Journal of Applied Phycology* 31(6): 3941–3948.

Babich, O., S. Sukhikh, V. Larina, O. Kalashnikova, E. Kashirskikh, A. Prosekov, S. Noskova, S. Ivanova, I. Fendri and S. Smaoui. 2022. Algae: study of edible and biologically active fractions, their properties and applications. *Plants* 11(6): 780.

Bachheti, R.K., L. Abate, A. Bachheti, A. Madhusudhan and A. Husen. 2021. Algae-, fungi-, and yeast-mediated biological synthesis of nanoparticles and their various biomedical applications. In: B. Kharisov and O. Kharissova (eds) *Handbook of Greener Synthesis of Nanomaterials and Compounds*. Amsterdam: Elsevier, 701–734.

Banskota, A.H., R. Stefanova, S. Speaker, S.P. Lall, J.S. Craigie, J.T. Hafting and A.T. Critchley. 2014. Polar lipids from the marine macroalga *Palmaria palmata* inhibit lipopolysaccharide-induced nitric oxide production in RAW264.7 macrophage cells. *Phytochemistry* 101: 101–108.

Belghit, I., J.D. Rasinger, S. Heesch, I. Biancarosa, N. Liland, B. Torstensen, R. Waagbø, E.J. Lock and C.G. Bruckner. 2017. In-depth metabolic profiling of marine macroalgae confirms strong biochemical differences between brown, red and green algae. *Algal Research* 26: 240–249.

Campos, A., C.B. Souza, C. Lhullier, M. Falkenberg, E.P. Schenkel, R.M. Ribeiro-do-Valle and J.M. Siqueira. 2012. Anti-tumour effects of elatol, a marine derivative compound obtained from red algae *Laurencia microcladia*. *Journal of Pharmacy and Pharmacology*, 64(8):1146–1154.

Carpena, M., C. Caleja, E. Pereira, C. Pereira, A. Ćirić, M. Soković, A. Soria-Lopez, M. Fraga-Corral, J. Simal-Gandara and I.C. Ferreira. 2021. Red seaweeds as a source of nutrients and bioactive compounds: optimization of the extraction. *Chemosensors* 9(6): 132.

Carpena, M., P. Garcia-Perez, P. Garcia-Oliveira, F. Chamorro, P. Otero, C. Lourenço-Lopes, H. Cao, J. Simal-Gandara and M. Prieto. 2022. Biological properties and potential of compounds extracted from red seaweeds. *Phytochemistry Reviews* 1–32. Advance online publication. https://doi.org/10.1007/s11101-022-09826-z

Černá, M. 2011. Seaweed proteins and amino acids as nutraceuticals. *Advances in Food Nutrition Research* 64: 297–312.

Chakraborty, K., D. Joseph and N.K. Praveen. 2015. Antioxidant activities and phenolic contents of three red seaweeds (Division: Rhodophyta) harvested from the Gulf of Mannar of Peninsular India. *Journal of Food Science Technology* 52(4): 1924–1935.

Chambers, L.D., C. Hellio, K.R. Strokes, S.P. Dennington, L.R. Goodes, R.J.K. Wood, and F.C. Walsh. 2011. Investigation of *Chondrus crispus* as a potential source of new antifouling agents. *International Biodeterioration and Biodegradation* 65(7): 939–946.

Chaves, R.P., S.R. da Silva, L.G.N. Neto, R.F. Carneiro, A.L.C. da Silva, A.H. Sampaio, B.L. de Sousa, M.G. Cabral, P.A. Videira and E.H. Teixeira. 2018. Structural characterization of two isolectins from the marine red alga *Solieria filiformis* (Kützing) PW Gabrielson and their anticancer effect on MCF-7 breast cancer cells. *International Journal of Biological Macromolecules* 107: 1320–1329.

Chen, K.J., C.K. Tseng, F.R. Chang, J.I. Yang, C.C. Yeh, W.C. Chen, S.F. Wu, H.W. Chang and J.C. Lee. 2013. Aqueous extract of the edible *Gracilaria tenuistipitata* inhibits hepatitis C viral replication via cyclooxygenase-2 suppression and reduces virus-induced inflammation. *PLoS One* 8(2): e57704.

Chen, Y.H., C.J. Tu and H.T. Wu. 2004. Growth-inhibitory effects of the red alga *Gelidium amansii* on cultured cells. *Biological Pharmaceutical Bulletin* 27(2): 180–184.

Cian, R.E., S.R. Drago, F. Sanchez de Medina and O. Martínez-Augustin. 2015. Proteins and carbohydrates from red seaweeds: evidence for beneficial effects on gut function and microbiota. *Marine Drugs* 13(8): 5358–5383.

Corey, P., J. Kim, D. Garbary, B. Prithiviraj and J. Duston. 2012. Bioremediation potential of *Chondrus crispus* (Basin Head) and *Palmaria palmata*: effect of temperature and high nitrate on nutrient removal. *Journal of Applied Phycology* 24(3): 441–448.

Costa, L.S., G.P. Fidelis, S.L. Cordeiro, R.M. Oliveira, D.A. Sabry, R.B.G. Câmara, L.T.D.B. Nobre, M.S.S.P. Costa, J. Almeida-Lima and E. Farias. 2010. Biological activities of sulfated polysaccharides from tropical seaweeds. *Biomedicine Pharmacotherapy* 64(1): 21–28.

Coura, C.O., R.B. Souza, J.A.G. Rodrigues, E. de S.O. Vanderlei, I.W.F. de Araújo, N.A. Ribeiro, A.F. Frota, K.A. Ribeiro, H.V. Chaves and K.M.A. Pereira. 2015. Mechanisms involved in the anti-inflammatory action of a polysulfated fraction from *Gracilaria cornea* in rats. *PLoS One* 10(3): e0119319.

Cui, M., J. Wu, S. Wang, H. Shu, M. Zhang, K. Liu and K. Liu. 2019. Characterization and anti-inflammatory effects of sulphated polysaccharide from the red seaweed *Gelidium pacificum okamure*. *Int J Biol Macromol* 127: 377–385.

Dayuti, S. 2018. Antibacterial activity of red algae (Gracilaria verrucosa) extract against Escherichia coli and Salmonella typhimurium. *IOP Conference Series: Earth and Environmental Science* 137: 012074.

De Almeida, C.L. Falcão, G.R. Lima, C.A. Montenegro, N.S. Lira, P.F. de Athayde-Filho, L.C. Rodrigues, M. de F.V. De Souza, J.M. Barbosa-Filho and L.M. Batista. 2011. Bioactivities from marine algae of the genus *Gracilaria*. *International Journal of Molecular Sciences* 12(7): 4550–4573.

De Araújo, I.W.F., E.D.S.O. Vanderlei, J.A.G. Rodrigues, C.O. Coura, A.L.G. Quinderé, B.P. Fontes, I.N.L. de Queiroz, R.J.B. Jorge, M.M. Bezerra, A.A. Rodrigues and H.V. Chaves. 2011. Effects of a sulfated polysaccharide isolated from the red seaweed Solieria filiformis on models of nociception and inflammation. *Carbohydrate polymers* 86(3): 1207–1215.

Diyana, A.F., A. Abdullah, Z.S. Hisham and K. Chan. 2015. Antioxidant activity of red algae *Kappaphycus alvarezii* and *Kappaphycus striatum*. *International Food Research Journal* 22(5): 1977.

El Baz, F.K., G.S. El Baroty, H.H.A. Abd El Baky, O.I. Abd El-Salam, and E.A. Ibrahim. 2013. View of structural characterization and biological activity of sulfolipids from selected marine algae. *Grasas y aceites* 64:561–571.

El Din, S.M.M. and A.M. El-Ahwany. 2016. Bioactivityand phytochemical constituents of marine red seaweeds (*Jania rubens, Corallina mediterranea* and *Pterocladia capillacea*). *Journal of Taibah University for Science* 10(4):471–484.

FAO. 2020. *El estado mundial de la pesca y la acuicultura 2020: La sostenibilidad en acción*. Rome: FAO.

Fidelis, G.P., R.B.G. Camara, M.F. Queiroz, M.S.S.P. Costa, P.C. Santos, H.A.O. Rocha and L.S. Costa. 2014. Proteolysis, NaOH and ultrasound-enhanced extraction of anticoagulant and antioxidant sulfated polysaccharides from the edible seaweed, *Gracilaria birdiae*. *Molecules* 19(11): 18511–18526.

Fleita, D., M. El-Sayed and D. Rifaat. 2015. Evaluation of the antioxidant activity of enzymatically-hydrolyzed sulfated polysaccharides extracted from red algae; *Pterocladia capillacea*. *LWT-Food Science Technology* 63(2): 1236–1244.

Gamero-Vega, G., M. Palacios-Palacios and V. Quitral. 2020. Nutritional composition and bioactive compounds of red seaweed: a mini-review. *Journal of Food and Nutrition Research* 8(8): 431–440.

Gheda, S., M. El-Sheekh and A. Abou-Zeid. 2018. In vitro anticancer activity of polysaccharide extracted from red alga *Jania rubens* against breast and colon cancer cell lines. *Asian Pacific Journal of Tropical Medicine* 11(10): 583.

Gómez-Ordóñez, E., A. Jiménez-Escrig and P. Rupérez. 2010. Dietary fibre and physicochemical properties of several edible seaweeds from the northwestern Spanish coast. *Food Research International* 43(9): 2289–2294.

Gómez-Ordóñez, E., A. Jiménez-Escrig and P. Rupérez. 2014. Bioactivity of sulfated polysaccharides from the edible red seaweed *Mastocarpus stellatus*. *Bioactive Carbohydrates Dietary Fibre* 3(1): 29–40.

Guérin-Deremaux, L., M. Pochat, C. Reifer, D. Wils, S. Cho and L.E. Miller. 2011. The soluble fiber NUTRIOSE induces a dose-dependent beneficial impact on satiety over time in humans. *Nutrition Research* 31(9): 665–672.

Hardouin, K., A.-S. Burlot, A. Umami, A. Tanniou, V. Stiger-Pouvreau, I. Widowati, G. Bedoux and N. Bourgougnon. 2014. Biochemical and antiviral activities of enzymatic hydrolysates from different invasive French seaweeds. *Journal of Applied Phycology* 26(2): 1029–1042.

He, X., A. Yamauchi, T. Nakano, T. Yamaguchi and Y. Ochiai. 2019. The composition and anti-inflammatory effect of polysaccharides from the red alga *Chondrus verrucosus*. *Fisheries Science* 85(5): 859–865.

Heo, S.-J., S.H. Cha, K.-W. Lee and Y.J. Jeon. 2006. Antioxidant activities of red algae from Jeju Island. *Algae* 21(1): 149–156.

Heo, S.-J., W.J. Yoon, K.N. Kim, G.N. Ahn, S.M. Kang, D.H. Kang, C. Oh, W.K. Jung and Y.J. Jeon. 2010. Evaluation of anti-inflammatory effect of fucoxanthin isolated from brown algae in lipopolysaccharide-stimulated RAW 264.7 macrophages. *Food Chemical Toxicology* 48(8–9): 2045–2051.

Hmani, I., L. Ktari, A. Ismail, C. M'dallel and M. El Bour. 2021. Assessment of the antioxidant and antibacterial properties of red algae (Rhodophyta) from the north coast of Tunisia. *Euro-Mediterranean Journal for Environmental Integration* 6(1): 13.

Hornicar, V.B., G. Parfene, A.K. Tyagi, D. Gottardi, R. Dinică, M.E. Guerzoni and G. Bahrim. 2014. Extraction and characterization of the volatile compounds and fatty acids from red and green macroalgae from the Romanian Black Sea in order to obtain valuable bio additatives and preservatives. *Journal of Applied Phycology* 26: 551–559.

Iliopoulou, D., N. Mihopoulos, C. Vagias, P. Papazafiri and V. Roussis. 2003. Novel cytotoxic brominated diterpenes from the red alga *Laurencia obtusa*. *Journal of Organic Chemistry*, 68(20):7667–7674.

Jang, H. and K. Park. 2020. Omega-3 and omega-6 polyunsaturated fatty acids and metabolic syndrome: a systematic review and meta-analysis. *Clinical Nutrition* 39(3): 765–773.

Jiang, Z., Y. Hama, K. Yamaguchi and T. Oda. 2012. Inhibitory effect of sulphated polysaccharide porphyran on nitric oxide production in lipopolysaccharide-stimulated RAW264.7 macrophages. *Journal of Biochemistry* 151(1): 65–74.

Kang, K.A., H.D. Bu, D.S. Park, G.M. Go, Y. Jee, T. Shin and J.W. Hyun. 2005. Antioxidant activity of ethanol extract of *Callophyllis japonica*. *Phytotherapy Research* 19(6): 506–510.

Kasanah, N., T. Triyanto, D.S. Seto, W. Amelia and A. Isnansetyo. 2015. Antibacterial compounds from red seaweeds (Rhodophyta). *Indonesian Journal of Chemistry* 15(2): 201–209.

Kazłowska, K., T. Hsu, C.-C. Hou, W.C. Yang and G.J. Tsai. 2010. Anti-inflammatory properties of phenolic compounds and crude extract from *Porphyra dentata*. *Journal of Ethnopharmacology* 128(1): 123–130.

Kim, C.O., Y.N. Kim and D.C. Lee. 2019. Effects of *Gelidium elegans* on weight and fat mass reduction and obesity biomarkers in overweight or obese adults: a randomized double-blinded study. *Nutrients* 11(7): 1513.

Kim, J.Y., M.Y. Yoon, M.-R. Cha, J.H. Hwang, E. Park, S.U. Choi, H.R. Park and Y.I. Hwang. 2007. Methanolic extracts of *Plocamium telfairiae* induce cytotoxicity and caspase-dependent apoptosis in HT-29 human colon carcinoma cells. *Journal of Medicinal Food* 10(4): 587–593.

Kim, M.-M. and S.K. Kim. 2010. Effect of phloroglucinol on oxidative stress and inflammation. *Food Chemical Toxicology* 48(10): 2925–2933.

Kim, S.K. and S. Himaya. 2011. Medicinal effects of phlorotannins from marine brown algae. *Advances in Food Nutrition Research* 64: 97–109.

Kiruba, N.J.M., M.A. Pradeep and S.J.B. Juliana. 2015. Study of phytoconstituents and antibacterial activity of *Kappaphycus alvarezii*. *International Journal of Current Microbiology Applied Sciences* 4: 1209–1217.

Kolanjinathan, K., P. Ganesh and M. Govindarajan. 2009. Antibacterial activity of ethanol extracts of seaweeds against fish bacterial pathogens. *European Review for Medical and Pharmacological Sciences* 13(3): 173–177.

König, G.M., A.D. Wright and A. Linden. 1999. *Plocamium hamatum* and its monoterpenes: chemical and biological investigations of the tropical marine red alga. *Phytochemistry* 52(6): 1047–1053.

Kumar, K.S., K. Ganesan and P.S. Rao. 2008. Antioxidant potential of solvent extracts of *Kappaphycus alvarezii* (Doty) Doty – an edible seaweed. *Food Chemistry* 107(1): 289–295.

Kumar, S., R. Gupta, G. Kumar, D. Sahoo and R.C. Kuhad. 2013. Bioethanol production from *Gracilaria verrucosa*, a red alga, in a biorefinery approach. *Bioresource Technology* 135: 150–156.

Lane, A.L., L. Nyadong, A.S. Galhena, T.L. Shearer, E.P. Stout, R.M. Parry, M. Kwasnik, M.D. Wang, M.E. Hay, F.M. Fernandez and J. Kubanek. 2009. Desorption electrospray ionization mass spectrometry reveals surface-mediated antifungal chemical defense of a tropical seaweed. *Proceedings of the National Academy of Sciences* 106(18):7314–7319.

Lavanya, R. and N. Veerappan. 2011. Antibacterial potential of six seaweeds collected from Gulf of Mannar of southeast coast of India. *Advances in Biological Research* 5(1): 38–44.

Leal, M.C., M.H. Munro, J.W. Blunt, J. Puga, B. Jesus, R. Calado, R. Rosa and C. Madeira. 2013. Biogeography and biodiscovery hotspots of macroalgal marine natural products. *Natural Product Reports* 30(11): 1380–1390.

Lee, D.S., W.S. Park, S.J. Heo, S.H. Cha, D. Kim, Y.J. Jeon, S.G. Park, S.K. Seo, J.S. Choi and S.J. Park. 2011. *Polyopes affinis* alleviates airway inflammation in a murine model of allergic asthma. *Journal of Biosciences* 36(5): 869–877.

Lee, X.J., H.C. Ong, Y.Y. Gan, W.H. Chen and T.M.I. Mahlia. 2020. State of art review on conventional and advanced pyrolysis of macroalgae and microalgae for biochar, bio-oil and bio-syngas production. *Energy Conversion Management* 210: 112707.

Liao, W., J. Lin, W. Shieh, W. Jeng and R. Huang. 2003. Antibiotic activity of lectins from marine algae against marine vibrios. *Journal of Industrial Microbiology Biotechnology* 30(7): 433–439.

Lim, C.S., D.Q. Jin, J.Y. Sung, J.H. Lee, H.G. Choi, I. Ha and J.S. Han. 2006. Antioxidant and anti-inflammatory activities of the methanolic extract of *Neorhodomela aculeate* in hippocampal and microglial cells. *Biological Pharmaceutical Bulletin* 29(6): 1212–1216.

Lins, K.O., D.P. Bezerra, A.P.N. Alves, N.M. Alencar, M.W. Lima, V.M. Torres, W.R. Farias, C. Pessoa, M.O. de Moraes and L.V. Costa-Lotufo. 2009. Antitumor properties of a sulfated polysaccharide from the red seaweed *Champia feldmannii* (Diaz-Pifferer). *Journal of Applied Toxicology* 29(1): 20–26.

Makkar, F. and K. Chakraborty. 2017. Antidiabetic and anti-inflammatory potential of sulphated polygalactans from red seaweeds *Kappaphycus alvarezii* and *Gracilaria opuntia*. *International Journal of Food Properties* 20: 1326–1337.

Makkar, F. and K. Chakraborty. 2018. Antioxidant and anti-inflammatory oxygenated meroterpenoids from the thalli of red seaweed *Kappaphycus alvarezii*. *Medicinal Chemistry Research* 27(8): 2016–2026.

Manefield, M., M. Welch, M. Givskov, G.P. Salmond and S. Kjelleberg. 2001. Halogenated furanones from the red alga, *Delisea pulchra*, inhibit carbapenem antibiotic synthesis and exoenzyme virulence factor production in the phytopathogen *Erwinia carotovora*. *FEMS Microbiology Letters* 205(1): 131–138.

Manilal, A., S. Sujith, J. Selvin, G.S. Kiran, C. Shakir and A.P. Lipton. 2015. Antimicrobial potential of marine organisms collected from the southwest coast of India against multiresistant human and shrimp pathogens. *Scientia Marina* 74(2): 287–296.

Maschek, J.A. and B.J. Baker. 2008. The chemistry of algal secondary metabolism. In: C.D. Amsler (ed) *Algal Chemical Ecology*. Berlin: Springer, 1–24.

Miranda, J.M., M. Trigo, J. Barros-Velázquez and S.P. Aubourg. 2022. Antimicrobial activity of red alga flour (*Gelidium* sp.) and its effect on quality retention of *Scomber scombrus* during refrigerated storage. *Foods* 11(7): 904.

Mišurcová, L. 2012. Chemical composition of seaweeds. In: S.K. Kim (ed) *Handbook of Marine Macroalgae: Biotechnology and Applied Phycology*. Chichester: John Wiley, 171–192.

Namvar, F., S. Mohamed, S.G. Fard, J. Behravan, N.M. Mustapha, N.B.M. Alitheen and F. Othman. 2012. Polyphenol-rich seaweed (*Eucheuma cottonii*) extract suppresses breast tumour via hormone modulation and apoptosis induction. *Food Chemistry* 130(2): 376–382.

Niu, J.F., G.C. Wang, B.C. Zhou, X.Z. Lin and C.S. Chen. 2007. Purification of R-phycoerythrin from *Porphyra haitanensis* (Bangiales, Rhodophyta) using expanded-bed absorption 1. *Journal of Phycology* 43(6): 1339–1347.

Onofrejová, L., J. Vašíčková, B. Klejdus, P. Stratil, L. Mišurcová, S. Kráčmar, J. Kopecký and J. Vacek. 2010. Bioactive phenols in algae: the application of pressurized-liquid and solid-phase extraction techniques. *Journal of Pharmaceutical Biomedical Analysis* 51(2): 464–470.

Øverland, M., L.T. Mydland and A. Skrede. 2019. Marine macroalgae as sources of protein and bioactive compounds in feed for monogastric animals. *Journal of the Science of Food Agriculture* 99(1): 13–24.

Palasí Mascarós, J.T. 2015. *Caracterización físico-química y nutricional de algas en polvo empleadas como ingrediente alimentario*. Universitat Politècnica de València.

Pangestuti, R. and S.K. Kim. 2015. Seaweed proteins, peptides, and amino acids. In: B.K. Tiwari and D.J. Toy (eds) *Seaweed Sustainability: Food and Non-Food Applications*. San Diego, CA: Academic Press, 125–140.

Panneer, J. and C.P. Balakrishnan. 2017. Phytoconstituents screening and in-vitro evaluation of total antioxidant activity of marine red algae *Gracilaria fergusonii*. *International Journal of Pharmaceutical and Phytopharmacological Research* 6: 9–13.

Pedersén, M., P. Saenger and L. Fries. 1974. Simple brominated phenols in red algae. *Phytochemistry* 13(10): 2273–2279.

Philippus, A.C., G.A. Zatelli, T. Wanke, M.G. de A. Barros, S.A. Kami, C. Lhullier, L. Armstrong, L.P. Sandjo and M. Falkenberg. 2018. Molecular networking prospection and characterization of terpenoids and C 15-acetogenins in Brazilian seaweed extracts. *RSC Advances* 8(52): 29654–29661.
Rafiquzzaman, S., M.U. Ahmad, J.M. Lee, E.Y. Kim, Y.O. Kim, D.G. Kim and I.S. Kong. 2016. Phytochemical composition and antioxidant activity of edible red alga *Hypnea musciformis* from Bangladesh. *Journal of Food Processing Preservation* 40(5): 1074–1083.
Rajasulochana, P., R. Dhamotharan, P. Krishnamoorthy and S. Murugesan. 2009. Antibacterial activity of the extracts of marine red and brown algae. *Journal of American Science* 5(3): 20–25.
Rhimou, B., R. Hassane, M. José and B. Nathalie. 2010. The antibacterial potential of the seaweeds (Rhodophyceae) of the Strait of Gibraltar and the Mediterranean Coast of Morocco. *African Journal of Biotechnology* 9(38): 6365–6372.
Rodrigues, D., S. Sousa, A. Silva, M. Amorim, L. Pereira, T.A. Rocha-Santos, A.M. Gomes, A.C. Duarte and A.C. Freitas. 2015. Impact of enzyme-and ultrasound-assisted extraction methods on biological properties of red, brown, and green seaweeds from the central west coast of Portugal. *Journal of Agricultural Food Chemistry* 63(12): 3177–3188.
Rosemary, T., A. Arulkumar, S. Paramasivam, A. Mondragon-Portocarrero and J.M. Miranda. 2019. Biochemical, micronutrient and physicochemical properties of the dried red seaweeds *Gracilaria edulis* and *Gracilaria corticata*. *Molecules* 24(12): 2225.
Rupérez, P. 2002. Mineral content of edible marine seaweeds. *Food Chemistry* 79: 23–26.
Saleh, B. and A. Al Mariri. 2017. Antimicrobial activity of the marine algal extracts against selected pathogens. *Journal of Agricultural Science and Technology* 19(5): 1067–1077.
Salhi, G., H. Zbakh, H. Moussa, M. Hassoun, V. Bochkov, C.J. Ciudad, V. Noé and H. Riadi. 2018. Antitumoral and anti-inflammatory activities of the red alga *Sphaerococcus coronopifolius*. *European Journal of Integrative Medicine* 18: 66–74.
Shankhadarwar, S. 2015. Phytochemical analysis of red alga *Acanthophora spicifera* (Vahl) collected from Mumbai, India. *Journal of Chemical Pharmaceutical Research* 7(12): 441–444.
Shao, P., X. Chen and P. Sun. 2013. In vitro antioxidant and antitumor activities of different sulfated polysaccharides isolated from three algae. *International Journal of Biological Macromolecules* 62: 155–161.
Shoeib, N.A., M.C. Bibby, G. Blunden, P.A. Linley, D.J. Swaine, R.T. Wheelhouse, and C.W. Wright. 2004. In-vitro Cytotoxic Activities of the Major Bromophenols of the Red Alga Polysiphonia anosmia and Some Novel Synthetic Isomers. *Journal of Natural Products*, 67(9):1445–1449.
Shu, M.-H., D. Appleton, K. Zandi and S. AbuBakar. 2013. Anti-inflammatory, gastroprotective and anti-ulcerogenic effects of red algae *Gracilaria change* (Gracilariales, Rhodophyta) extract. *BMC Complementary Alternative Medicine* 13(1): 1–13.
Škrovánková, S. 2011. Seaweed vitamins as nutraceuticals. *Advances in Food Nutrition Research* 64: 357–369.
Sørensen, L.E., P.B. Jeppesen, C.B. Christiansen, K. Hermansen and S. Gregersen. 2019. Nordic seaweed and diabetes prevention: exploratory studies in KK-Ay mice. *Nutrients* 11(6): 1435.
Souza, R.B., A.F. Frota, J. Silva, C. Alves, A.Z. Neugebauer, S. Pinteus, J.A.G. Rodrigues, E.M.S. Cordeiro, R.R. de Almeida and R. Pedrosa. 2018. In vitro activities of kappa-carrageenan isolated from red marine alga *Hypnea musciformis*: antimicrobial, anticancer and neuroprotective potential. *International Journal of Biological Macromolecules* 112: 1248–1256.
Sudhakar, K., R. Mamat, M. Samykano, W. Azmi, W. Ishak and T. Yusaf. 2018. An overview of marine macroalgae as bioresource. *Renewable Sustainable Energy Reviews* 91: 165–179.
Sugahara, T., Y. Ohama, A. Fukuda, M. Hayashi, A. Kawakubo and K. Kato. 2001. The cytotoxic effect of *Eucheuma serra* agglutinin (ESA) on cancer cells and its application to molecular probe for drug delivery system using lipid vesicles. *Cytotechnology* 36(1): 93–99.
Suwal, S., V. Perreault, A. Marciniak, É. Tamigneaux, É. Deslandes, L. Bazinet, H. Jacques, L. Beaulieu and A. Doyen. 2019. Effects of high hydrostatic pressure and polysaccharides on the extraction of antioxidant compounds from red macroalgae, *Palmaria palmata* and *Solieria chordalis*. *Journal of Food Engineering* 252: 53–59.
Talyshinsky, M.M., Y.Y. Souprun and M.M. Huleihel. 2002. Antiviral activity of red microalgal polysaccharides against retroviruses. *Cancer Cell International* 2(1): 1–7.
Torres, M.D., N. Flórez-Fernández and H. Domínguez. 2019. Integral utilization of red seaweed for bioactive production. *Marine Drugs* 17(6): 314.

Wang, T., R. Jónsdóttir, H.G. Kristinsson, G.O. Hreggvidsson, J.Ó. Jónsson, G. Thorkelsson and G. Ólafsdóttir. 2010. Enzyme-enhanced extraction of antioxidant ingredients from red algae *Palmaria palmata*. *LWT-Food Science Technology* 43(9): 1387–1393.

Wen, X., C. Peng, H. Zhou, Z. Lin, G. Lin, S. Chen and P. Li. 2006. Nutritional composition and assessment of *Gracilaria lemaneiformis* Bory. *Journal of Integrative Plant Biology* 48(9): 1047–1053.

Widowati, I., D. Lubac, M. Puspita and N. Bourgougnon. 2014. Antibacterial and antioxidant properties of the red alga *Gracilaria verrucosa* from the north coast of Java, Semarang, Indonesia. *International Journal of Latest Research in Science Technology* 3(3): 179–185.

Wijesinghe, W.A.J.P., E.A. Kim, M.C. Kang, W.W. Lee, H.S. Lee, C.S. Vairappan, and Y.J. Jeon. 2014. Assessment of anti-inflammatory effect of 5β-hydroxypalisadin B isolated from red seaweed Laurencia snackeyi in zebrafish embryo in vivo model. *Environmental toxicology and pharmacology*, 37(1):110–117.

Xu, W., H. Qin, H. Tian, L. Liu, J. Gao, F. Peng and Y. Tu. 2022. Biohybrid micro/nanomotors for biomedical applications. *Applied Materials Today* 27: 101482.

Yang, J.I., C.C. Yeh, J.C. Lee, S.C. Yi, H.W. Huang, C.N. Tseng and H.W. Chang. 2012. Aqueous extracts of the edible *Gracilaria tenuistipitata* are protective against H2O2-induced DNA damage, growth inhibition, and cell cycle arrest. *Molecules* 17(6): 7241–7254.

Yeh, C.C., C.N. Tseng, J.I. Yang, H.W. Hwang, Y. Fang, J.Y. Tang, F.R. Chang and H.W. Chang. 2012a. Antiproliferation and induction of apoptosis in Ca9-22 oral cancer cells by ethanolic extract of *Gracilaria tenuistipitata*. *Molecules* 17(9): 10916–10927.

Yeh, C.C., J.I. Yang, J.C. Lee, C.N. Tseng, Y.C. Chan, Y.C. Hseu, J.Y. Tang, L.Y. Chuang, H.W. Hwang, F.R. Chang and H.W. Chang. 2012b. Anti-proliferative effect of methanolic extract of *Gracilaria tenuistipitata* on oral cancercells involves apoptosis, DNA damage, and oxidative stress. *BMC Complementary and Alternative medicine* 12(1): 1–9.

Yanshin, N., A. Kushnareva, V. Lemesheva, C. Birkemeyer and E. Tarakhovskaya. 2021. Chemical composition and potential practical application of 15 red algal species from the White Sea Coast (the Arctic Ocean). *Molecules* 26(9): 2489.

Yoon, W.J., S.J. Heo, S.C. Han, H.J. Lee, G.J. Kang, H.K. Kang, J.W. Hyun, Y.S. Koh and E.S. Yoo. 2012. Anti-inflammatory effect of sargachromanol G isolated from *Sargassum siliquastrum* in RAW 264.7 cells. *Archives of Pharmacal Research* 35(8): 1421–1430.

Yuan, Y.V., M.F. Carrington and N.A. Walsh. 2005. Extracts from dulse (*Palmaria palmata*) are effective antioxidants and inhibitors of cell proliferation in vitro. *Food Chemical Toxicology* 43(7): 1073–1081.

Zandi, K., S. Tajbakhsh, I. Nabipour, Z. Rastian, F. Yousefi, S. Sharafian and K. Sartavi. 2010a. In vitro antitumor activity of *Gracilaria corticate* (a red alga) against Jurkat and molt-4 human cancer cell lines. *African Journal of Biotechnology* 9(40): 6787–6790.

Zandi, K., S. Ahmadzadeh, S. Tajbakhsh, Z. Rastian, F. Yousefi, F. Farshadpour and K. Sartavi. 2010b. Anticancer activity of *Sargassum oligocystum* water extract against human cancer cell lines. *European Review for Medical Pharmacological Sciences* 14: 669–673.

Zhou, C., X. Yu, Y. Zhang, R. He and H. Ma. 2012. Ultrasonic degradation, purification and analysis of structure and antioxidant activity of polysaccharide from *Porphyra yezoensis* Ueda. *Carbohydrate Polymers* 87(3): 2046–2051.

Index

Note: Page locators in **bold** and *italics* represents tables and figures, respectively.

C

D

E

F

G

H

I

For Product Safety Concerns and Information please contact our EU
representative GPSR@taylorandfrancis.com
Taylor & Francis Verlag GmbH, Kaufingerstraße 24, 80331 München, Germany

www.ingramcontent.com/pod-product-compliance
Ingram Content Group UK Ltd.
Pitfield, Milton Keynes, MK11 3LW, UK
UKHW052220190926
475702UK00006B/1116
* 9 7 8 1 0 3 2 1 8 5 4 0 8 *